Statistics
AND
Econometrics

Statistics
AND
Econometrics

Charles R. Frank, Jr.
PRINCETON UNIVERSITY

HOLT, RINEHART and WINSTON, Inc.

NEW YORK CHICAGO SAN FRANCISCO ATLANTA
DALLAS MONTREAL TORONTO LONDON SYDNEY

To Pat

Preface

This book is intended as a basic introductory text on statistics and econometrics for students at the advanced undergraduate or graduate level. It is the product of my lecture notes for an undergraduate course at Yale University. These notes were expanded and presented with a good deal more mathematical sophistication to a graduate workshop in econometrics at Princeton University. Student reaction in these courses contributed greatly to the organization and method of presentation of this volume.

The book was designed so that a student with no mathematical preparation beyond elementary college algebra could master the material in the main text. Students with more formal mathematical training will find additional topics and derivations in the appendices which require knowledge of elementary calculus and matrix algebra. Although a formal mathematical background is not a prerequisite for the interested student, mathematical aptitude is essential for careful perusal of all thirteen chapters.

The first eight chapters (Parts 1–3) develop elementary statistical theory and its applications. Here selectivity was exercised to highlight those subjects necessary for comprehension of the last five chapters (Parts 4 and 5), which deal with topics often offered in courses on econometrics. Many of the subjects discussed in these latter chapters are usually reserved for advanced graduate courses.

In order to cover a large range of topics in a book of reasonable length, many theorems and propositions have been presented without proofs. However, the author has attempted to maintain rigor and precision in the statement of theorems and other propositions. Implications and applications of theorems are explained and illustrated in the text. The student interested in theorem proofs should consult the references at the end of each part.

In order to maintain the flow and continuity of the text, the development of simple mathematical tools which are often not covered in courses in college algebra and lengthy or complicated derivations are placed in appendices.

The author has designed the chapter sequence so that each succeeding chapter represents an increasing level of difficulty. Thus, simple and multiple regression and correlation are presented in Chapters 3 and 4; the more difficult concepts of probability and probability distribution are in Chapters 5 and 6; and regression analysis as a problem in statistical inference is included in Chapter 9. This order of treatment enables the less mathematically inclined student or the student with limited time to concentrate on Part 1 and gain an appreciation of the most commonly used statistical measures. Thus, the five parts of the book may be combined in various ways to suit the ability, background, and interests of the reader.

The author is extremely grateful for a careful and extraordinarily thorough reading of the manuscript by Charles Bischoff of Yale University. Mr. Bischoff also taught undergraduate statistics and econometrics at Yale and offered voluminous comments concerning organization, style, clarity, and mathematical rigor. Most of his suggestions were incorporated into the final version of the manuscript. Mrs. Diane Sliney also read through the entire manuscript and offered useful suggestions on style and organization. Mrs. Sliney proofread for typing errors and checked all the tables and examples for mistakes. Her help is very much appreciated. Mr. Takatoshi Kato and Mr. David Wheeler also read parts of the manuscript to check for errors. Mr. Wheeler suggested many of the problems at the end of each chapter and worked out the answers.

I am indebted to the Literary Executor of the late Sir Ronald R. Fisher, F.R.S., and to Oliver & Boyd, Edinburgh, for their permission to reprint Table C from their book, *Statistical Methods for Research Workers*.

Mrs. Manya Vas typed nearly all of the manuscript in an extremely skillful and patient manner. Mrs. Jane Huyeck typed some of the first chapters and Mrs. Dorothy Reiger assisted in typing odd bits and some of the revisions.

The author assumes responsibility for all remaining errors and earnestly hopes they are few.

—C. R. F.

January 1971
Princeton, New Jersey

Contents

Chapter 4

Part II · Probability   97

Chapter 5

Chapter 6

Part III · Statistical Inference   157

Chapter 7

Chapter 8

Part IV · Linear Models   205

Part V · Problems in Econometrics   271

Contents

Chapter 12

Chapter 13

Statistics
AND
Econometrics

Chapter 1

Introduction

Econometrics is the application of the methods of mathematical statistics to problems of economics. The econometrician (one who practices econometrics) therefore must have a fairly good grounding in the fundamentals of mathematical statistics. The necessity for a knowledge of the fundamentals of mathematical statistics, however, extends beyond the practicing econometrician. Since much of what is written in the field of economics and in other social sciences involves statistical analysis, any serious student of economics must know these fundamentals whether he expects to use econometric techniques himself or not.

Econometrics involves the measurement of economic variables and economic relationships. There are two fundamentally different approaches to this kind of measurement. The first involves *descriptive statistics*. Descriptive statistics summarize characteristics of large amounts of data or transform data into more readily comprehensible forms. The techniques of *statistical inference* enable one to make probabilistic statements concerning the parameters of theoretical models of economic processes.

Descriptive statistics include measures of central tendency of data, such as the mean, the mode, or the median, and measures of the degree of dispersion of data, such as the variance, the range, or the standard deviation. Measures of the nature and strength of relationships among variables such as those associated with regression and correlation analysis may also be treated as descriptive statistics. Methods of descriptive statistics enable one to calculate growth rates of variables over time, determine price and quantity index numbers, remove seasonal and trend influences from time-series data, or determine the periodicity of cyclical changes in economic aggregates.

Methods of statistical inference are grounded in the theory of probability. The use of probabilistic models is particularly suitable in the measurement of economic relationships that are difficult to measure exactly because of a number of sources of error. Errors arise in economic data and relationships for a number of reasons. First, there are errors of measurement. For example, a country's exports or imports are unlikely to be measured with complete accuracy, as mistakes will be made on invoices and bills of lading, customs officials will often incorrectly record prices and quantities of goods moving into or out of customs houses, smuggled items will escape detection, and mistakes will be made in adding up individual items to get an export or import total.

Second, errors are introduced when economic data are compiled on a sampling basis. For example, unemployment and employment data are usually based on monthly samples. That is, only a small fraction of the total number of individuals in the work force and only a small fraction of employers will be sampled to determine overall unemployment or employment rates. The reported rate of unemployment or employment in the sample will not always accurately reflect the overall rates. This will be particularly true if the sample is "biased" in such a way as to include a proportionately large number of people likely to have or not have access to jobs.

Third, errors in economic relationships arise from random shocks such as war, natural disaster, or variations in the weather. Finally, errors may arise because of the influence of variables not included or not measured when estimating a particular relationship. For example, the supply of feed grains is not only a function of the price of feed grains but perhaps also the price of meat, the price of nonfeed grains and other farm products, the availability of farm credit, and the amount of storage facilities available. If one wishes to estimate a supply relationship that depends only on the price of feed grains, large errors will be introduced by the effects of these other variables.

Since errors in the data and in relationships are so prevalent, it is useful to be able to hedge or surround econometric measurements with a degree of uncertainty. The techniques of statistical inference enable one to say such things as: The elasticity of feed-grain supply lies within the interval from 0.9 to 1.2 with a 95 percent degree of confidence; or the unemployment rate among males under age 25 is significantly greater than the overall employment rate. The terms *95 percent degree of confidence* and *significantly greater* are used to hedge statistics, but they have a very precise technical meaning in terms of probability and probability distributions.

In order to apply the methods of statistical inference, first one must formulate a mathematical model of an economic process. The model may be very simple and consist of only one equation, such as a supply equation. For example, the model may simply relate the supply of cotton (S) to its price (P) in a linear equation:

$$S = a + b \cdot P, \tag{1.1}$$

where a and b are constants—that is, parameters of the model. In order to apply the techniques of statistical inference, a random error term (μ) must enter the model. The error term usually is simply added to the relationship to make the relationship *stochastic* or random.

$$S = a + b \cdot P + \mu. \tag{1.2}$$

Another simple model is the Keynesian consumption function

$$C = c + d \cdot Y + \mu, \tag{1.3}$$

where C is consumption, Y is income, μ is a random-error term, and c and d are parameters of the model (d is the marginal propensity to consume).

A model may involve several equations instead of just one. For example, one might wish not only to specify a supply function for cotton as in (1.2), but to add a demand equation to the model as well:

$$D = \alpha + \beta \cdot P + \gamma \cdot Y + v, \tag{1.4}$$

where D is quantity of cotton demanded, P is price, Y is income, v is an error term, and α, β, and γ are parameters of the demand equation. In order to complete the model, an equilibrium condition must be added—that is, demand equals supply:

$$D = S. \tag{1.5}$$

The three equations, (1.2), (1.4), and (1.5), constitute a stochastic model of a cotton market.

Once a model has been constructed, the parameters of the model are estimated from the data on the variables of the model. For example, the parameters of the supply equation (1.2) can be estimated from a series of data on quantity supplied (S) and price (P). The techniques of statistical inference may then be used to *infer* certain properties of the model. For example, one might *infer* that the price elasticity of supply of cotton is significantly less than unity. As stated above, these inferences about the parameters of a model are always interpreted in terms of probabilities.

The set of parameters of a model are often termed the *structure* of the model. It is often impossible, however, to make inferences concerning the structure of a model, either because the data are imperfect or because the structure is inherently incapable of being estimated. If the structure of the model is inherently incapable of being estimated, we say that the model is *nonidentifiable*. Models can often be made *identifiable* in a theoretical sense by adding variables to the model or by making a priori assumptions concerning *some* of the parameters of the model. It may not be crucial, however, to identify the structure of a model. It all depends on the purpose to which the model is put.

For purposes of prediction or forecasting, it is not always necessary that the structure of a model be identified. For example, total advertising expenditures are closely tied to retail sales. This is really the result of a two-way relationship. Advertising budgets are often based on estimates of retail sales. At the same time retail sales depend to a certain extent on the amount of advertising. In order to estimate the *structure* of a retail-sales model, we would have to be able to estimate both of these functional relationships. It may not be possible, however, to identify the structure of both. Nevertheless, if one simply wants to forecast retail sales taking advertising expenditure into account, these may be based on a single estimated relationship between these variables even though this single relationship results from the combined effect of two separate functional relationships.

On the other hand, there are instances when a good knowledge of the structure of relationships is absolutely crucial. For example, if one wants to know the effect of placing a government-imposed limit on the amount of advertising expenditure allowed by a firm, one of the relationships between advertising expenditures will be fundamentally changed or destroyed. In order to make any reasonable estimate of the impact of such restrictions, the structure of the model must be identified. If one wants to know the effect of an excise tax on sales of a commodity or the effect of a devaluation on the balance of payments, it is essential to know the structure of the economic processes involved. That is, one must know the supply and demand elasticities for the commodity being taxed, or the supply and demand elasticities for internationally traded goods for the impact of a devaluation.

The next three chapters of this book concentrate on the descriptive approach to statistics. Chapters 5 and 6 introduce the notion of probability and probability distributions. Sampling and estimation procedures for probabilistic (stochastic) models are discussed in Chapter 7. Chapter 8 introduces statistical inference in the form of hypothesis testing and the placing of confidence limits about estimated parameters. Chapters 9 and 10 develop a very general linear model and discuss estimation and hypothesis testing procedures for this model. Chapter 11 focuses on problems that arise when the assumptions of the general model are not satisfied and suggests modified estimation and hypothesis testing procedures. Chapter 12 discusses the problem of identification of econometric models with several structural equations and methods of estimating the structure of such models. Chapter 13 contains some supplementary remarks referring to methods that are used by econometricians but do not fit easily into the organizational structure of the preceding chapters. The treatment of these supplementary topics is not very complete but should provide the reader with a brief introduction and a stimulus to further reading.

Part I

Descriptive Statistics

Chapter 2

Central
Tendency
and
Dispersion

Large quantities of data are often difficult to comprehend. One way to derive meaning from a large mass of data is to use a *statistic* to describe certain properties of the data. A statistic is a number computed from a set of data using some formula. In this chapter, we shall limit our discussion to statistics that describe two different properties of data: (1) those which measure central tendency, and (2) those which measure spread or dispersion.

Central Tendency

Let $Y_1, Y_2, \ldots, Y_n$ stand for n numbers or n pieces of data. A common measure of central tendency is the *average* or *mean* value of the numbers $Y_1, Y_2, \ldots, Y_n$. The mean is written

$$\bar{Y} = \frac{Y_1 Y_2 + \cdots + Y_n}{n} = \frac{\sum\limits_{i=1}^{n} Y_i}{n}. \tag{2.1}$$

Example 2.1. Table 2.1 is drawn from the *Statistical Abstract of the United States* and lists foreign aid commitments from 1949 to 1965. Suppose we wish to calculate the average total economic and military assistance. Then from Table 2.1, column 1, we add the numbers

Y_1	5,026
Y_2	3,670
Y_3	3,603
Y_4	3,466
Y_5	6,119
Y_6	5,524
Y_7	4,217
Y_8	4,434
Y_9	3,712
Y_{10}	3,984
Y_{11}	4,026
Y_{12}	3,584
Y_{13}	3,386
Y_{14}	3,956
Y_{15}	4,107
Y_{16}	3,633
Y_{17}	3,336

$$\sum_{i=1} Y_i = 69,783$$

$$\bar{Y} = 4,105$$

The mean annual foreign aid bill of the United States has been $4,105 million from 1949 to 1965. The mean amount of economic assistance has been $2,281 million and the mean military assistance $1,824 million.

An alternative measure of central tendency is the *median*. The median splits any set of data into two groups. Half of the numbers in a set of data $Y_1, Y_2, \ldots, Y_n$ lie below the median, and half lie above the median.

When the number n of pieces of data is odd, the median may be defined precisely by ordering the data so that

$$Y_{i+1} \geqq Y_i \qquad \text{for } i = 1, 2, \ldots, n - 1. \tag{2.2}$$

The median then is the number

$$Y_{(n+1)/2} \qquad \text{for } n \text{ odd.} \tag{2.3}$$

Example 2.2.   Let us determine the median amount of economic assistance given by the United States from Table 2.1, column 2. There are 17 years

TABLE 2.1 United States foreign assistance commitments under economic and military assistance programs (millions of dollars)

	TOTAL ECONOMIC AND MILITARY ASSISTANCE (1)	ECONOMIC ASSISTANCE			MILITARY ASSISTANCE		
		(2) TOTAL	(3) GRANTS	(4) LOANS	(5) TOTAL	(6) GRANTS	(7) LOANS
1949	5,026	5,026	4,093	933	—	—	—
1950	3,670	3,614	3,451	163	56	56	—
1951	3,603	2,622	2,577	45	980	980	—
1952	3,466	1,985	1,784	201	1,481	1,481	—
1953	6,119	1,960	1,934	26	4,159	4,159	—
1954	5,524	2,228	2,114	114	3,296	3,296	—
1955	4,217	1,821	1,624	197	2,396	2,396	—
1956	4,434	1,506	1,298	208	2,928	2,920	8
1957	3,712	1,627	1,305	322	2,085	2,078	7
1958	3,984	1,620	1,203	417	2,363	2,325	39
1959	4,026	1,916	1,291	626	2,110	2,050	60
1960	3,584	1,866	1,302	564	1,718	1,697	21
1961	3,386	2,012	1,305	707	1,374	1,344	30
1962	3,956	2,508	1,180	1,329	1,448	1,427	21
1963	4,107	2,297	954	1,343	1,809	1,765	44
1964	3,633	2,136	808	1,328	1,498	1,415	83
1965	3,336	2,026	904	1,122	1,310	1,237	73

Source: Statistical Abstract of the United States, 1966, Table No. 1245, p. 852.

between 1949 and 1965 so that the number n of pieces of data is odd. Let us order the data from column 2 as indicated by (2.2).

$Y_1 = 1,506$ $Y_{10} = 2,026$

$Y_2 = 1,620$ $Y_{11} = 2,136$

$Y_3 = 1,627$ $Y_{12} = 2,228$

$Y_4 = 1,821$ $Y_{13} = 2,297$

$Y_5 = 1,866$ $Y_{14} = 2,508$

$Y_6 = 1,916$ $Y_{15} = 2,622$

$Y_7 = 1,960$ $Y_{16} = 3,614$

$Y_8 = 1,985$ $Y_{17} = 5,026$

$Y_9 = 2,012$

Now $n + 1 = 17 + 1 = 18$ and $(n + 1)/2 = 18/2 = 9$. Thus the median is

$Y_9 = 2,012$

The median amount of economic assistance is $2,012 million.

When the number n of pieces of data is even, the median must be defined in some other way, since $(n + 1)/2$ is not a whole number. One way of defining the median for n even is to split the difference between the pieces of data $Y_{n/2}$ and $Y_{(n+2)/2}$. That is, if the data are ordered according to (2.2), the median is

$$\tfrac{1}{2}(Y_{n/2} + Y_{(n+2)/2}) \qquad \text{for } n \text{ even.} \tag{2.4}$$

Example 2.3.   Let us determine the median amount of military assistance given by the United States from Table 2.1, column 5, exclusive of the year 1949, in which no military aid was given. There are 16 years between 1950 and 1965, so that the number n of pieces of data is even. Let us order the data.

$Y_1 =$	56	$Y_9 =$	1,809
$Y_2 =$	980	$Y_{10} =$	2,085
$Y_3 =$	1,310	$Y_{11} =$	2,110
$Y_4 =$	1,374	$Y_{12} =$	2,363
$Y_5 =$	1,448	$Y_{13} =$	2,396
$Y_6 =$	1,481	$Y_{14} =$	2,928
$Y_7 =$	1,498	$Y_{15} =$	3,296
$Y_8 =$	1,718	$Y_{16} =$	4,159

Here $n = 16$, so that $n/2 = 8$ and $(n + 2)/2 = 18/2 = 9$. Thus the median is

$$\tfrac{1}{2}(Y_8 + Y_9) = \tfrac{1}{2}(1,718 + 1,809) = 1,763.5.$$

The median amount of military assistance is $1,763.5 million.

The mean and median generally give different numbers as measures of central tendency. If the data are symmetrical about the median, then the mean and the median are the same. Data are symmetrical about the median when the sum of the absolute value of the divergences above the median is equal to the sum below the median. That is, if Y_I stands for the median and the data are ordered according to (2.2), then

$$D(+) = \sum_{i=I+1}^{n} |Y_i - Y_I| \tag{2.5}$$

is the sum above the median and

$$D(-) = \sum_{i=1}^{I-1} |Y_i - Y_I| \tag{2.6}$$

is the sum below the median. The bars $|\ \ |$ about a number indicate its absolute value. For example, $|3.2| = 3.2$ and $|-4.8| = 4.8$.

Example 2.4.   Consider the following pieces of hypothetical data:

Y_1	Y_2	Y_3	Y_4	Y_5
4	7	8	10	11

These data are symmetrical. The median is $Y_3 = 8$ and

$$D(+) = |10 - 8| + |11 - 8| = 2 + 3 = 5,$$
$$D(-) = |7 - 8| + |4 - 8| = 1 + 4 = 5.$$

The mean is also equal to 8, as the reader may verify for himself.

When $D(+)$ is greater than $D(-)$, the data are said to be skewed positively. When $D(+)$ is less than $D(-)$, the data are negatively skewed. With positively skewed data, the mean is greater than the median; for negatively skewed data, the mean is less than the median.

Example 2.5. Let us compare the median and mean amount of United States economic assistance from column 2 of Table 2.1. The median amount of aid calculated in Example 2.2 was $2,012 million. The mean amount of economic assistance calculated in Example 2.1 was $2,281 million, or $269 million higher than the median. The difference is largely due to the large amounts of aid in 1949 and 1950, during the latter stages of United States Marshall Plan aid to western Europe. The large amounts of aid in those two years skew the data in a positive direction, so that the mean is greater than the median. If the aid given in 1949 and 1950 were halved, the median amount of aid would not change. The mean, however, would be very much affected, since the large amounts of aid in 1949 and 1950 carry a heavy weight in the calculation of the mean.

Another measure of central tendency is the *mode*. The mode is applicable to data that are grouped into various classes. Each *class* of data is assigned a frequency. The mode is the class with the greatest frequency.

Example 2.6. Let us group the data on United States economic assistance in Column 2 of Table 2.1. The grouping can be done most easily by reference to the ordered data in Example 2.2.

CLASS		
FROM	TO	FREQUENCY
0–	499	0
500–	999	0
1,000–	1,499	0
1,500–	1,999	8
2,000–	2,499	5
2,500–	2,999	2
3,000–	3,499	0
3,500–	3,999	1
4,000–	4,499	0
4,500–	4,999	0
5,000–	5,499	1

The mode is the class from $1,500 million to $1,999 million. If one wishes to pick a single number to represent the mode, one might use the middle of the interval between $1,500 million and $1,999 million or a mode of approximately $1,750 million.

A fourth measure of central tendency is the weighted mean. A weighted mean is applied to a set of data $Y_1, Y_2, \ldots, Y_n$, which have been assigned weights $w_1, w_2, \ldots, w_n$. The weighted mean is given by the formula

$$Y_w = \frac{\sum\limits_{i=1}^{n} w_i \cdot Y_i}{\sum\limits_{i=1}^{n} w_i}. \tag{2.7}$$

Each piece of data Y_i is multiplied by its weight. The sum of the weighted data is then divided by the sum of the weights.

Example 2.7. The weighted mean may be used to calculate an approximate value of the mean for grouped data such as that in Example 2.6. The numbers Y_i are the middle of the class intervals and the weights the frequency of each class. Using the grouped data in Example 2.6, we have:

$Y_1 = 250$	$w_1 = 0$	$w_1 \cdot Y_1 = 0$
$Y_2 = 750$	$w_2 = 0$	$w_2 \cdot Y_2 = 0$
$Y_3 = 1,250$	$w_3 = 0$	$w_3 \cdot Y_3 = 0$
$Y_4 = 1,750$	$w_4 = 8$	$w_4 \cdot Y_4 = 14,000$
$Y_5 = 2,250$	$w_5 = 5$	$w_5 \cdot Y_5 = 11,250$
$Y_6 = 2,750$	$w_6 = 2$	$w_6 \cdot Y_6 = 5,500$
$Y_7 = 3,250$	$w_7 = 0$	$w_7 \cdot Y_7 = 0$
$Y_8 = 3,750$	$w_8 = 1$	$w_8 \cdot Y_8 = 3,750$
$Y_9 = 4,250$	$w_9 = 0$	$w_9 \cdot Y_9 = 0$
$Y_{10} = 4,750$	$w_{10} = 0$	$w_{10} \cdot Y_{10} = 0$
$Y_{11} = 5,250$	$w_{11} = 1$	$w_{11} \cdot Y_{11} = 5,250$

$$\sum_{i=1}^{11} w_{11} = 17 \qquad \sum_{i=1}^{11} w_i \cdot Y_i = 39,750$$

The weighted mean is

$$\frac{39,750}{17} = 2,338.$$

The weighted mean of $2,338 million from the grouped data is an approximation of the true mean of $2,281 million, which we calculated in Example 2.1 from the *raw* or ungrouped data.

Example 2.8. Table 2.2 lists the Gross Domestic Product (GDP) per capita for several Latin American countries. If one took a simple average of the per capita GDP, one might obtain a very misleading impression. The countries with low populations would tend to pull the simple average down out of proportion to their importance. A weighted mean, using

TABLE 2.2 Per capita gross domestic product in United States dollars and population for selected Latin American countries in 1963

	PER CAPITA GROSS DOMESTIC PRODUCT	POPULATION (MILLIONS)
Argentina	616	21.7
Bolivia	125	3.6
Brazil	215	76.4
Chile	457	8.2
Colombia	268	16.9
Ecuador	183	4.7
Guatemala	274	4.2
Guyana	259	.6
Mexico	390	38.4
Nicaragua	256	1.5
Paraguay	192	1.9
Peru	247	11.0
Uruguay	518	2.6
Venezuela	881	8.1

Sources: United Nations Yearbook of National Accounts Statistics, 1965, Tables 9A and 9B, pp. 494 and 500; United Nations Demographic Yearbook, 1965, Table 4, pp. 131 and 133.

population as weights, however, would give a more meaningful average per capita GDP. The weighted average is

$$Y_w = \frac{67,611}{199.8} = 338.4.$$

The numerator is the sum of each country's per capita GDP multiplied by population (total GDP), and the denominator is the total population of all 14 Latin American countries listed in Table 2.2. The simple average per capita GDP of \$349.1 can be compared with the weighted average of \$338.4. The weighted average is smaller because some of the countries with a larger-than-average population have a small per capita GDP and several of the smaller countries have a large per capita GDP. For example, Venezuela and Uruguay with a large per capita GDP have smaller-than-average populations and thus receive relatively small weight. On the other hand, Brazil with a very large population has a relatively small per capita income that receives a large weight.

Two other measures of central tendency are the *geometric mean* and the *harmonic mean*. These are used relatively infrequently. The geometric mean is given by

$$Y_G = \sqrt[n]{Y_1 \cdot Y_2 \cdot \ldots \cdot Y_n}. \tag{2.8}$$

Example 2.9.   The geometric mean is sometimes used with ratios, rates of change, and percentages. For example, suppose an astute investor in stock-exchange securities is able to double his money in one year. The next year he does even better and quadruples his money. He then runs into a streak of bad luck the next year and makes nothing but loses nothing either. If his original investment is $1,000, then the following illustrates its growth:

YEAR	MONEY INVESTED (1)	MONEY INVESTED PLUS CAPITAL GAIN (2)	MULTIPLE (2) ÷ (1)
1	$1,000	$2,000	2
2	$2,000	$8,000	4
3	$8,000	$8,000	1

If we take the geometric mean of the last column, we get

$$\sqrt[3]{2 \cdot 4 \cdot 1} = \sqrt[3]{8} = 2.$$

We may say that on the average, this astute investor doubles his money every year.

The harmonic mean is given by the formula

$$Y_H = \frac{n}{\displaystyle\sum_{i=1}^{n} \frac{1}{Y_i}}. \tag{2.9}$$

Example 2.10.   The harmonic mean is used in the following type of problem. Suppose a motorist buys $5 worth of gasoline at $.30 a gallon and then later buys $5 worth at $.36 a gallon. To calculate his average cost per gallon we take the harmonic mean of the two different prices paid.

$$\frac{2}{\dfrac{1}{\$.30} + \dfrac{1}{\$.36}} = \$.327$$

Thus, on the average, the motorist paid 32.7 cents per gallon.

Measures of Dispersion

Two sets of data with the same mean may not be alike at all. The one set of data may have a wide dispersion of values on either side of the mean, while the other set may be grouped closely about the mean. The most common measure of dispersion is the *standard deviation*. The standard deviation of a set of data $Y_1, Y_2, \ldots, Y_n$ is denoted by s_y and is given by the formula

$$s_y = \sqrt{\frac{1}{n} \sum_{i=1}^{n} (Y_i - \bar{Y})^2}. \tag{2.10}$$

To calculate the standard deviation, one first determines the squared difference between each piece of data and the mean. The squared quantities are summed and divided by n, the number of pieces of data, to obtain s_y^2. The number s_y^2 is called the variance. The square root of the variance is s_y, the standard deviation.

Example 2.11. Let us calculate the standard deviation for the amount of United States economic assistance in Table 2.1. The mean amount of economic assistance calculated in Example 2.1 was

$$\bar{Y} = 2{,}281.$$

Y_i	$Y_i - \bar{Y}$	$(Y_i - \bar{Y})^2$
5,026	2,745	7,535,025
3,614	1,333	1,776,889
2,622	341	116,281
1,985	−296	87,616
1,960	−321	103,041
2,228	−53	2,809
1,821	−460	211,600
1,506	−775	600,625
1,627	−654	427,716
1,620	−661	436,921
1,916	−365	133,225
1,866	−415	172,225
2,012	−269	72,361
2,508	227	51,529
2,297	16	256
2,136	−145	21,025
2,026	−255	65,025

Summing the items in the last column, we get

$$\sum_{i=1}^{n} (Y_i - \bar{Y})^2 = 11{,}814{,}169,$$

$$\frac{1}{n} \sum_{i=1}^{n} (Y_i - \bar{Y})^2 = \frac{11{,}814{,}169}{17} = 694{,}951,$$

$$s_y = \sqrt{694{,}951} = 834.$$

The standard deviation of United States economic assistance is \$834. The standard deviation for total United States economic and military aid is only \$752. The larger variation in economic aid is due largely to the two atypical years 1949 and 1950. The squared deviation from the mean $(Y_i - \bar{Y})^2$ is quite large for those years. For total economic and military aid, none of the deviations from the mean is so large.

Other less frequently used measures of variation are the *range*, the *interquartile range*, and the *mean absolute deviation*. The range is the difference between the largest item Y_n and the smallest item Y_1.

Example 2.12.   The range of the per capita incomes in Table 2.2 is

\$881 − \$125 = \$756.

This is the difference in per capita incomes between Venezuela, the wealthiest country in per capita terms, and Bolivia, the poorest country.

The mean deviation is the average of the absolute deviations from the mean:

$$\text{Mean deviation} = \frac{\sum_{i=1}^{n} |Y_i - \bar{Y}|}{n}. \tag{2.11}$$

Example 2.13.   Let us calculate the mean deviation for the per capita incomes in Table 2.2. Here $\bar{Y} = 349$.

Y_i	$Y_i - \bar{Y}$	$|Y_i - \bar{Y}|$
616	267	267
125	−224	224
215	−134	134
457	108	108
268	−81	81
183	−166	166
274	−75	75
259	−90	90
390	41	41
256	−93	93
192	−157	157
247	−102	102
518	169	169
881	532	532
		$\Sigma |Y_i - \bar{Y}| = 2239$

$$\text{Mean deviation} = \frac{2{,}239}{14} = 159.9.$$

The interquartile range is obtained by first dividing the data into four groups, or quartiles, approximately equal in number. The groups are ordered so that any data in the first quartile are less in value than those in the second, those in the second less than those in the third, and so on. The interquartile range is the range of the data in the middle two quartiles. quartiles.

Example 2.14. Let us compute the interquartile range for the data on total United States military assistance in Table 2.1. The data have been ordered by size in Example 2.3. These data may be divided into quartiles as follows:

$$
1 \quad \begin{cases} Y_1 = & 56 \\ Y_2 = & 980 \\ Y_3 = 1{,}310 \\ Y_4 = 1{,}374 \end{cases}
$$

$$
2 \quad \begin{cases} Y_5 = 1{,}448 \\ Y_6 = 1{,}481 \\ Y_7 = 1{,}498 \\ Y_8 = 1{,}718 \end{cases}
$$

$$
3 \quad \begin{cases} Y_9 = 1{,}809 \\ Y_{10} = 2{,}085 \\ Y_{11} = 2{,}110 \\ Y_{12} = 2{,}363 \end{cases}
$$

$$
4 \quad \begin{cases} Y_{13} = 2{,}396 \\ Y_{14} = 2{,}928 \\ Y_{15} = 3{,}296 \\ Y_{16} = 4{,}159 \end{cases}
$$

The semiinterquartile range is $Y_{12} - Y_5 = 2363 - 1448 = 915$.

Frequency Distributions

A convenient way of summarizing a large amount of data graphically is through the use of a histogram or a frequency diagram. In order to draw a histogram, the data must be grouped into classes. The frequency of each class is measured on the vertical axis of the histogram diagram and the class interval along the horizontal axis.

Example 2.15. In Example 2.6 we grouped the data on U.S. economic assistance into classes. In Fig. 2.1 a bar is drawn above each class interval. The height of the bar depends on the frequency. Figure 2.1 is called a histogram or a digaram of a frequency distribution.

It is often convenient to represent a frequency distribution by drawing a smooth curve through the tops of the bars in the histogram, as is done in Fig 2.2. Using these curves, it is then possible to compare different

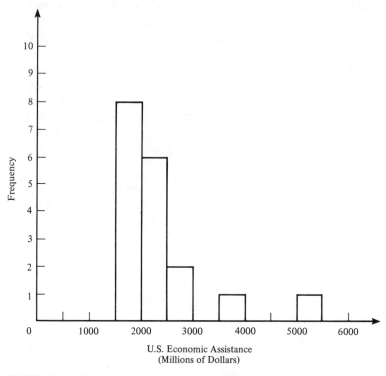

FIGURE 2.1   Histogram for U.S. Economic Assistance

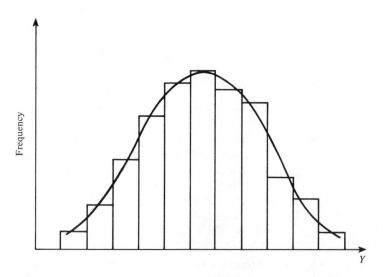

FIGURE 2.2   Histogram and Frequency Distribution

characteristics of frequency distributions. For example, the frequency distribution in Fig. 2.3(a) has less dispersion than that in Fig. 2.3(b). The variance or standard deviation is smaller in Fig. 2.3(a). The frequency distribution in Fig. 2.3(c) is positively skewed. The mode is represented by the highest point of the frequency distribution. The median is to the right of the mode and the mean is greater than the median. The frequency distribution in Fig. 2.3(d) is negatively skewed. The mean is less than the median, which in turn is less than the mode. The frequency distributions in Figs. 2.3(a) and 2.3(b) are both symmetric. The mean, the mode, and the median coincide.

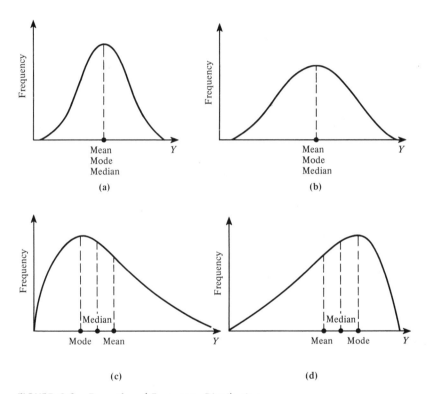

FIGURE 2.3   Examples of Frequency Distributions

APPENDIX   Use of the Summation Sign

The summation sign Σ is the Greek capital letter *sigma*. Associated with a summation sign are an index i and an upper and lower limit of the index. Thus suppose there are three numbers labeled Y_1, Y_2, and Y_3. The sum of these numbers is written

$$\sum_{i=1}^{3} Y_i = Y_1 + Y_2 + Y_3.$$

The index is i; the lower limit of the index is 1 and the upper limit is 3. If there are n numbers Y_1, Y_2, $\ldots$, Y_n, then the sum of these numbers is written

$$\sum_{i=1}^{n} Y_i = Y_1 + Y_2 + \cdots + Y_n.$$

Certain rules with regard to the use of summation signs can be derived from the basic definition of a sum.

Suppose there are two sets of numbers X_1, X_2, $\ldots$, X_n and Y_1, Y_2, $\ldots$, Y_n. Then let us denote the sum of each pair of numbers (X_i, Y_i) by $Z_i = X_i + Y_i$. That is, $Z_1 = X_1 + Y_1$, $Z_2 = X_2 + Y_2$, and so on. The sum of the Z's is written

$$\sum_{i=1}^{n} Z_i = \sum_{i=1}^{n} (X_i + Y_i).$$

Rule 1.   The sum

$$\sum_{i=1}^{n} (X_i + Y_i) = \left(\sum_{i=1}^{n} X_i \right) + \left(\sum_{i=1}^{n} Y_i \right).$$

Thus if $n = 3$, we have

$$\sum_{i=1}^{3} Z_i = \sum_{i=1}^{3} (X_i + Y_i) = (X_1 + Y_1) + (X_2 + Y_2) + (X_3 + Y_3)$$
$$= (X_1 + X_2 + X_3) + (Y_1 + Y_2 + Y_3)$$
$$= \left(\sum_{i=1}^{3} X_i \right) + \left(\sum_{i=1}^{3} Y_i \right).$$

Let a be any constant term that does not change with the index. Let Y_1, Y_2, $\ldots$, Y_n be a set of numbers that do vary with the index.

Rule 2. The sum

$$\sum_{i=1}^{n} (aY_i) = a\left(\sum_{i=1}^{n} Y_i\right).$$

Thus if $n = 3$, we have

$$\sum_{i=1}^{3} (aY_i) = aY_1 + aY_2 + aY_3 = a(Y_1 + Y_2 + Y_3) = a\left(\sum_{i=1}^{3} Y_i\right).$$

Rule 3.

$$\sum_{i=1}^{n} (Y_i + a) = \left(\sum_{i=1}^{n} Y_i\right) + n \cdot a.$$

For $n = 3$, we have

$$\sum_{i=1}^{3} (Y_i + a) = Y_1 + a + Y_2 + a + Y_3 + a$$

$$= (Y_1 + Y_2 + Y_3) + a + a + a$$

$$= \left(\sum_{i=1}^{3} Y_i\right) + 3 \cdot a.$$

The algebraic manipulations performed with sums as illustrated in the three rules above are used very often in deriving statistical formulae.

One common misunderstanding among students refers to the notation for sums of squares as opposed to the square of a sum.

Sum of squares: $\quad \sum_{i=1}^{n} Y_i^2 = Y_1^2 + Y_2^2 + \cdots + Y_n^2.$

Square of sum: $\quad \left(\sum_{i=1}^{n} Y_i\right)^2 = (Y_1 + Y_2 + \cdots + Y_n)^2.$

The two are not equivalent, as an example for $n = 2$ will show:

$$\sum_{i=1}^{2} Y_i^2 = Y_1^2 + Y_2^2,$$

$$\left(\sum_{i=1}^{2} Y_i\right)^2 = Y_1^2 + Y_2^2 + 2Y_1Y_2.$$

The difference between the two is the cross product $2Y_1Y_2$.

Sometimes data may be classified in more than one way. For example, the table below gives a two-way classification of data by industry and by country. Let Y_{ij} be the average wage in the ith industry in the jth country. Suppose there are four countries and three industries. The data may be arranged in the following table.

	INDIA (1)	JAPAN (2)	FRANCE (3)	USSR (4)
Steels (1)	Y_{11}	Y_{12}	Y_{13}	Y_{14}
Chemicals (2)	Y_{21}	Y_{22}	Y_{23}	Y_{24}
Textiles (3)	Y_{31}	Y_{32}	Y_{33}	Y_{34}

Then, for example, Y_{23} stands for the average wage in the second industry (chemicals) in the third country (France). Instead of one index, there are two indices i and j. One may sum over either index or both. Thus

$$\sum_{i=1}^{3} Y_{i2} = Y_{12} + Y_{22} + Y_{32}$$

is the sum of the average wages in all three industries in the second country (Japan). This is the sum of all industries down the second column of the table. Also

$$\sum_{j=1}^{4} Y_{3j} = Y_{31} + Y_{32} + Y_{33} + Y_{34}$$

is the sum across all four countries for the third industry (textiles). This is a sum across the third row of the table.

A double summation is written

$$\sum_{i=1}^{n} \sum_{j=1}^{n} Y_{ij}.$$

This consists of the sum of all Y_{ij} variables, where i and j, the indices, assume all possible combinations between their respective lower and upper limits, 1 and m for the index i and 1 and n for the index j. Using the data format in the above table, the double summation may be viewed as the sum of row sums. That is,

$$\sum_{i=1}^{3} \sum_{j=1}^{4} Y_{ij} = \sum_{i=1}^{3} \left(\sum_{j=1}^{4} Y_{ij} \right) = \sum_{j=1}^{4} Y_{1j} + \sum_{j=1}^{4} Y_{2j} + \sum_{j=1}^{4} Y_{3j}.$$

Often, one may reverse the order of summation and view the double summation as a sum of column sums:

$$\sum_{i=1}^{3} \sum_{j=1}^{4} Y_{ij} = \sum_{j=1}^{4} \sum_{i=1}^{3} Y_{ij} = \sum_{j=1}^{4} \left(\sum_{i=1}^{3} Y_{ij} \right)$$

$$= \sum_{i=1}^{3} Y_{i1} + \sum_{i=1}^{3} Y_{i2} + \sum_{i=1}^{3} Y_{i3} + \sum_{i=1}^{4} Y_{i4}.$$

PROBLEMS

1. (a) Determine the median amount of U.S. grants for economic assistance (column 3 of Table 2.1) and the median amount of grants from 1960 to 1965 for military assistance (column 6 of Table 2.1). (b) Calculate the mean amount of grants for economic assistance and the mean amount of grants for military assistance.

2. (a) Are the data on U.S. loans for economic assistance (column 4 of Table 2.1) positively skewed or negatively skewed? (b) Calculate $D(+)$ and $D(-)$. (c) Which is larger—the mean U.S. loan for economic assistance or the median loan?

3. (a) Group the data in column 4 of Table 2.1 into six classes of $250 million from zero to $1,500 million. (b) Calculate the weighted mean. (c) How does the weighted mean compare with the mean computed from the raw data?

4. A high school physics class contains one genius. On the final examination, four students score 25, four score 40, four score 30, and the genius scores 100. Calculate the mean and the median.

5. Given the following sets of grades by a class of 40:

GRADES	WEIGHTS	PROPORTIONAL WEIGHTS	PERCENTAGE WEIGHTS
50	10	.25	25
60	10	.25	25
75	20	.50	50

Calculate the weighted mean grade, using proportional weights, percentage weights, and weights. Are they the same? Why?

6. (a) What is the geometric mean of the following data: 2, 4, 3, 6? (b) If a manufacturing plant doubles its output every year for 36 years, what is the geometric mean of its yearly growth?

7. If one buys $500 worth of stock at $44/share and $500 worth of stock at $24/share, what is the average cost of his combined holdings per share?

8. (a) Calculate the variance in per capita gross domestic product for the countries listed in Table 2.2. (b) What is the standard deviation?

9. Assume each Latin American country in Table 2.2 was given $500 million in economic assistance by the United States. Determine the range and the mean deviation of per capita assistance.

10. Determine the interquartile range of the populations of the Latin American countries listed in Table 2.2.

11. Assume that the population of a country has grown geometrically. In 1900 it was 40 million, and in 1960 it was 80 million. What was the population in 1930?

Chapter 3

Simple Regression and Correlation

Often there are strong associations between two or more sets of data. For example, the exports of a country each year may be closely related to national income for that year. The income of a parent may bear a close association to the future income of a child. The thickness of the earth's crust at any point is related to elevation above sea level. The imports of a country may be related to both total income and investment. Air travel to and from an urban center is associated with both population and average income level in that center.

When two or more sets of data are closely related, one often wants to know both the form of the association or relationship *and* the strength of that relationship. The measurement of the form and strength of relationships between variables is called *regression* and *correlation* analysis.

Simple Linear Regression

Suppose there are two sets of data $Y_1, Y_2, \ldots, Y_n$ and $X_1, X_2, \ldots, X_n$. One often assumes the variables are related in a linear fashion

$$Y = a + b \cdot X, \tag{3.1}$$

where a is the intercept on the vertical or Y-axis and b is the slope of the line, as shown in Fig. 3.1.

Example 3.1. Table 3.1 contains data on the total exports (column 2) and the Gross Domestic Product (column 4) of Kenya from 1955 to 1964. The two magnitudes are obviously very closely related. By methods which we will describe below, one may obtain the following linear relationship:

$$Y = -22.6 + .278 \cdot X \qquad (a = -22.6; b = .278), \tag{3.2}$$

24

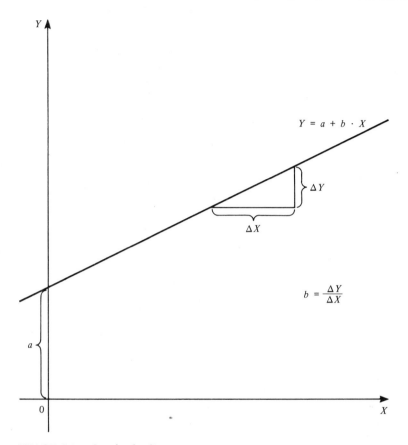

FIGURE 3.1   Graph of a Line

where Y stands for exports and X stands for Gross Domestic Product. This relationship is called an estimated linear equation or linear regression. For any given value of Gross Domestic Product, say for 1960, an estimated level of exports is given by substituting the 1960 Gross Domestic Product into equation (3.2).

$$Y = -22.6 + (.278)(225.5) = 40.1. \tag{3.3}$$

This value of Y is an estimate, because it is not the exact value of exports for 1960, but it is a good approximation. The actual value of exports from Table 3.1 is £40.2 million, while the estimated value from (3.2) is £40.1 million.

The intercept in equation (3.2) indicates that when Gross Domestic Product is projected back to zero, the estimated level of exports is equal

TABLE 3.1   Kenya exports, imports, gross capital formation, and Gross Domestic Product, 1955-1964 (£ million)

	IMPORTS (1)	EXPORTS (2)	GROSS CAPITAL FORMATION (3)	GROSS DOMESTIC PRODUCT (4)
1955	71.5	28.0	43.8	180.5
1956	69.8	33.0	45.7	193.2
1957	72.0	31.2	45.6	205.9
1958	60.9	33.2	40.0	208.1
1959	61.5	38.4	40.3	214.8
1960	70.1	40.2	41.4	225.5
1961	68.9	41.7	31.9	224.7
1962	69.5	45.1	33.3	244.1
1963	73.7	51.0	30.4	259.1
1964	76.6	53.5	32.8	277.7

Source: Republic of Kenya Statistical Abstract 1965, Table 33(a), p. 22; Table 130, p. 102; and Table 129(a), p. 101.

to the intercept. The slope of equation (3.2) implies that for each £1-million increase in Gross Domestic Product, the estimated increase in exports is £.278 million.

The data do not normally give rise to an exact linear relationship. For a given value X_i, the value of the Y variable given by an estimated linear relationship will generally differ from the actual value Y_i by an amount u_i, called a residual. This may be expressed by writing

$$Y_i = a + bX_i + u_i \tag{3.4}$$

or

$$u_i = Y_i - (a + bX_i) = Y_i - \hat{Y}_i, \tag{3.5}$$

where Y_i is the actual value of the Y variable and $(a + bX_i) = \hat{Y}_i$ is the estimated value of the Y variable.

Example 3.2.   Given the estimated relationship (3.2) from the previous example, the predicted value of Y for 1960 is given by (3.3). This differs from the actual value Y_i for 1960 by an amount $u_i = 40.2 - 40.1 = 0.1$.

The error term u_i for each year may be shown graphically by means of a scatter diagram as shown in Fig. 3.2. Each pair of values (X_i, Y_i) is plotted on this diagram to obtain a "scatter" of points. The line (3.2) is plotted as a solid line on this diagram. The error term u_i for each point

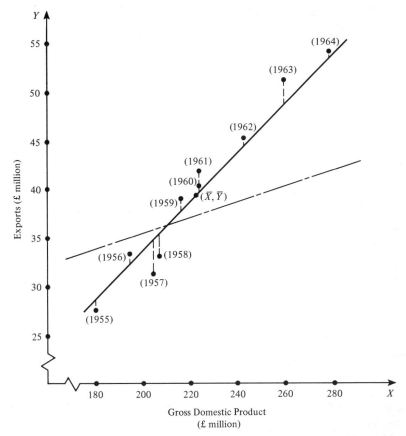

FIGURE 3.2 Error Terms in Regression of Kenya Exports on Gross Domestic Product

(X_i, Y_i) is the vertical distance between the point and the solid line. These distances are shown by the dotted vertical lines in Fig. 3.2.

One way of determining the estimated linear relationship between two variables Y and X is to find a line that minimizes the sum of the squared residuals—that is, minimizes the sum of the squared vertical distances between each point on a scatter diagram and the line of relationship. A line that satisfies this property is called a *least-squares regression line*. The sum of the squared residuals is denoted by S and is determined by the formula

$$S = \sum_{i=1}^{n} u_i^2 = \sum_{i=1}^{n} (Y_i - a - bX_i)^2 = \sum_{i=1}^{n} (Y_i - \hat{Y}_i)^2. \tag{3.6}$$

Example 3.3. Note that in Fig. 3.2 there are two lines. The equation for the solid line is given by (3.2). The equation for the dashed line is

$$Y = -5.0 + .200 \cdot X. \tag{3.7}$$

Let us compare the sum of the squared error terms that one gets from (3.2) as opposed to (3.7). For (3.7), we have:

Y_i	X_i	$b \cdot X_i$ ($b = .200$)	$\hat{Y}_i = a + bX_i$ ($a = -5.0$)	$Y_i - \hat{Y}_i$	$(Y_i - \hat{Y}_i)^2$
28.0	180.5	36.10	31.10	−3.10	9.61
33.0	193.2	38.64	33.64	−0.64	0.41
31.2	205.9	41.18	36.18	−4.98	24.80
33.2	208.1	41.62	36.62	−3.42	11.70
38.4	214.8	42.96	37.96	0.44	0.19
40.2	225.5	45.10	40.10	0.10	0.01
41.7	224.7	44.94	39.94	1.76	3.10
45.1	244.1	48.82	43.82	1.28	1.64
51.0	259.1	51.82	46.82	4.18	17.47
53.5	277.7	55.54	50.54	2.96	8.76
					77.69

Taking the sum of the entries in the last column, we get

$S = 77.69$

as the sum of the squared error terms for equation (3.7).
    For equation (3.2), the calculations are as follows:

$b \cdot X_i$ ($b = .278$)	$\hat{Y}_i = a + bX_i$ ($a = -22.6$)	$Y_i - \hat{Y}_i$	$(Y_i - \hat{Y}_i)^2$
50.18	27.58	.42	.18
53.71	31.11	1.89	3.57
57.24	34.64	−3.44	11.83
57.85	35.25	−2.05	4.20
59.71	37.11	1.29	1.66
62.69	40.09	0.11	.01
62.47	39.87	1.83	3.35
67.86	45.26	−.16	.03
72.03	49.43	1.57	2.46
77.20	54.60	−1.10	1.21
			28.50

Taking the sum of the entries in the last column, we obtain

$$S = 28.50$$

for the sum of the squared error terms of equation (3.2). From the least-squares criterion, equation (3.2) is much more satisfactory than equation (3.7).

The problem then is to find a value for a, the intercept, and b, the slope, that minimizes S in (3.6) when the values of the data

$$Y_1, Y_2, \ldots, Y_n \quad \text{and} \quad X_1, X_2, \ldots, X_n$$

are used. It is well known that these values for a and b are obtained by solving the following linear equations for a and b (see Appendix 2 to this chapter for a derivation):

$$a + \frac{1}{n}\left(\sum_{i=1}^{n} X_i\right)\cdot b = \frac{1}{n}\left(\sum_{i=1}^{n} Y_i\right),$$

$$\frac{1}{n}\left(\sum_{i=1}^{n} X_i\right)\cdot a + \frac{1}{n}\left(\sum_{i=1}^{n} X_i^2\right)\cdot b = \frac{1}{n}\left(\sum_{i=1}^{n} Y_i X_i\right).$$

(3.8)

These equations are called the *normal* equations. The solution to this set of equations can be obtained by using the method of determinants. First, however, let us rewrite equations (2.19) using a simpler notation:

$$a\cdot m_{00} + b\cdot m_{01} = m_{0y},$$
$$a\cdot m_{10} + b\cdot m_{11} = m_{1y},$$

(3.9)

where we have set

$$m_{00} = 1,$$

$$m_{01} = m_{10} = \frac{1}{n}\left(\sum_{i=1}^{n} X_i\right) = \bar{X},$$

$$m_{11} = \frac{1}{n}\left(\sum_{i=1}^{n} X_i^2\right),$$

(3.10)

$$m_{0y} = \frac{1}{n}\left(\sum_{i=1}^{n} Y_i\right) = \bar{Y},$$

$$m_{1y} = \frac{1}{n}\left(\sum_{i=1}^{n} Y_i X_i\right).$$

Now let us apply the method of determinants (see Appendix 1 to this chapter) to the form (3.9) of the normal equations.

$$a = \frac{\begin{vmatrix} m_{0y} & m_{01} \\ m_{1y} & m_{11} \end{vmatrix}}{\begin{vmatrix} m_{00} & m_{01} \\ m_{10} & m_{11} \end{vmatrix}} = \frac{m_{0y}m_{11} - m_{01}m_{1y}}{m_{00}m_{11} - m_{01}m_{10}},$$

$$b = \frac{\begin{vmatrix} m_{00} & m_{0y} \\ m_{10} & m_{1y} \end{vmatrix}}{\begin{vmatrix} m_{00} & m_{01} \\ m_{10} & m_{11} \end{vmatrix}} = \frac{m_{0y}m_{11} - m_{01}m_{1y}}{m_{00}m_{11} - m_{01}m_{10}}.$$

$$(3.11)$$

Equations (3.11) give the values for a and b that minimize S in (3.6).

One important thing to note about a least-squares line is that it passes through the point of means. That is, if we plot the point $(\bar{X}, \bar{Y})$ on a graph and draw the least-squares line, it will go through this point. This can be seen from the first normal equation in (3.8). Since

$$\bar{X} = \frac{1}{n}\left(\sum_{i=1}^{n} X_i\right) \quad \text{and} \quad \bar{Y} = \frac{1}{n}\left(\sum_{i=1}^{n} Y_i\right),$$

the first normal equation can be written

$$a + b\bar{X} = \bar{Y}. \tag{3.12}$$

Example 3.4. Let us calculate the least-squares regression line using Kenya exports as the Y variable and Gross Domestic Product as the X variable.

Y_i	X_i	$(X_i)^2$	$X_i \cdot Y_i$
28.0	180.5	32,580.25	5,054.00
33.0	193.2	37,326.24	6,375.60
31.2	205.9	42,394.81	6,424.08
33.2	208.1	43,305.61	6,908.92
38.4	214.8	46,139.04	8,248.32
40.2	225.5	50,850.25	9,065.10
41.7	224.7	50,490.09	9,369.99
45.1	244.1	59,584.81	11,008.91
51.0	259.1	67,132.81	13,214.10
53.5	277.7	77,117.29	14,856.95
395.3	2,233.6	506,921.20	90,525.97

Next add each column to get

$$\sum_{i=1}^{10} Y_i = 395.3,$$

$$\sum_{i=1}^{10} X_i = 2,233.6,$$

$$\sum_{i=1}^{10} X_i^2 = 506,921.20,$$

$$\sum_{i=1}^{10} X_i \cdot Y_i = 90,525.97.$$

Using (3.10), we calculate

$m_{00} = 1,$

$m_{01} = m_{10} = \frac{1}{10}(2,233.6) = 223.36 = \bar{X},$

$m_{11} = \frac{1}{10}(506,921.20) = 50,692.12,$

$m_{0y} = \frac{1}{10}(395.3) = 39.53 = \bar{Y},$

$m_{1y} = \frac{1}{10}(90,525.97) = 9,052.60.$

Next use (3.11) to calculate a and b:

$$a = \frac{\begin{vmatrix} 39.53 & 223.36 \\ 9,052.60 & 50,692.12 \end{vmatrix}}{\begin{vmatrix} 1 & 223.36 \\ 223.36 & 50,692.12 \end{vmatrix}} = \frac{(39.53)(50,692.12) - (223.36)(9,052.60)}{50,692.12 - (223.36)^2},$$

$$b = \frac{\begin{vmatrix} 1 & 39.53 \\ 223.36 & 9,052.60 \end{vmatrix}}{\begin{vmatrix} 1 & 223.36 \\ 223.36 & 50,692.12 \end{vmatrix}} = \frac{9,052.60 - (39.53)(223.36)}{50,692.12 - (223.36)^2}$$

or

$$a = \frac{-18,129.2324}{802.43} = -22.6,$$

$$b = \frac{223.18}{802.43} = .278.$$

Thus the least-squares regression line is

$$Y = -22.6 + .278X.$$

This is the same as the linear relationship (3.2) that we discussed in Example 3.1. Note that if we plot the point $(\bar{X}, \bar{Y}) = (224.16, 39.53)$ in Fig. 3.2, it lies on the least-squares regression line—that is, the solid line in Fig. 3.2.

It is rarely most advantageous to compute the coefficients or parameters a and b using equations (3.10) and (3.11). An easier method of calculation is to determine b first by the equation (see Appendix 2 for a derivation):

$$b = \frac{\sum_{i=1}^{n} X_i Y_i - n \cdot \bar{X} \cdot \bar{Y}}{\sum_{i=1}^{n} X_i^2 - n \cdot (\bar{X})^2} \tag{3.13}$$

and then substitute this value into the following equation [derived from (3.12)] to obtain the value of a:

$$a = \bar{Y} - b\bar{X}. \tag{3.14}$$

Example 3.5.   Let us calculate the regression line from Example 3.4 using the formulae (3.13) and (3.14). From Example 3.4, we have

$$n \cdot \bar{X} \cdot \bar{Y} = 10(223.36)(39.53) = 88{,}294.21,$$
$$n \cdot (\bar{X})^2 = 10(223.36)^2 = 498{,}896.90,$$
$$\Sigma X_i \cdot Y_i = 90{,}525.97$$
$$\Sigma X_i^2 = 506{,}921.20,$$
$$b = \frac{90{,}525.97 - 88{,}294.21}{506{,}921.20 - 498{,}896.90} = .278,$$
$$a = \bar{Y} - b\bar{X} = 39.53 - (.278)(223.36),$$
$$a = 39.53 - 62.09 = -22.6.$$

The answer is the same as that obtained in Example 3.4 but the calculations are less involved.

Simple Linear Correlation

So far we have concerned ourselves only with the exact *form* of a linear association between two variables Y and X. The least-squares principle was used to find the linear relationship. The strength of a relationship or association can, however, be inferred from the calculation of a least-squares line. The reasoning is that the better the scatter points "fit" the least-squares regression line, the stronger is the relationship between two variables. Three commonly used statistics measure the "goodness of fit": the *standard error of the estimate*, the *coefficient of determination*, and the *correlation coefficient*. As one might expect, all are closely related to the sum of the squared error terms.

The *standard error of the estimate* is determined by calculating the actual sum of the squared error terms, dividing by n, the total number of pieces of data, and taking the square root. (The number n is often called the number of observations.) That is, one first calculates the slope b and the intercept a. Let $\hat{a}$ and $\hat{b}$ stand for these estimated values of a and b. Then, substituting these estimated values of a and b into (3.5), we get an estimated residual u_i for each of the n observations.

$$u_i = Y_i - (\hat{a} + \hat{b} \cdot X_i) = Y_i - \hat{Y}_i \qquad \text{for } i = 1, \ldots, n. \qquad (3.15)$$

The standard error of the estimate is denoted by s_u, where s_u^2 is given by the formula

$$s_u^2 = \frac{\sum_{i=1}^{n} (u_i)^2}{n} = \frac{\sum [Y_i - (\hat{a} + \hat{b} \cdot X_i)]^2}{n} = \frac{\sum_{i=1}^{n} (Y_i - \hat{Y}_i)^2}{n}. \qquad (3.16)$$

Example 3.6.   The sum of the squared error terms for the least-squares regression line (3.2) was calculated in Example 3.3 as the sum $S = 28.50$. The standard error of the estimate is

$$\sqrt{\tfrac{1}{10}(28.50)} = \sqrt{2.850} = 1.69 \quad \text{or} \quad \pounds 1.69 \text{ million.}$$

Note that in Table 3.1 exports and Gross Domestic Product were measured in £ million. If we measure exports, the Y variable, in terms of £ thousand, then the standard error of estimate changes and would be the square root of £1690 thousand. On the other hand, change in the units of measurement of the X variable, Gross Domestic Product, while it changes the units of a and b, does not affect the units of the standard error of the estimate.

In the example above we noted that the standard error of the estimate is a measure of goodness of fit that depends on the units of measurement of the Y variable. The *coefficient of determination* does not depend on any units of measurement. The rationale behind this statistic is that the variance of the variable Y can be divided into two parts: the *explained variation* and the *unexplained variation*. From Chapter 2 [equation (2.10)], we can write the variance of the Y variable, s_Y^2, as follows:

$$s_Y^2 = \frac{\sum_{i=1}^{n} (Y_i - \bar{Y})^2}{n}.$$

The variance is the sum of the squared deviations about the mean value of Y divided by n. This may be viewed as the *total variation* of the variable Y. The standard error of the estimate as given by (3.16) is the sum of the squared deviations "about the regression line." We may view this as the *unexplained variation* of the variable Y—that is, the variation of Y left unexplained by the regression line. In Fig. 3.3 we have drawn a regression line and a scatter of points. The vertical distance is drawn from each

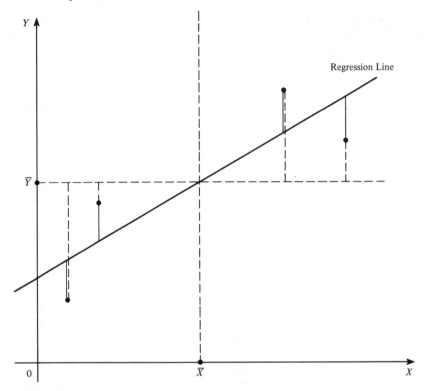

FIGURE 3.3   **Explained and Unexplained Variation from Regression Line**

point to the regression line by means of a solid line and to the horizontal line $Y = \bar{Y}$ by means of a dotted line. The sum of squared solid-line differences is the unexplained variation. The sum of the squared dotted-line distances is the total variation. The total variation is always larger than the unexplained variation for a least-squares regression line, since by definition the regression line minimizes the unexplained variation. The *explained variation* of the variable Y is the total variation s_Y^2 less the unexplained variation s_u^2:

$$\begin{bmatrix} \text{explained} \\ \text{variation} \end{bmatrix} \text{ equals } \begin{bmatrix} \text{total variation } (s_Y^2) \\ \text{minus} \\ \text{unexplained variation} \\ (s_u^2) \end{bmatrix}.$$

The coefficient of determination (R^2) is the explained variation divided by the total variation.

$$R^2 = \frac{s_Y^2 - s_u^2}{s_Y^2} = \frac{\text{explained variation}}{\text{total variation}}. \tag{3.17}$$

The coefficient of determination may vary from zero (no explained variation) to one for a "perfect fit" (all the variation of Y explained by the regression line). The coefficient of determination R^2 may be interpreted as the percentage of variation of the variable Y "explained" by the regression line.

The formula (3.17) is rarely used in actual computation. Two alternative formulae are the following:

$$R^2 = \frac{\hat{b}\left(\sum_{i=1}^{n} X_i Y_i - n \cdot \bar{X} \cdot \bar{Y}\right)}{\left[\sum_{i=1}^{n} Y_i^2 - n(\bar{Y})^2\right]}, \tag{3.18}$$

$$R^2 = \frac{\left(\sum_{i=1}^{n} X_i Y_i - n \cdot \bar{X} \cdot \bar{Y}\right)^2}{\left[\sum_{i=1}^{n} X_i^2 - n \cdot (\bar{X})^2\right]\left[\sum_{i=1}^{n} Y_i^2 - n \cdot (\bar{Y})^2\right]}. \tag{3.19}$$

These formulae are equivalent to (3.17) but often easier to compute (see Appendix 2 for a derivation).

Example 3.7.   In Example 3.6 we calculated s_u, the standard error of the estimate, to be

$$s_u = 1.69 \quad \text{and} \quad s_u^2 = 2.850.$$

Using Kenya exports in Table 3.1 as the Y variable, we can calculate

$$\tfrac{1}{10}\left[\sum_{i=1}^{10} (Y_i - \bar{Y})^2\right] = \tfrac{1}{10}(649.22) = 64.922.$$

The coefficient of determination is

$$R^2 = \frac{64.922 - 2.850}{64.922} = \frac{62.072}{64.922},$$

$$R^2 = .956.$$

That is, about 95.6 percent of the variation in exports is "explained" by variations in Gross Domestic Product.

In addition to the standard error of the estimate (3.16) and the coefficient of determination (3.17), the *correlation coefficient* R is used as a measure of goodness of fit or strength of a relationship between two variables. The correlation coefficient R is the square root of the coefficient of determination R^2; that is,

$$R = \frac{\sum_{i=1}^{n} X_i \cdot Y_i - n \cdot \bar{X} \cdot \bar{Y}}{\sqrt{\left[\sum_{i=1}^{n} X_i^2 - n(\bar{X})^2\right]\left[\sum_{i=1}^{n} Y_i^2 - n(\bar{Y})^2\right]}}. \tag{3.20}$$

The correlation coefficient differs from the coefficient of determination in that it can be positive or negative. If the correlation coefficient is negative, it indicates an inverse relationship between the variables Y and X. That is, as Y goes down, X tends to go up, and vice versa. (This inverse relationship is also indicated by a negative slope b of the estimated regression line.) The correlation coefficient can vary from -1 to $+1$. If $R = -1$, then the regression line fits perfectly and has a negative slope. If $R = +1$, then the regression line fits perfectly and the slope b is positive. If $R = 0$, then there is no correlation or association between the variables Y and X.

Example 3.8.   Let us consider the data presented in Table 2.1 of the last chapter. Let the Y variable be total U.S. economic assistance, column 2 of the table, and the X variable be total U.S. military assistance, column 5. Using these data, we can calculate

$$\sum_{i=1}^{17} X_i^2 = 74{,}183{,}577,$$

$$\sum_{i=1}^{17} Y_i^2 = 100{,}232{,}572,$$

$$\sum_{i=1}^{17} X_i Y_i = 60{,}853{,}577,$$

$$\bar{X} = 1{,}824.2,$$

$$\bar{Y} = 2{,}280.6,$$

$$n \cdot \bar{X} \cdot \bar{Y} = 17(1{,}824.2)(2{,}280.6) = 70{,}724{,}599,$$

$$n \cdot (\bar{X})^2 = 56{,}570{,}996,$$

$$n \cdot (\bar{Y})^2 = 88{,}419{,}318,$$

$$\Sigma X_i Y_i - n \cdot \bar{X} \cdot \bar{Y} = -9{,}871{,}022,$$

$$\Sigma X_i^2 - n \cdot (\bar{X})^2 = 17{,}612{,}581,$$

$$\Sigma Y_i^2 - n \cdot (\bar{Y})^2 = 11{,}813{,}254.$$

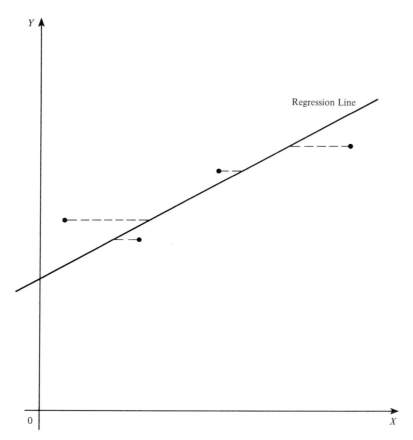

FIGURE 3.4   Horizontal Deviations from Regression Line

Using equation (3.20), we get

$$R = \frac{-9,871,022}{\sqrt{(17,612,581)(11,813,254)}} = \frac{-9,871}{14,426} = -.684.$$

This is a high negative correlation between the amount of economic assistance given by the United States and the amount of military assistance. That is, there is a fairly strong tendency for economic aid to go down when military aid goes up and vice versa.

Linear Regression—Variables Reversed

The curious reader at this point may well ask, why does one minimize the vertical sum of squared distances (see Fig. 3.2) in order to obtain an estimated linear relationship between two variables? One might just as well minimize the sum of the horizontal squared distances (see Fig. 3.4) or the sum of a combination of horizontal and vertical squared distances. Furthermore, why minimize the sum of the *squared* distances? Why not minimize the sum of the distances themselves or the sum of the cubed distances? There is no clear answer to these questions unless one makes some specific assumptions about the nature of the variables Y_i and X_i. These will be discussed in later chapters. At this point, however, we shall consider two other commonly used methods of obtaining a linear relationship, without discussing the rationale for their use.

First, let us consider the method by which one minimizes the horizontal sum of squares as shown in Fig. 3.4, which is a scatter diagram with the horizontal distances marked by dotted lines. To minimize the sum of the squared horizontal distances, we simply reverse the roles played by Y and X in minimizing the sum of the squared vertical distances. Thus the regression equation is

$$X = a^* + b^* \cdot Y, \tag{3.21}$$

where the asterisks are used to distinguish the slope b^* and intercept a^* from a and b, the intercept and slope in equation (3.1). Solve (3.21) for Y in terms of X.

$$Y = \left(\frac{-a^*}{b^*}\right) + \left(\frac{1}{b^*}\right) X. \tag{3.22}$$

To obtain the form (3.1), we set

$$a = \frac{-a^*}{b^*} \quad \text{and} \quad b = \frac{1}{b^*}. \tag{3.23}$$

To determine a^* and b^*, we can reverse the roles of Y and X in (3.8) and use the normal equations

$$a^* + \frac{1}{n}\left(\sum_{i=1}^{n} Y_i\right) \cdot b^* = \frac{1}{n}\left(\sum_{i=1}^{n} X_i\right),$$

$$\frac{1}{n}\left(\sum_{i=1}^{n} Y_i\right) \cdot a^* + \frac{1}{n}\left(\sum_{i=1}^{n} Y_i^2\right) \cdot b^* = \frac{1}{n}\left(\sum_{i=1}^{n} Y_i \cdot X_i\right). \tag{3.24}$$

We can use the method of determinants to solve (3.24) for a^* and b^*. Alternatively, one may use the following formula to solve for b^*.

$$b^* = \frac{\sum_{i=1}^{n} Y_i \cdot X_i - n \cdot \bar{Y} \cdot \bar{X}}{\sum_{i=1}^{n} Y_i^2 - n \cdot (\bar{Y})^2} .$$

(3.25)

Then a^* may be solved by substituting the value of b^* obtained from (3.25) into

$$a^* b^* - \bar{X} = \cdot \bar{Y}.$$

(3.26)

The equations (3.25) and (3.26) are the same as (3.13) and (3.14) with the roles of X and Y interchanged.

Finally, the values of a and b that minimize the horizontal sum of squares can be determined by substituting the values obtained for a^* and b^* into (3.23).

Example 3.9.   Let us recalculate the regression in Example 3.5, this time however, reversing the roles of Y and X. We can calculate

$n \cdot \bar{X} \cdot \bar{Y} = 88{,}294.21,$

$n \cdot (\bar{Y})^2 = 10(39.53)^2 = 15{,}626.209,$

$\Sigma \, X_i Y_i = 90{,}525.97,$

$\Sigma \, Y_i^2 = 16{,}275.43.$

Using (3.25), we get

$$b^* = \frac{90{,}525.97 - 88{,}294.21}{16{,}275.43 - 15{,}626.209} = \frac{2{,}231.76}{649.22} = 3.438,$$

$a^* = \bar{X} - b^* \cdot \bar{Y} = 223.36 - (3.438)(39.53) = 223.36 - 135.90 = 87.46.$

To convert to least-squares estimates of a and b, we use (3.23)

$$a = \frac{-a^*}{b^*} = \frac{-87.46}{3.438} = -25.44,$$

$$b = \frac{1}{b^*} = \frac{1}{3.438} = .291.$$

Note that the value of b obtained by minimizing the horizontal sum of squares is larger than the b obtained in minimizing the vertical sum of squares in Examples 3.4 and 3.5. On the other hand, the value of a is smaller.

The goodness of fit of a regression line that minimizes the horizontal sum of squared distances may also be measured by a standard error of the estimate, a coefficient of determination, and a correlation coefficient. These measures are defined as in the case of minimizing the vertical sum of squares merely by reversing the roles of Y and X. The standard error of the estimate, however, depends on the units in which we measure the X variable. In the vertical-sum-of-squares case, the standard error of estimate depends on the units of measurement of the Y variable.

The coefficient of determination and the correlation coefficient are exactly the same whether one minimizes the vertical or horizontal sum of squares. This may be seen by examination of the formulae in (3.19) and (3.20), where reversing the roles of Y and X does not change the value of R^2 or R. One interesting thing to note is that if one multiplies the value of b obtained from minimizing the vertical sum of squares [equation (3.13)] by the value of b^* obtained by minimizing the horizontal sum of squares [equation (3.25)], one obtains the coefficient of determination [equation (3.19)]. That is,

$$R^2 = b \cdot b^*, \tag{3.27}$$

where b is given by (3.13), b^* by (3.25), and R^2 by (3.19). Since $b = 1/b^*$ is the slope of the regression line when minimizing the horizontal sum of squares, we see that if $R^2 = 1$, then $b' \cdot 1/b'' = 1$ or $b' = b''$, where b' is the slope of the regression line when minimizing the vertical sum of squares and b'' is the slope when minimizing the horizontal sum of squares. In other words, we obtain the same regression line when the fit is perfect.

Example 3.10.   Let us multiply the value of b obtained in Example 3.5 by the value of b^* in Example 3.9.

$b = .278,$
$b^* = 3.438,$
$R^2 = b \cdot b^* = .956.$

This is the same as the value of R^2 obtained in Example 3.7, which was computed by using equation (3.17).

Orthogonal Regression

We have seen how one can minimize either the vertical or horizontal sum of squares. No rationale was presented for preferring one method over the other. The absolute value of the slope b determined by minimizing the vertical sum of squares is usually smaller than the absolute value of b determined by minimizing the horizontal sum of squares. Instead of minimizing the vertical or horizontal sum of squares, one could minimize the sum of squares at an angle to the regression line (see Fig. 3.5). The result of minimizing an angular sum of squares is a coefficient b that lies

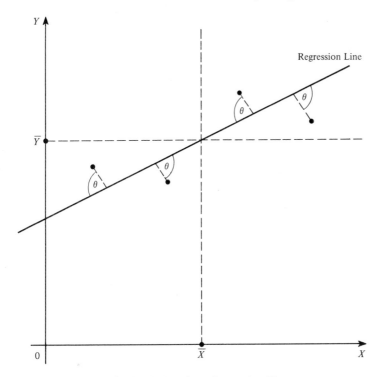

FIGURE 3.5   Angular Deviations from Regression Line

between the two b's produced by minimizing the vertical and horizontal sum of squares. Thus minimizing the vertical or horizontal sum of squares can be viewed as limiting extreme cases of minimizing the sum of squared distances from scatter points to a line of regression.

It might seem natural to compromise between the two extremes of horizontal and vertical distances and minimize the sum of the squared distances between each scatter point and a line drawn perpendicular to the line of regression. (In Fig. 3.5 the dotted line from the scatter point to the line of regression is perpendicular when the angle θ is 90 degrees or a right angle.) This is called orthogonal regression.

The algebra necessary to derive the coefficients a and b from an orthogonal regression becomes quite messy. The computation of orthogonal regression coefficients is more difficult than in the horizontal or vertical case. The coefficients a and b that minimize the orthogonal sum of squares are determined by first solving for b using the following formulae.

$$b = \frac{B_1 + \sqrt{B_1^2 + 4B_2^2}}{2B_2},$$ (3.28)

where

$$B_1 = \Sigma\, Y_i^2 - \Sigma\, X_i^2 - n[(\bar{Y})^2 - (\bar{X})^2],$$
$$B_2 = \Sigma\, Y_i X_i - n \cdot \bar{Y} \cdot \bar{X}.$$

Once b is determined using (3.28), we determine a, the intercept, according to (3.29):

$$a = \bar{Y} - b \cdot \bar{X}. \tag{3.29}$$

Example 3.11. Let us calculate the orthogonal regression for Kenya exports (Y) and Gross Domestic Product (X). Here

$$\Sigma\, Y_i^2 - n \cdot (\bar{Y})^2 = 649.22,$$
$$\Sigma\, X_i^2 - n \cdot (\bar{X})^2 = 8{,}024.30,$$
$$B_2 = \Sigma\, Y_i X_i - n \cdot \bar{Y} \cdot \bar{X} = 2{,}231.76,$$
$$B_1 = 649.22 - 8{,}024.30 = -7{,}375.08,$$
$$B_1^2 = 54{,}391{,}805,$$
$$B_2^2 = (2{,}231.76)^2 = 4{,}980{,}753,$$
$$\sqrt{B_1^2 + 4B_2^2} = \sqrt{74{,}314{,}817} = 8{,}621.$$

Thus from (3.28)

$$b = \frac{-7{,}375 + 8{,}621}{2(2{,}232)} = \frac{1{,}246}{4{,}464},$$
$$b = .279.$$

We see that the value of b obtained from minimizing the orthogonal sum of squares lies between the values $b = .278$ and $b = .291$ obtained by minimizing the vertical and horizontal sum of squares. The value of a is, from (3.29),

$$a = \bar{Y} - b \cdot \bar{X} = 39.53 - (.279)(223.36),$$
$$a = -22.8.$$

Note that as in the case of horizontal and vertical regression, the regression line passes through the point of means $(\bar{X}, \bar{Y})$.

Conclusion

In this chapter we showed how data on two variables could be used to determine a linear relationship between the variables by minimizing the vertical, horizontal, or angular sum of squared residuals. A single measure of the association between two variables is the correlation coefficient. The square of the correlation coefficient, the coefficient of determination, is a measure of the strength of a linear relationship between two variables.

The correlation coefficient or coefficient of determination is independent of whether one minimizes the vertical, horizontal, or angular sum of squared residuals.

In the next chapter we consider various extensions of simple regression and correlation analysis to cover relationships among several variables and nonlinear relationships among variables.

APPENDIX 1 Solving Systems of Linear Equations

Consider the following system of linear equations:

$$
\begin{aligned}
x_{00}a_0 + x_{01}a_1 + \cdots + x_{0n}a_n &= y_0 \\
x_{10}a_0 + x_{11}a_1 + \cdots + x_{1n}a_n &= y_1 \\
x_{20}a_0 + x_{21}a_1 + \cdots + x_{2n}a_n &= y_2 \\
&\;\;\vdots \\
x_{n0}a_0 + x_{n1}a_1 + \cdots + x_{nn}a_n &= y_n
\end{aligned}
\tag{3.30}
$$

The x's and y's are assumed to be constants and the a's are unknowns. There are $n + 1$ equations and $n + 1$ unknowns $a_0, a_1, a_2, \ldots, a_n$. In this appendix we shall consider methods of solving the linear equations to determine the value of the unknowns. We shall assume that there is a unique set of values for $a_0, a_1, a_2, \ldots, a_n$ that solve these equations.

Matrices

A matrix is any array of numbers in rows and columns. A matrix can be illustrated best by some examples.

Example 1

$$
\begin{bmatrix} 5 & 3 & 1 \\ 4 & 2 & 8 \end{bmatrix}
$$

This is a matrix with two rows and three columns. The first row is (5, 3, 1). The second row is (4, 2, 8). The second column is (3, 2).

Example 2

$$
\begin{bmatrix} 8 & 4 & 19.3 & 2.7 \end{bmatrix}
$$

This matrix has one row and four columns.

Example 3

$$
\begin{bmatrix} 3 \\ 2 \\ 1 \end{bmatrix}
$$

This is a matrix with three rows and one column.

Example 4

$$X = \begin{bmatrix} x_{00} & x_{01} & x_{02} & \cdots & x_{0n} \\ x_{10} & x_{11} & x_{12} & \cdots & x_{1n} \\ \cdot & \cdot & \cdot & & \cdot \\ \cdot & \cdot & \cdot & & \cdot \\ \cdot & \cdot & \cdot & & \cdot \\ x_{k0} & x_{k1} & x_{k2} & \cdots & x_{kn} \end{bmatrix}$$

X is a matrix with $k + 1$ columns and $n + 1$ rows.

Example 5

$$C = \begin{bmatrix} c_{11} & c_{12} & \cdots & c_{1n} \\ c_{21} & c_{22} & \cdots & c_{2n} \\ \cdot & \cdot & & \cdot \\ \cdot & \cdot & & \cdot \\ \cdot & \cdot & & \cdot \\ c_{n1} & c_{n2} & \cdots & c_{nn} \end{bmatrix}$$

C is a matrix with n rows and n columns. A matrix with the same number of rows as columns is called a *square matrix*. The number of rows or columns n is called the *order* of the square matrix.

Determinants

Every square matrix has a determinant. We can define the determinant of a square matrix by what is called the recursive method of definition. For a matrix of order one, the determinant is written:

$$\det |c| = |c| = c. \tag{3.31}$$

Thus for the matrix $|10.2|$ of order one (one row and one column)

$$\det |10.2| = |10.2| = 10.2.$$

For a matrix of order n, the determinant is

$$\det C = \det \begin{bmatrix} c_{11} & c_{12} & \cdots & c_{1n} \\ c_{21} & c_{22} & \cdots & c_{2n} \\ \cdot & \cdot & & \cdot \\ \cdot & \cdot & & \cdot \\ \cdot & \cdot & & \cdot \\ c_{n1} & c_{n2} & \cdots & c_{nn} \end{bmatrix}$$

$$= c_{11}C_{11} - c_{21}C_{21} + \cdots + (-1)^{n+1}c_{n1}C_{n1}. \tag{3.32}$$

C_{11} is the determinant of the square matrix formed by striking out the first row and first column of C. C_{21} is the determinant of the square matrix formed by striking out the second row and first column of C. Each of the terms in (3.32) alternate in sign. If n is odd, then the last term is positive. That is, $(-1)^{n+1} = +1$ for n odd. For example, if $n = 3$, then

$$
\begin{aligned}
(-1)^{n+1} &= (-1)^4 \\
&= (-1)(-1)[(-1)(-1)] \\
&= [(-1)(-1)](+1) = (+1)(+1) = (+1).
\end{aligned}
$$

For n even, the last term is negative. For example, if $n = 4$, then $(-1)^5 = -1$.

Let us evaluate the determinant of

$$
C = \begin{bmatrix} c_{11} & c_{12} \\ c_{21} & c_{22} \end{bmatrix} = \begin{bmatrix} 5 & 2 \\ 3 & 4 \end{bmatrix}
$$

The determinant is

$$
\det C = c_{11}C_{11} - c_{21}C_{21}, \tag{3.33}
$$

where C_{11} is the determinant of the matrix formed by striking out the first row and first column of C—that is,

$$
C_{11} = \det \begin{bmatrix} \cancel{c_{11}} & \cancel{c_{12}} \\ \cancel{c_{21}} & c_{22} \end{bmatrix} = \det [c_{22}] = c_{22}. \tag{3.34}
$$

Similarly

$$
C_{21} = \det [c_{12}] = c_{12}. \tag{3.35}
$$

Thus from (3.33)

$$
\det C = c_{11}c_{22} - c_{21}c_{12} = 5 \cdot 4 - 3 \cdot 2 = 20 - 6 = 14. \tag{3.36}
$$

The expression (3.36) holds for any square matrix of order 2. The determinant is evaluated by multiplying the two diagonal elements c_{11} and c_{22}, and subtracting the product of the two off-diagonal elements c_{21} and c_{12}.

Next let us evaluate the determinant of the following matrix of order 3:

$$
\begin{aligned}
\det \begin{bmatrix} 7 & 5 & 1 \\ 3 & 4 & 3 \\ 2 & 2 & 5 \end{bmatrix} &= \begin{vmatrix} 7 & 5 & 1 \\ 3 & 4 & 3 \\ 2 & 2 & 5 \end{vmatrix} \\
&= 7 \begin{vmatrix} 4 & 3 \\ 2 & 5 \end{vmatrix} - 3 \begin{vmatrix} 5 & 1 \\ 2 & 5 \end{vmatrix} + 2 \begin{vmatrix} 5 & 1 \\ 4 & 3 \end{vmatrix} \\
&= 7(20 - 6) - 3(25 - 2) + 2(15 - 4) \\
&= 7 \cdot 14 - 3 \cdot 23 + 2 \cdot 11 \\
&= 98 - 69 + 22 = 51.
\end{aligned}
$$

The expression (3.32) is called the evaluation of the determinant by expansion along the first column. We could also evaluate by expanding along the second column:

$$\det C = -c_{12}C_{12} + c_{22}C_{22} + \cdots + (-1)^{n+2}c_{n2}C_{n2}. \tag{3.37}$$

Note that when expanding along the second column, although the terms alternate in sign, they start with a minus, in contrast to (3.32) whose terms begin with a plus. In general, if one expands along an odd (respectively even) column, the terms begin with a minus (respectively plus). The expression for a general expansion along an arbitrary column, say the jth column, is

$$\det C = (-1)^{1+j}c_{1j}C_{1j} + (-1)^{2+j}c_{2j}C_{2j} + \cdots + (-1)^{n+j}c_{nj}C_{nj}. \tag{3.38}$$

Thus if j is odd, then $(-1)^{1+j}$ is positive $(+1)$ and the first term is a plus. If j is even, then $(-1)^{1+j}$ is negative (-1) and the first term is a minus.

The general term c_{ij} is the element of the ith row and jth column of the matrix C. The general term C_{ij} is the determinant of the matrix formed by striking out the ith row and jth column of C. C_{ij} is called the *minor* of the ith row and jth column. When multiplied by $(-1)^{i+j}$, it is called the *cofactor* of the ith row and jth column. Thus

$$C_{ij}^* = (-1)^{i+j}C_{ij} \tag{3.39}$$

is the cofactor of the ith row and the jth column. Thus (3.38) can be written

$$\det C = c_{1j}C_{1j}^* + c_{2j}C_{2j}^* + \cdots + c_{nj}C_{nj}^*. \tag{3.40}$$

Let us evaluate the determinant of

$$C = \begin{bmatrix} c_{11} & c_{12} & c_{13} \\ c_{21} & c_{22} & c_{23} \\ c_{31} & c_{32} & c_{33} \end{bmatrix} = \begin{bmatrix} 7 & 3 & 2 \\ 5 & 4 & 2 \\ 1 & 3 & 5 \end{bmatrix}$$

by expanding along the third column:

$$\det C = (-1)^{1+3} \cdot 2 \cdot \begin{vmatrix} 5 & 4 \\ 1 & 3 \end{vmatrix} + (-1)^{2+3} \cdot 2 \cdot \begin{vmatrix} 7 & 3 \\ 1 & 3 \end{vmatrix} + (-1)^{3+3} \cdot 5 \cdot \begin{vmatrix} 7 & 3 \\ 5 & 4 \end{vmatrix}$$

$$= +2 \cdot (15 - 4) - 2 \cdot (21 - 3) + 5 \cdot (28 - 15)$$
$$= 2 \cdot 11 - 2 \cdot 18 + 5 \cdot 13$$
$$= 22 - 36 + 65 = 51.$$

In terms of cofactors

$$C_{13}^* = (-1)^{1+3}C_{13} = + \begin{vmatrix} 5 & 4 \\ 1 & 3 \end{vmatrix} = 11,$$

$$C_{23}^* = (-1)^{2+3}C_{23} = - \begin{vmatrix} 7 & 3 \\ 1 & 3 \end{vmatrix} = -18,$$

$$C_{33}^* = (-1)^{3+3}C_{33} = + \begin{vmatrix} 7 & 3 \\ 5 & 4 \end{vmatrix} = +13.$$

Then

$$\det C = c_{13}C_{13}^* + c_{23}C_{23}^* + c_{33}C_{33}^*$$
$$= 2(11) + 2(-18) + 5(13)$$
$$= 22 - 36 + 65 = 51.$$

Determinant Method of Solution for Systems of Linear Equations

Consider the following three linear equations with three unknowns a_0 a_1, and a_2:

$$\begin{aligned} x_{00}a_0 + x_{01}a_1 + x_{02}a_2 &= y_0, \\ x_{10}a_0 + x_{11}a_1 + x_{12}a_2 &= y_1, \\ x_{20}a_0 + x_{21}a_1 + x_{22}a_2 &= y_2. \end{aligned} \tag{3.41}$$

The constant x's can be written in the form of a matrix

$$X = \begin{bmatrix} x_{00} & x_{01} & x_{02} \\ x_{10} & x_{11} & x_{12} \\ x_{20} & x_{21} & x_{22} \end{bmatrix} \tag{3.42}$$

The constant column on the right-hand side of (3.41) can be substituted for each one of the columns of X in turn to obtain three more matrices:

$$X^0 = \begin{bmatrix} y_0 & x_{01} & x_{02} \\ y_1 & x_{11} & x_{12} \\ y_2 & x_{21} & x_{22} \end{bmatrix} \tag{3.43}$$

$$X^1 = \begin{bmatrix} x_{00} & y_0 & x_{02} \\ x_{10} & y_1 & x_{12} \\ x_{20} & y_2 & x_{22} \end{bmatrix} \tag{3.44}$$

$$X^2 = \begin{bmatrix} x_{00} & x_{01} & y_0 \\ x_{10} & x_{11} & y_1 \\ x_{20} & x_{21} & y_2 \end{bmatrix} \tag{3.45}$$

Then the solution to the equations (3.41) is

$$a_0 = \frac{\det X^0}{\det X}, \qquad a_1 = \frac{\det X^1}{\det X}, \qquad a_2 = \frac{\det X^2}{\det X}. \tag{3.46}$$

Consider the following example: Solve for a_0, a_1, and a_2 given the system of linear equations

$$\begin{aligned} 7a_0 + 5a_1 + 1a_2 &= 3, \\ 3a_0 + 4a_1 + 3a_2 &= 2, \\ 2a_0 + 2a_1 + 5a_2 &= 4. \end{aligned} \tag{3.47}$$

Then

$$\det X = \det \begin{bmatrix} 7 & 5 & 1 \\ 3 & 4 & 3 \\ 2 & 2 & 5 \end{bmatrix} = 51, \tag{3.48}$$

$$a_0 = \frac{\det X^0}{\det X} = \frac{\begin{vmatrix} 3 & 5 & 1 \\ 2 & 4 & 3 \\ 4 & 2 & 5 \end{vmatrix}}{51} = \frac{40}{51},$$

$$a_1 = \frac{\det X^1}{\det X} = \frac{\begin{vmatrix} 7 & 3 & 1 \\ 3 & 2 & 3 \\ 2 & 4 & 5 \end{vmatrix}}{51} = \frac{-33}{51},$$

(3.49)

$$a_2 = \frac{\det X^2}{\det X} = \frac{\begin{vmatrix} 7 & 5 & 3 \\ 3 & 4 & 2 \\ 2 & 2 & 4 \end{vmatrix}}{51} = \frac{38}{51}.$$

The determinant of X^0 may be expanded along the first column, X^1 along the second column, and X^2 along the third column. From (3.43), (3.44), and (3.45) we have

$$\det X^0 = y_0 X_{11}^* + y_1 X_{21}^* + y_2 X_{31}^*,$$
$$\det X^1 = y_0 X_{12}^* + y_1 X_{22}^* + y_3 X_{32}^*,$$
$$\det X^2 = y_0 X_{13}^* + y_1 X_{23}^* + y_3 X_{33}^*,$$

(3.50)

where X_{ij}^* is the cofactor of the ith row and jth column of X.

Using these expressions (3.50) for the determinants of X^0, X^1, and X^2, we can rewrite (3.49) as follows:

$$a_0 = \frac{y_0 X_{11}^* + y_1 X_{21}^* + y_2 X_{31}^*}{\det X},$$

$$a_1 = \frac{y_0 X_{12}^* + y_1 X_{22}^* + y_2 X_{32}^*}{\det X},$$

(3.51)

$$a_2 = \frac{y_0 X_{13}^* + y_1 X_{23}^* + y_2 X_{33}^*}{\det X}.$$

The coefficients of the y's on the right-hand side of (3.51) can be written in the form of a matrix:

$$X^{-1} = \begin{bmatrix} \dfrac{X_{11}^*}{\det X} & \dfrac{X_{21}^*}{\det X} & \dfrac{X_{31}^*}{\det X} \\ \dfrac{X_{12}^*}{\det X} & \dfrac{X_{22}^*}{\det X} & \dfrac{X_{32}^*}{\det X} \\ \dfrac{X_{13}^*}{\det X} & \dfrac{X_{23}^*}{\det X} & \dfrac{X_{33}^*}{\det X} \end{bmatrix}$$

(3.52)

This matrix is denoted by X^{-1} and is called the inverse of X. In general, if X is a matrix of the form

$$X = \begin{bmatrix} x_{11} & x_{12} & \cdots & x_{1n} \\ x_{21} & x_{22} & \cdots & x_{2n} \\ \cdot & \cdot & & \cdot \\ \cdot & \cdot & & \cdot \\ \cdot & \cdot & & \cdot \\ x_{n1} & x_{n2} & \cdots & x_{nn} \end{bmatrix} \qquad (3.53)$$

the inverse of X is

$$X^{-1} = \begin{bmatrix} \dfrac{X_{11}^*}{\det X} & \dfrac{X_{21}^*}{\det X} & \cdots & \dfrac{X_{n1}^*}{\det X} \\ \dfrac{X_{12}^*}{\det X} & \dfrac{X_{22}^*}{\det X} & \cdots & \dfrac{X_{n2}^*}{\det X} \\ \cdot & \cdot & & \cdot \\ \cdot & \cdot & & \cdot \\ \cdot & \cdot & & \cdot \\ \dfrac{X_{1n}^*}{\det X} & \dfrac{X_{2n}^*}{\det X} & \cdots & \dfrac{X_{nn}^*}{\det X} \end{bmatrix} \qquad (3.54)$$

where X_{ij}^* is the cofactor of the ith row and jth column of X.

The inverse of a matrix X is formed by replacing each element of X, say x_{ij}, by its cofactor X_{ij}^* to obtain the matrix

$$\begin{bmatrix} X_{11}^* & X_{12}^* & \cdots & X_{1n}^* \\ X_{21}^* & X_{22}^* & \cdots & X_{2n}^* \\ \cdot & \cdot & & \cdot \\ \cdot & \cdot & & \cdot \\ \cdot & \cdot & & \cdot \\ X_{n1}^* & X_{n2}^* & \cdots & X_{nn}^* \end{bmatrix} \qquad (3.55)$$

This matrix is then *transposed*—that is, the rows and columns are interchanged:

$$\begin{bmatrix} X_{11}^* & X_{21}^* & \cdots & X_{n1}^* \\ X_{12}^* & X_{22}^* & \cdots & X_{n2}^* \\ \cdot & \cdot & & \cdot \\ \cdot & \cdot & & \cdot \\ \cdot & \cdot & & \cdot \\ X_{1n}^* & X_{2n}^* & \cdots & X_{nn}^* \end{bmatrix} \qquad (3.56)$$

Each element of this transposed matrix then is divided by $\det X$ to arrive at the inverse given by (3.52).

Suppose

$$X = \begin{bmatrix} 7 & 5 & 1 \\ 3 & 4 & 3 \\ 2 & 2 & 5 \end{bmatrix} \qquad (3.57)$$

This is the matrix of coefficients on the left-hand side of (3.47). Replace each element of X by its cofactor:

$$X_{11}^* = (-1)^{1+1} \begin{vmatrix} 4 & 3 \\ 2 & 5 \end{vmatrix} = +(20 - 6) = 14,$$

$$X_{12}^* = (-1)^{1+2} \begin{vmatrix} 3 & 3 \\ 2 & 5 \end{vmatrix} = -(15 - 6) = -9,$$

$$X_{13}^* = (-1)^{1+3} \begin{vmatrix} 3 & 4 \\ 2 & 2 \end{vmatrix} = +(6 - 8) = -2,$$

$$X_{21}^* = (-1)^{2+1} \begin{vmatrix} 5 & 1 \\ 2 & 5 \end{vmatrix} = -(25 - 2) = -23,$$

$$X_{22}^* = (-1)^{2+2} \begin{vmatrix} 7 & 1 \\ 2 & 5 \end{vmatrix} = +(35 - 2) = 33,$$

$$X_{23}^* = (-1)^{2+3} \begin{vmatrix} 7 & 5 \\ 2 & 2 \end{vmatrix} = -(14 - 10) = -4,$$

$$X_{31}^* = (-1)^{3+1} \begin{vmatrix} 5 & 1 \\ 4 & 3 \end{vmatrix} = +(15 - 4) = 11,$$

$$X_{32}^* = (-1)^{3+2} \begin{vmatrix} 7 & 1 \\ 3 & 3 \end{vmatrix} = -(21 - 3) = -18,$$

$$X_{33}^* = (-1)^{3+3} \begin{vmatrix} 7 & 5 \\ 3 & 4 \end{vmatrix} = +(28 - 15) = 13,$$

and we get

$$\begin{bmatrix} 14 & -9 & -2 \\ -23 & 33 & -4 \\ 11 & -18 & 13 \end{bmatrix}$$

Interchange the rows and columns:

$$\begin{bmatrix} 14 & -23 & 11 \\ -9 & 33 & -18 \\ -2 & -4 & 13 \end{bmatrix}$$

From (3.48), we have det $X = 51$. Thus the inverse of X is

$$X^{-1} = \begin{bmatrix} \dfrac{14}{51} & \dfrac{-23}{51} & \dfrac{11}{51} \\ \dfrac{-9}{51} & \dfrac{33}{51} & \dfrac{-18}{51} \\ \dfrac{-2}{51} & \dfrac{-4}{51} & \dfrac{13}{51} \end{bmatrix} \tag{3.58}$$

Let us use the inverse to solve for a_0, a_1, and a_2 in (3.47). Here $y_0 = 3$, $y_1 = 2$, and $y_2 = 4$. From (3.51), we have

$$a_0 = 3\left(\frac{14}{51}\right) + 2\left(\frac{-23}{51}\right) + 4\left(\frac{11}{51}\right) = \frac{40}{51},$$

$$a_1 = 3\left(\frac{-9}{51}\right) + 2\left(\frac{33}{51}\right) + 4\left(\frac{-18}{51}\right) = \frac{-33}{51},$$

$$a_2 = 3\left(\frac{-2}{51}\right) + 2\left(\frac{-4}{51}\right) + 4\left(\frac{13}{51}\right) = \frac{38}{51}.$$

APPENDIX 2   Derivation of Formulae

In this appendix we shall derive the formulae in Chapter 3 for which no derivations are presented in the text. In some cases the derivations depend on a knowledge of the differential calculus, especially the theory concerned with maximization or minimization of functions of several variables.

In nearly every case, the derivations are less messy if one works with deviations from means rather than with the variables X and Y themselves. The deviations from means are denoted by lower-case letters:

$$x_i = X_i - \bar{X}, \qquad y_i = Y_i - \bar{Y}. \tag{3.59}$$

If one takes the sum of squared deviations x_i^2 from the mean, the result is

$$\sum_{i=1}^{n} x_i^2 = \sum (X_i - \bar{X})^2$$

$$= \sum [X_i^2 - 2 \cdot \bar{X} \cdot X_i + (\bar{X})^2]$$

$$= \sum X_i^2 - 2 \cdot \bar{X} \cdot \sum X_i + n \cdot (\bar{X})^2$$

$$= \sum X_i^2 - 2 \cdot \bar{X} \cdot n \sum \frac{X_i}{n} + n \cdot (\bar{X})^2 \tag{3.60}$$

$$= \sum X_i^2 - 2 \cdot n \cdot (\bar{X})^2 + n \cdot (\bar{X})^2,$$

$$\sum_{i=1}^{n} x_i^2 = \sum_{i=1}^{n} X_i^2 - n \cdot (\bar{X})^2.$$

Similarly, for the sum of the y_i^2 we have

$$\sum_{i=1}^{n} y_i^2 = \sum_{i=1}^{n} Y_i^2 - n \cdot (\bar{Y})^2. \tag{3.61}$$

These two relationships (3.60) and (3.61) between sums of squares will be used several times.

Also of use in later derivations is the sum of the product $x_i \cdot y_i$ of deviations from the mean:

$$\sum_{i=1}^{n} x_i y_i = \sum_{i=1}^{n} (X_i - \bar{X}) \cdot (Y_i - \bar{Y})$$

$$= \sum (X_i \cdot Y_i - \bar{X} \cdot Y_i - \bar{Y} \cdot X_i + \bar{X} \cdot \bar{Y})$$

$$= \sum X_i \cdot Y_i - \bar{X} \cdot \sum Y_i - \bar{Y} \cdot \sum X_i + n \cdot \bar{X} \cdot \bar{Y} \qquad (3.62)$$

$$= \sum X_i \cdot Y_i - \bar{X} \cdot n \cdot \bar{Y} - \bar{Y} \cdot n \cdot \bar{X} + n \cdot \bar{X} \cdot \bar{Y},$$

$$\sum_{i=1}^{n} x_i y_i = \sum X_i Y_i - n \cdot \bar{X} \cdot \bar{Y}.$$

The Normal Equations

The normal equations (3.8) are derived by minimizing the sum of squared vertical deviations from the regression line as given by (3.6), which is rewritten here for convenience.

$$S = \sum_{i=1}^{n} (Y_i - a - bX_i)^2. \qquad (3.63)$$

Let us take the partial derivatives with respect to a and b and set equal to zero.

$$\frac{\partial S}{\partial a} = \sum_{i=1}^{n} 2(Y_i - a - bX_i)(-1) = 0,$$

$$\qquad (3.64)$$

$$\frac{\partial S}{\partial b} = \sum_{i=1}^{n} 2(Y_i - a - bX_i)(-X_i) = 0.$$

Using the rules for summation set out in the Appendix to Chapter 2, we may write

$$-2 \sum_{i=1}^{n} Y_i + 2a \cdot n + 2b \cdot \sum_{i=1}^{n} X_i = 0,$$

$$\qquad (3.65)$$

$$-2 \sum_{i=1}^{n} Y_i X_i + 2a \cdot \sum_{i=1}^{n} X_i + 2b \cdot \sum_{i=1}^{n} X_i^2 = 0.$$

If we add $2 \sum_{i=1}^{n} Y_i$ to both sides of the first equation and $2 \sum_{i=1}^{n} Y_i X_i$ to both sides of the second equation, we obtain

$$2a \cdot n + 2b \cdot \sum_{i=1}^{n} X_i = 2 \cdot \sum_{i=1}^{n} Y_i,$$

$$\tag{3.66}$$

$$2a \cdot \sum_{i=1}^{n} X_i + 2b \cdot \sum_{i=1}^{n} X_i^2 = 2 \cdot \sum_{i=1}^{n} Y_i \cdot X_i.$$

Next divide both sides of these equations by $2n$.

$$a + \frac{1}{n} \left(\sum_{i=1}^{n} X_i \right) b = \frac{1}{n} \left(\sum_{i=1}^{n} Y_i \right),$$

$$\tag{3.67}$$

$$\frac{1}{n} \left(\sum_{i=1}^{n} X_i \right) a + \frac{1}{n} \left(\sum_{i=1}^{n} X_i^2 \right) b = \frac{1}{n} \left(\sum_{i=1}^{n} Y_i \cdot X_i \right).$$

These are the normal equations (3.8) in the text.

Let us solve the normal equations (3.67) by the method of elimination of variables. The first of these equations may be written

$$a + b \cdot \bar{X} = \bar{Y}. \tag{3.68}$$

Solving for a, we obtain

$$a = \bar{Y} - b \cdot \bar{X}. \tag{3.69}$$

Substitute this value for a into the second equation of (3.67).

$$\bar{X} \cdot (\bar{Y} - b\bar{X}) + \frac{1}{n} \left(\sum_{i=1}^{n} X_i^2 \right) b = \frac{1}{n} \left(\sum_{i=1}^{n} X_i Y_i \right).$$

Solving for b, we have

$$b \left[\frac{1}{n} \left(\sum_{i=1}^{n} X_i^2 \right) - (\bar{X})^2 \right] = \left[\frac{1}{n} \left(\sum_{i=1}^{n} Y_i \cdot X_i \right) - \bar{X} \cdot \bar{Y} \right]$$

or

$$b = \frac{\left[\displaystyle\sum_{i=1}^{n} Y_i \cdot X_i - n \cdot \bar{X} \cdot \bar{Y} \right]}{\left[\displaystyle\sum_{i=1}^{n} X_i^2 - n \cdot (\bar{X})^2 \right]}. \tag{3.70}$$

This expression for b is the same as that given in the text in (3.13). We may use (3.70) to calculate the value of b and substitute this value into (3.69) [the same as (3.14) in the text] to get a value for a.

Using the expressions (3.60) and (3.62), we may rewrite (3.70) as follows:

$$b = \frac{\sum_{i=1}^{n} x_i y_i}{\sum_{i=1}^{n} x_i^2}. \tag{3.71}$$

Coefficient of Determination

The definition of R^2 in (3.18) is

$$R^2 = \frac{s_y^2 - s_u^2}{s_y^2}. \tag{3.72}$$

Using the definitions of s_u^2 (3.16) and s_y^2 (3.17), we may write

$$R^2 = \frac{\frac{\sum (Y_i - \bar{Y})^2}{n} - \frac{\sum (Y_i - \hat{a} - \hat{b} X_i)^2}{n}}{\frac{\sum (Y_i - \bar{Y})^2}{n}}, \tag{3.73}$$

where the "hats" over the a and b indicate that these have been estimated by solving the normal equations.

Using the expression (2.11), we have

$$\hat{a} = \bar{Y} - \hat{b} \cdot \bar{X}. \tag{3.74}$$

Substitute this value of a into (3.73) and multiply both the numerator and denominator in (3.73) by n.

$$R^2 = \frac{\sum (Y_i - Y)^2 - \sum (Y_i - \bar{Y} + \hat{b} \cdot \bar{X} - \hat{b} \cdot X_i)^2}{\sum (Y_i - \bar{Y})^2}.$$

Using the notation of (3.59), we obtain

$$R^2 = \frac{\sum y_i^2 - \sum (y_i - \hat{b} \cdot x_i)^2}{\sum y_i^2}.$$

Expand the squared terms in the numerator.

$$R^2 = \frac{\sum y_i^2 - \sum (y_i - 2\hat{b} \cdot x_i y_i + \hat{b}^2 \cdot x_i^2)}{\sum y_i^2},$$

$$R^2 = \frac{2\hat{b} \cdot \sum x_i y_i - \hat{b}^2 \cdot \sum x_i^2}{\sum y_i^2}.$$

From (3.71), we have

$$\hat{b} = \frac{\sum_{i=1}^{n} x_i y_i}{\sum_{i=1}^{n} x_i^2}.$$

Substitute this value into the last expression for R^2.

$$R^2 = \frac{\dfrac{2(\Sigma\, x_i y_i)(\Sigma\, x_i y_i)}{\Sigma\, x_i^2} - \dfrac{(\Sigma\, x_i y_i)^2(\Sigma\, x_i^2)}{(\Sigma\, x_i^2)^2}}{\Sigma\, y_i^2},$$

$$R^2 = \frac{(\Sigma\, x_i y_i)(\Sigma\, x_i y_i)}{(\Sigma\, x_i^2)(\Sigma\, y_i^2)} = \frac{(\Sigma\, x_i y_i)^2}{(\Sigma\, x_i^2)(\Sigma\, y_i^2)}.$$

(3.75)

Using the expressions (3.60), (3.61), (3.62), and (3.71), we obtain

$$R^2 = \frac{\hat{b}\cdot\left[\displaystyle\sum_{i=1}^{n} X_i Y_i - n\cdot\bar{X}\cdot\bar{Y}\right]}{\displaystyle\sum_{i=1}^{n} Y_i^2 - n\cdot(\bar{Y})^2}$$

(3.76)

and

$$R^2 = \frac{\left[\displaystyle\sum_{i=1}^{n} X_i Y_i - n\cdot\bar{X}\cdot\bar{Y}\right]^2}{\left[\displaystyle\sum_{i=1}^{n} X_i^2 - n\cdot(\bar{X})^2\right]\left[\displaystyle\sum_{i=1}^{n} Y_i^2 - n\cdot(\bar{Y})^2\right]}.$$

(3.77)

This is the derivation of the expressions (3.18) and (3.19) in the text.

Orthogonal Regression

Figure 3.6, shows a scatter point and a regression line. The vertical distance to the regression line is labeled u, the horizontal distance v, and the perpendicular (orthogonal) distance w. The lines u and v intersect with the regression line to form a right triangle. The line w cuts the hypotenuse of this right triangle into two parts c_u and c_v and also divides the right triangle into two smaller right triangles. The elementary theory of geometry tells us that the sum of the squared sides of a right triangle is equal to the square of the hypotenuse. Applying this analysis to the three triangles in Fig. 3.6, we get

$$c_v^2 + w^2 = v^2,$$

(3.78)

$$c_u^2 + w^2 = u^2,$$

(3.79)

$$(c_v + c_u)^2 = c_v^2 + c_u^2 + 2\cdot c_v c_u = u^2 + v^2.$$

(3.80)

Add the first two equations and subtract from the third.

$$2\cdot c_v\cdot c_u - 2w^2 = 0$$

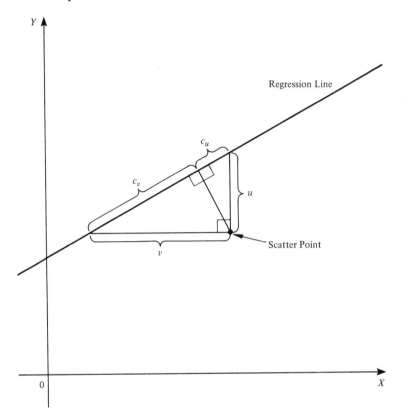

FIGURE 3.6   Orthogonal Regression

or

$$(w^2)^2 = c_v^2 \cdot c_u^2. \tag{3.81}$$

Solve (3.78) and (3.79) for c_v^2 and c_u^2, and substitute into (3.81).

$$(w^2)^2 = (v^2 - w^2)(u^2 - w^2). \tag{3.82}$$

Expanding the right-hand side of (3.82), we get

$$(w^2)^2 = v^2 u^2 - v^2 w^2 - u^2 w^2 + (w^2)^2$$

or

$$w^2(u^2 + v^2) = u^2 \cdot v^2. \tag{3.83}$$

Let b stand for the slope of the line in Fig. 3.6. By definition

$$b = \frac{u}{v} \tag{3.84}$$

or

$$v^2 = \frac{u^2}{b^2}.$$ (3.85)

Substitute (3.85) into (3.83).

$$w^2\left(u^2 + \frac{u^2}{b^2}\right) = u^2 \cdot \frac{u^2}{b^2}$$

or

$$w^2(b^2 + 1) = u^2$$

and

$$w^2 = \frac{u^2}{(1 + b^2)}.$$ (3.86)

Now u is the vertical deviation. The formula for the ith vertical deviation u_i is given by (3.5) and is $Y_i - a - bX_i$. Thus the sum of the squared orthogonal deviations is

$$S_w = \Sigma\, w_i^2 = \frac{\Sigma\, u_i^2}{1 + b^2} = \frac{\Sigma\, (Y_i - a - bX_i)^2}{1 + b^2}.$$ (3.87)

The sum S_w is a function of the parameters a and b. To minimize S_w one differentiates with respect to a and b. First, with respect to a,

$$\frac{\partial S_w}{\partial a} = \frac{2 \sum\limits_{i=1}^{n} (Y_i - a - bX_i)(-1)}{(1 + b^2)} = 0.$$ (3.88)

Divide both sides of (3.88) by $2n/(1 + b^2)$ and add $\Sigma\, Y_i$ to both sides to get

$$\frac{n \cdot a + b\,\Sigma\, X_i}{n} = \frac{\Sigma\, Y_i}{n}$$

or

$$a = \bar{Y} - b\bar{X}.$$ (3.89)

Thus we see that with orthogonal regression, the regression line goes through the point $(\bar{X},\, \bar{Y})$ of means.

Rather than differentiate S_w directly with respect to b, let us substitute the value of a from (3.89) into (3.87). We get

$$S_w = \frac{\sum\limits_{i=1}^{n} (Y_i - \bar{Y} + b\bar{X} - bX_i)^2}{1 + b^2}$$ (3.90)

or, using the notation in (3.59),

$$S_w = \frac{\displaystyle\sum_{i=1}^{n}(y_i - bx_i)^2}{1 + b^2}. \tag{3.91}$$

This more compact expression for S_w is differentiated with respect to b.

$$\frac{\partial S_w}{\partial b} = \frac{\displaystyle\sum_{i=1}^{n}2(y_i - bx_i)(-x_i)}{1 + b^2} - \frac{\displaystyle\sum_{i=1}^{n}(y_i - bx_i)^2(2b)}{(1 + b^2)^2} = 0. \tag{3.92}$$

Divide both sides of this expression by $2/(1 + b^2)$ and expand the summation terms.

$$[-\Sigma\, y_i x_i + b\,\Sigma\, x_i^2] - \frac{[b \cdot \Sigma\, y_i^2 - 2b^2 \cdot \Sigma\, x_i y_i + b^3 \cdot \Sigma\, x_i^2]}{1 + b^2} = 0.$$

Multiply both sides by $(1 + b^2)$ and collect terms.

$$[-\Sigma\, y_i x_i + b \cdot \Sigma\, x_i^2] + b^2 \cdot [-\Sigma\, y_i x_i + b \cdot \Sigma\, x_i^2] - b \cdot \Sigma\, y_i^2$$
$$+ 2b^2 \cdot \Sigma\, x_i y_i - b^3 \cdot \Sigma\, x_i^2 = 0$$

or

$$-\Sigma\, y_i x_i + b \cdot [\Sigma\, x_i^2 - \Sigma\, y_i^2] + b^2 \cdot (\Sigma\, x_i y_i) = 0$$

or

$$b^2 \cdot (\Sigma\, x_i y_i) + b \cdot (\Sigma\, x_i^2 - \Sigma\, y_i^2) + (-\Sigma\, x_i y_i) = 0. \tag{3.93}$$

The expression (3.87) is a quadratic function of b. The usual quadratic formula may be used to solve for b.

$$b = \frac{(-\Sigma\, x_i^2 + \Sigma\, y_i^2) \pm \sqrt{[(\Sigma\, x_i^2) - (\Sigma\, y_i)^2]^2 + 4(\Sigma\, x_i y_i)^2}}{2(\Sigma\, x_i y_i)}. \tag{3.94}$$

The equation (3.87) has two roots—that is, two possible values of b. We may show, by taking the second derivatives of S_w with respect to b, that the larger root makes S_w a minimum. Thus

$$b = \frac{B_1 + \sqrt{B_1^2 + 4B_2^2}}{2B_2}, \tag{3.95}$$

where from (3.94) and (3.60), (3.61), and (3.62)

$$B_1 = \Sigma\, x_i^2 + \Sigma\, y_i^2 = \Sigma\, Y_i^2 - \Sigma\, X_i^2 - n[(\bar{Y})^2 - (\bar{X})^2]$$
$$B_2 = \Sigma\, x_i y_i = \Sigma\, X_i Y_i - n \cdot \bar{X} \cdot \bar{Y}.$$

The expression (3.94) is the same as (3.28) in the text.

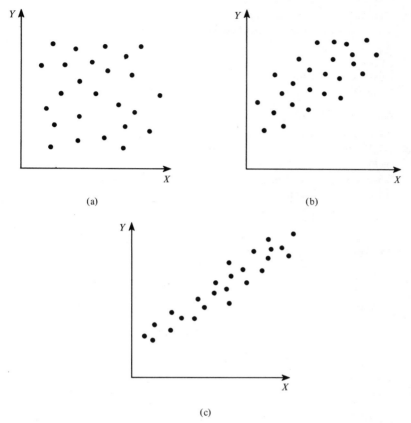

(a)

(b)

(c)

FIGURE 3.7

PROBLEMS

1. Generally speaking, which of the scatter diagrams in Fig. 3.7 indicates the highest correlation? Why?
2. Which scatter diagram in Fig. 3.7 indicates that the correlation coefficient will be about zero? Why?
3. The following results have been computed from data on years of service (X) and weekly salary for 30 employees of a retail store.

$$n = 30, \qquad\qquad \Sigma\,X^2 = 510,$$
$$\Sigma\,X = 100 \text{ (years)}, \qquad \Sigma\,Y^2 = 82{,}130,$$
$$\Sigma\,Y = 2{,}000 \text{ (dollars)}, \qquad \Sigma\,XY = 5{,}780.$$

Compute (a) the line of regression and (b) the coefficient of correlation.

4. Data on two variables Y and X are used to determine a regression line by minimizing the vertical sum of squares. The result is

$Y = 35.2 - 14.3X.$

Next the same data are used to estimate the horizontal sum of squares. The result is

$X = 1.8 - .06Y.$

What is the coefficient of determination?

5. Which shows a stronger relationship—a correlation coefficient of $-.72$ or a coefficient of determination of 0.64? Why?

6. Given the following tables of employee experience (in years) and productivity (in output/man-hour):

X	Y
EXPERIENCE	PRODUCTIVITY
3	13
6	24
10	32
2	8
5	23
4	23
8	29
1	5

(a) Find the least-squares regression line. (b) What is the sum of the squared residuals from the least-squares regression line? (c) Calculate the sum of the squared residuals from the line $Y = 10.00 + 2.0X$. (d) What is the coefficient of determination for this line and how does it compare with the coefficient of determination from the least-squares regression line? (e) Calculate the standard error of the estimate of the least-squares regression line.

7. Given the following data, calculate the orthogonal regression of consumption on income.

X	Y
INCOME	CONSUMPTION
300	220
400	280
500	350
700	500
1000	800

*8. Evaluate the determinant of

$$\begin{bmatrix} 6 & 2 & 7 \\ 8 & 4 & 3 \\ 3 & 5 & 1 \end{bmatrix}$$

By expanding along the first column; the second column.

*9. Solve the following system of linear equations:

$$6a_0 + 3a_1 + 4a_2 = 3,$$
$$a_0 + 3a_1 + 7a_2 = 5,$$
$$2a_0 + a_1 + 9a_2 = 10.$$

*10. Form the inverse of the matrix:

$$\begin{bmatrix} 6 & 2 & 7 \\ 8 & 4 & 3 \\ 3 & 5 & 1 \end{bmatrix}$$

---

* Problems relating to Appendix 1 of this chapter.

Chapter 4

Multiple and Nonlinear Regression and Correlation

Frequently, one wishes to associate more than two variables. For example, suppose air travel to and from an urban center (Y), population of the center (X), and average income level in the center (Z) are all related.

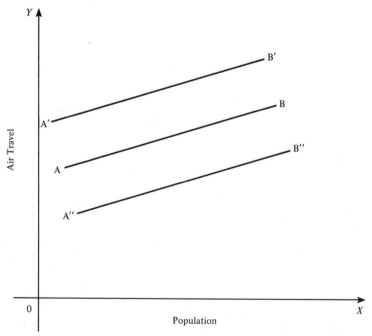

FIGURE 4.1   Relation between Air Travel and Population at Different Income Levels

Looking at the relationship between air travel (Y) and population (X) alone, a linear relationship might be found such as the line AB shown in Fig. 4.1. This relationship may hold, however, only if the income level is assumed to be held fixed. If the income level rises, the relationship may be shifted to $A'B'$ in Fig. 4.1. Similarly if the income level is lower, the relationship $A''B''$ may hold. Such a three-variable relationship may be illustrated as a plane in three-dimensional space, as shown in Fig. 4.2.

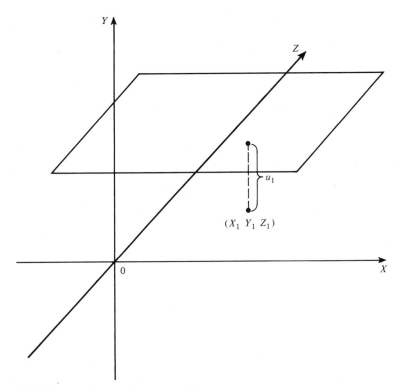

FIGURE 4.2   Multiple Regression Plane

Multiple Linear Regression

Let Y, X, and Z be three different variables and Y_1, Y_2, ... , Y_n, X_1, X_2, ... , X_n, and Z_1, Z_2, ... , Z_n be n observations or pieces of data relating to the three variables. Instead of a line of regression, one attempts to fit a regression plane in these dimensions. The equation of the plane is

$$Y = a + bX + cZ. \tag{4.1}$$

Figure 4.2 shows a three-dimensional space. The Y-axis and the X-axis are in the plane of the page and the Z-axis is perpendicular to the plane of the page. The regression plane $Y = a + bX + cZ$ is also shown in Fig. 4.2. The point (X_1, Y_1, Z_1) is plotted by finding the point X_1 on the X-axis and the point Y_1 on the Y-axis and going a distance Z_1, perpendicular to the plane of the page. The distance u_1 is represented by a line dropped from the point (X_1, Y_1, Z_1) to the regression plane. The line from u_1 to the regression plane is parallel to the Y-axis. The points (X_2, Y_2, Z_2), (X_3, Y_3, Z_3), $\ldots$, (X_n, Y_n, Z_n) may also be plotted and the distance along a line parallel to the Y-axis calculated. The least-squares regression plane is the plane that minimizes the sum of these squared distances. In other words, we must determine the parameters a, b, and c that minimize the sum of these squared distances. (We might also minimize the sum of the squared distances parallel to the X-axis, the Z-axis, or perpendicular to the regression plane, as was done in the two-variable case. For the sake of brevity, we shall confine ourselves to the minimization of the squared distances parallel to the Y-axis.)

The value of the variable Y_i estimated by the regression plane is

$$\hat{Y}_i = a + \hat{b}X_i + \hat{c}Z_i. \tag{4.2}$$

This estimated value of Y_i is subtracted from actual value of Y_i to obtain the estimated residual $\hat{u}_i$.

$$\hat{u}_i = Y_i - (\hat{a} + \hat{b}X_i + \hat{c}Z_i) = Y_i - \hat{Y}_i. \tag{4.3}$$

The residual u_i is the distance from the point Y_i to the regression plane. The sum of the squared deviations is

$$S = \sum_{i=1}^{n} u_i^2 = \sum_{i=1}^{n} (Y_i - a - bX_i - cZ_i)^2 \tag{4.4}$$

To find the coefficients a, b, and c, we solve the following normal equations for a, b, and c.

$$\begin{aligned}
a + (\Sigma\, X_i)b + (\Sigma\, Z_i)c &= (\Sigma\, Y_i), \\
(\Sigma\, X_i)a + (\Sigma\, X_i^2)b + (\Sigma\, Z_iX_i)c &= (\Sigma\, Y_iX_i), \\
(\Sigma\, Z_i)a + (\Sigma\, X_iZ_i)b + (\Sigma\, Z_i^2)c &= (\Sigma\, Y_iZ_i).
\end{aligned} \tag{4.5}$$

Example 4.1. Table 3.1 of the previous chapter gives Kenya imports (column 1), gross capital formation (column 3), and Gross Domestic Product (column 4). Let imports be the Y variable, gross capital formation the X variable, and Gross Domestic Product the Z variable.

Y_i (1)	X_i (2)	Z_i (3)	Y_iX_i (4)	Y_iZ_i (5)
71.5	43.8	180.5	3,131.70	12,905.75
69.8	45.7	193.2	3,189.86	13,485.36
72.0	45.6	205.9	3,283.20	14,824.80
60.9	40.0	208.1	2,436.00	12,673.29
61.5	40.3	214.8	2,478.45	13,210.20
70.1	41.4	225.5	2,902.14	15,807.55
68.9	31.9	224.7	2,197.91	15,481.83
69.5	33.3	244.1	2,314.35	16,964.95
73.7	30.4	259.1	2,240.48	19,095.67
76.6	32.8	277.7	2,512.48	21,271.82

X_iZ_i (6)	Y_i^2 (7)	X_i^2 (8)	Z_i^2 (9)
7,905.90	5,112.25	1,918.44	32,580.25
8,829.24	4,872.04	2,088.49	37,326.24
9,389.04	5,184.00	2,079.36	42,394.81
8,324.00	3,708.81	1,600.00	43,305.61
8,656.44	3,782.25	1,624.09	46,139.04
9,335.70	4,914.01	1,713.96	50,850.25
7,167.93	4,747.21	1,017.61	50,490.09
8,128.53	4,830.25	1,108.89	59,584.81
7,876.64	5,431.69	924.16	67,132.81
9,108.56	5,867.56	1,075.84	77,117.29

Sum all the numbers in each of the nine columns to obtain

$\Sigma Y_i = 694.5,$
$\Sigma X_i = 385.2,$
$\Sigma Z_i = 2,233.6,$
$\Sigma Y_iX_i = 26,686.57,$
$\Sigma Y_iZ_i = 155,721.22,$
$\Sigma X_iZ_i = 84,721.98,$
$\Sigma Y_i^2 = 48,450.07,$
$\Sigma X_i^2 = 15,150.84,$
$\Sigma Z_i^2 = 506,921.20.$

The normal equations are given by

$$a + (385.2)b + (2,233.6)c = 694.5,$$
$$(385.2)a + (15,150.84)b + (84,721.98)c = 26,686.57,$$
$$(2,233.6)a + (84,721.98)b + (506,921.20)c = 155,721.22.$$

$$a = \frac{\begin{vmatrix} 694.5 & 385.2 & 2{,}233.6 \\ 26{,}686.57 & 15{,}150.84 & 84{,}721.98 \\ 155{,}721.22 & 84{,}721.98 & 506{,}921.20 \end{vmatrix}}{\begin{vmatrix} 1 & 385.2 & 2{,}233.6 \\ 385.2 & 15{,}150.84 & 84{,}721.98 \\ 2{,}233.6 & 84{,}721.98 & 506{,}921.20 \end{vmatrix}}.$$

By evaluating the determinants in the numerator and denominator and dividing the first by the latter, we obtain

$a = 48.866.$

Similarly,

$b = .44494,$

$c = .14748.$

Thus for each £1-million increase in Gross Domestic Product (Z), imports (Y) rise by about £.15 million or 15 percent of the rise in Gross Domestic Product with gross capital formation (X) held constant. If Gross Domestic Product (Z) is constant, a £1-million rise in gross capital formation (X) results in about a £.44-million rise in imports (Y).

Once the parameters a, b, and c are determined, one may calculate the various measures of goodness of fit as in the two-variable case. The standard error of the estimate s_u is the sum of the squared residuals, given by

$$s_u^2 = \frac{\sum_{i=1}^{n} (\hat{u}_i)^2}{n} = \frac{\sum_{i=1}^{n} (Y_i - \hat{a} - \hat{b} \cdot X_i - \hat{c} \cdot Z_i)^2}{n} = \frac{\sum_{i=1}^{n} (Y_i - \hat{Y}_i)^2}{n}, \qquad (4.6)$$

where $\hat{a}$, $\hat{b}$, and $\hat{c}$ are the estimated values of the parameters, a, b, and c—that is, those obtained by solving the normal equations (4.5). The coefficient of determination is

$$R^2 = \frac{s_Y^2 - s_u^2}{s_Y^2}, \qquad (4.7)$$

where

$$s_Y^2 = \frac{1}{n} \sum_{i=1}^{n} (Y_i - \bar{Y})^2.$$

As in the two-variable case, the coefficient of determination measures the variation in the Y variable "explained" by variations in X and Z as a percentage of the total variation in the Y variable.

More convenient formulae to use in the actual calculations of s_u^2 and R^2 are the following:

$$s_u^2 = \frac{\Sigma \, Y_i^2 - n(\bar{Y})^2 - \hat{b}[\Sigma \, X_iY_i - n \cdot \bar{X} \cdot \bar{Y}] - \hat{c}[\Sigma \, Z_iY_i - n \cdot \bar{Z} \cdot \bar{Y}]}{n} \qquad (4.8)$$

and

$$R^2 = \frac{\hat{b}[\Sigma \, X_iY_i - n \cdot \bar{X} \cdot \bar{Y}] + \hat{c}[\Sigma \, Z_iY_i - n \cdot \bar{Y} \cdot \bar{Z}]}{\Sigma \, Y_i^2 - n \cdot (\bar{Y})^2} , \qquad (4.9)$$

where $\hat{b}$ and $\hat{c}$ are the least-squares estimates of the coefficients b and c.

Example 4.2. Let us calculate the coefficient of determination for the regression in the previous example.

$$\Sigma \, X_iY_i = 26{,}686.57,$$

$$n \cdot \bar{X} \cdot \bar{Y} = 10 \left(\frac{694.5}{10} \right) \left(\frac{385.2}{10} \right) = 26{,}752.140,$$

$$\Sigma \, X_iY_i - n \cdot \bar{X} \cdot \bar{Y} = -65.570,$$

$$\Sigma \, Z_iY_i = 155{,}721.220,$$

$$n \cdot \bar{Z} \cdot \bar{Y} = 10 \left(\frac{2{,}233.6}{10} \right) \left(\frac{694.5}{10} \right) = 155{,}123.520,$$

$$\Sigma \, Z_iY_i - n \cdot \bar{Z} \cdot \bar{Y} = 597.700,$$

$$\Sigma \, Y_i^2 = 48{,}450.070,$$

$$n \cdot (\bar{Y})^2 = 10 \left(\frac{694.5}{10} \right)^2 = 48{,}233.025,$$

$$\Sigma \, Y_i^2 - n(\bar{Y})^2 = 217.045,$$

$$R^2 = \frac{(.44494)(-65.570) + (.14748)(597.700)}{217.045} ,$$

$$R^2 = .27171.$$

That is, about 27 percent of the variation in imports (Y) are explained by variations in capital formation (X) and domestic product (Z).

Regression and correlation analysis can be generalized to cover any number of variables. Suppose we have the variables $Y, X_1, X_2, \ldots , X_k$, and that there are n pieces of data for each variable. Then the Y data are $Y_1, Y_2, \ldots , Y_n$. The X_1 data are denoted by $X_{11}, X_{12}, X_{13}, \ldots , X_{1n}$. The X_2 data are $X_{21}, X_{22}, X_{23}, \ldots , X_{2n}$, and so on.

The regression plane is

$$Y = a_0 + a_1X_1 + a_2X_2 + \cdots + a_kX_k. \qquad (4.10)$$

The estimated residual $\hat{u}_i$ is denoted by

$$\hat{u}_i = Y_i - (\hat{a}_0 + \hat{a}_1X_{1i} + \hat{a}_2X_{2i} + \cdots + \hat{a}_kX_{ki}) \qquad (4.11)$$

or

$$\hat{u}_i = Y_i - \hat{Y}_i, \tag{4.12}$$

where $\hat{a}_0, \hat{a}_1, \ldots, \hat{a}_k$ are the estimated values of $a_0, a_1, \ldots, a_k$. The sum of the squared residuals is

$$S = \sum_{i=1}^{n} \hat{u}_i^2$$

$$= \sum_{i=1}^{n} [Y_i - (\hat{a}_0 + \hat{a}_1 X_{1i} + \hat{a}_2 X_{2i} + \cdots + \hat{a}_k X_{ki})]^2 \tag{4.13}$$

$$= \sum_{i=1}^{n} (Y_i - \hat{Y}_i)^2.$$

The normal equations are

$$n \cdot a_0 + \left(\sum_{i=1}^{n} X_{1i}\right) \cdot a_1 + \cdots + \left(\sum_{i=1}^{n} X_{ki}\right) \cdot a_k = \left(\sum_{i=1}^{n} Y_i\right)$$

$$\left(\sum_{i=1}^{n} X_{1i}\right) a_0 + \left(\sum_{i=1}^{n} X_{1i}^2\right) \cdot a_1 + \cdots + \left(\sum_{i=1}^{n} X_{ki} X_{1i}\right) \cdot a_k = \left(\sum_{i=1}^{n} Y_i X_{1i}\right)$$

$$\left(\sum_{i=1}^{n} X_{ki}\right) a_0 + \left(\sum_{i=1}^{n} X_{1i} X_{ki}\right) \cdot a_1 + \cdots + \left(\sum_{i=1}^{n} X_{ki}^2\right) \cdot a_k = \left(\sum_{i=1}^{n} Y_i X_{ki}\right)$$

$$\tag{4.14}$$

To find the regression parameters, one must solve this system of linear equations for the parameters $a_0, a_1, \ldots, a_k$. The standard error of the estimate s_u is the square root of

$$s_u^2 = \frac{\sum_{i=1}^{n} (\hat{u}_i)^2}{n}. \tag{4.15}$$

The coefficient of determination is

$$R^2 = \frac{s_Y^2 - s_u^2}{s_Y^2}, \tag{4.16}$$

where s_Y^2 is given by

$$s_Y^2 = \frac{\sum\limits_{i=1}^{n} (Y_i - \bar{Y})^2}{n}. \tag{4.17}$$

Partial Correlation

When there is an association between three or more variables, partial relationships between variables become important. For example, suppose the variables Y, X, and Z are related. If one computes the simple correlation coefficient between Y and X, the result may be highly misleading.

Example 4.3. In Table 3.1, Kenya's imports (Y) and gross capital formation (X) seem to move in opposite directions. The calculation of the simple correlation coefficient verifies this. The correlation between these two variables is $-.2618$. This relationship is not expected, since one would expect imports to rise as capital formation rises because of Kenya's dependence on imported capital goods. This negative association may be misleading, however, under the following circumstances. Suppose gross capital formation (X) and imports (Y) tend to have a positive association whenever Gross Domestic Product (Z) is constant. Gross Domestic Product (Z) and imports (Y) have a strong positive association when gross capital formation (X) is constant. If Gross Domestic Product (Z) goes up whenever gross capital formation (X) goes down, then imports (Y) can go up or down, since the effect of the Z variable is to push Y up while the effect of the X variable is a tendency for Y to go down. If the influence of the Z variable is stronger, then Y will go up. The pattern that one sees in the data is a tendency for X to go down when Y goes up, and the simple correlation between these variables will be negative. This negative correlation, however, is due to the overriding influence of the Z variable, which pushes Y up when X goes down. If the influence of the Z variable could be somehow removed, the "true" positive association between X and Y would emerge.

The *partial correlation coefficient* between X and Y is the correlation between X and Y obtained by removing (in a special way) the effects of the Z variable. The partial correlation coefficient may be defined in terms of the sum of squared residuals from least-squares regression lines. Let us determine the regression line using the data for the X, Y, and Z variables.

$$Y = a_{YXZ} + b_{YXZ}X + c_{YXZ}Z. \tag{4.18}$$

Next run the regression on the data for the Y and Z variables alone.

$$Y = a_{YZ} + c_{YZ}Z. \tag{4.19}$$

By solving the normal equations in each case, we are able to arrive at estimated values $\hat{a}_{YXZ}$, $\hat{b}_{YXZ}$, and $\hat{c}_{YXZ}$ for the parameters of the YXZ regression and $\hat{a}_{YZ}$ and $\hat{c}_{YZ}$ for the YZ regression. Let s_{YXZ} stand for the standard error of the estimate of the YXZ regression and s_{YZ} be the standard error of the estimate for the YZ regression. Then

$$s_{YXZ}^2 = \text{variation of } Y \text{ unexplained by } X \text{ and } Z,$$
$$s_{YZ}^2 = \text{variation of } Y \text{ unexplained by } Z,$$
$$s_{YZ}^2 - s_{YXZ}^2 = \text{reduction in the unexplained variation of } Y \text{ by introducing the } X \text{ variable; or}$$
$$s_{YZ}^2 - s_{YXZ}^2 = \text{additional variation in } Y \text{ explained by introduction of the } X \text{ variable.}$$

Then $R_{XY \cdot Z}^2$ is the coefficient of *partial* determination between X and Y, where

$$R_{YX \cdot Z}^2 = \frac{s_{YZ}^2 - s_{YXZ}^2}{s_{YZ}^2}, \tag{4.20}$$

or

$R_{YX \cdot Z}^2 = $ additional variation in Y explained by X relative to the variation left unexplained by Z.

The square root of the partial coefficient of determination is the partial correlation coefficient $R_{YX \cdot Z}$. A handy formula to use in calculating the partial correlation coefficient is

$$R_{YX \cdot Z} = \frac{R_{YX} - R_{YZ} \cdot R_{XZ}}{\sqrt{1 - R_{XZ}^2} \sqrt{1 - R_{YZ}^2}}, \tag{4.21}$$

where R_{XY}, R_{YZ}, and R_{XZ} are the simple correlations between each pair of variables.

Example 4.4.   The simple correlation between Kenya imports (Y) in Table 3.1 and capital formation X is

$$R_{YX} = -.2516,$$

which indicates, contrary to what one would expect, an inverse relationship. Using formula (4.71), let us calculate the partial correlation coefficient $R_{YX \cdot Z}$, where the effects of the Z variable (Gross Domestic Product) are taken into account. We have

$$R_{YZ} = .4529,$$
$$R_{XZ} = -.8307,$$
$$R_{YZ}^2 = .20512,$$
$$R_{XZ}^2 = .69006,$$
$$R_{YX \cdot Z} = \frac{-.2516 - (.4529)(-.8307)}{\sqrt{(1 - .69006)} \sqrt{(1 - .20512)}},$$

or

$$R_{YX \cdot Z} = +.1907.$$

A positive association between imports and capital formation emerges when domestic production is taken into account.

In general, partial coefficients of determination can be determined for any subset of a large group of variables. Suppose we have the $k + 1$ variables $Y, X_1, X_2, \ldots, X_k$. We wish to find the partial coefficient of determination between $Y, X_1, X_2, \ldots, X_{k*}$, removing the effects of variables $X_{k*+1}, X_{k*+2}, \ldots, X_k$. Let $s_{0,1,2,\ldots,k}$ be the standard error of the estimate obtained by finding a least-squares regression plane for $Y, X_1, X_2, \ldots, X_k$ and $s_{0k*+1,\ldots,k}$ be the standard error of the estimate by regressing $Y, X_{k*+1}, X_{k*+2}, \ldots, X_k$. Then the partial coefficient of determination among $Y, X_1, X_2, \ldots, X_{k*}$ is

$$R^2_{0,1,2,\ldots,k* \cdot k*+1,\ldots,k} = \frac{s^2_{0,k*+1,\ldots,k} - s^2_{0,1,2,\ldots,k}}{s^2_{0,k*+1,\ldots,k}}. \tag{4.22}$$

Nonlinear Regression and Correlation

Up to this point, we have been dealing only with *linear* regression relationships between variables. At times we may want to specify a nonlinear relationship. The following are some commonly used examples of nonlinear relationships.

Simple quadratic:

$$Y = a + bX + cX^2.$$

Two-variable quadratic:

$$Y = a + b_1X + b_2Z + c_1X^2 + c_2Z^2 + c_3X \cdot Z.$$

Double logarithmic:

$$\log_e Y = a + b \log_e X.$$

Logarithmic in dependent variable:

$$\log_e Y = a + bX.$$

Logarithmic in independent variable:

$$Y = a + b \cdot \log_e Y.$$

All of these types of relationships can be fitted to the data using the least-squares criteria by simple transformation of the data. For example, the two-variable quadratic relationship is transformed into a linear relationship by defining five different variables as follows:

$$X_1 = X, \quad X_2 = Z, \quad X_3 = X^2, \quad X_4 = Z^2, \quad X_5 = X \cdot Z. \tag{4.23}$$

Then the two-variable quadratic relationship can be written

$$Y = a + b_1 X_1 + b_2 X_2 + c_1 X_3 + c_2 X_4 + c_3 X_5. \tag{4.24}$$

This new relationship is linear in the transformed variables. The least-squares formula for multiple regression and correlation can be applied to the data if it is transformed as indicated by (4.23).

Example 4.5.   Let us indicate how to fit a two-variable quadratic form to the variables Y, X, and Z, which represent Kenya imports, capital formation, and domestic product, respectively. The variable X_1 corresponds to the data in column (2) of the calculations done in Example 4.1. The variable X_2 corresponds to the data in column (3). Similarly X_3, X_4, and X_5 correspond to the columns (8), (9), and (6), respectively. We treat the data in each of these columns as new data, and from (4.14) we derive the normal equations to solve for the parameters a $(= a_0)$, b_1 $(= a_1)$, b_2 $(= a_2)$, c_1 $(= a_3)$, c_2 $(= a_4)$, and c_3 $(= a_5)$. This gives us a least-squares regression plane of the form (4.10).

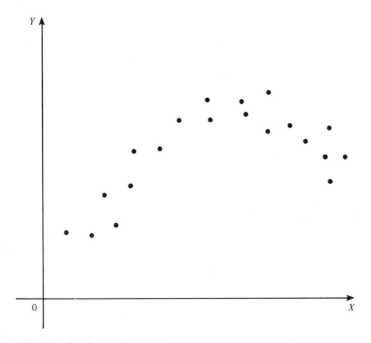

FIGURE 4.3   Scatter Diagram

Whether one should use a simple linear relationship as discussed in previous sections or one of the nonlinear forms suggested here in order to determine a least-squares relationship among various variables is largely a matter of judgment. The choice of the type of relationship is often governed by (a) the "shape" of the scatter diagram or (b) a priori theoretical considerations.

Example 4.6. If the scatter diagram for the data on two variables X and Y looks like that in Fig. 4.3, then one is tempted to try a simple quadratic function and obtain a least-squares fit.

Example 4.7. Suppose an agronomist has data on various amounts of fertilizer application and wheat yields on presumably similar plots of land. Past experience indicates that small fertilizer applications give large increases in yield but larger and larger fertilizer applications give successively smaller increments in yield. This situation is depicted by a yield curve, shown in Fig. 4.4. Very large fertilizer applications might actually have a negative effect on yields. In Fig. 4.4 this is shown as a decreasing yield curve for fertilizer applications beyond X^*. In a case such as this, the agronomist might want to fit a simple quadratic relationship. The c coefficient in the simple quadratic relationship will presumably turn out to be negative, so that the cX^2 term reduces yields more and more below what one would get using a simple linear relationship. (See Fig. 4.4.)

Example 4.8. The two-variable quadratic relationship might be used whenever there is a suspicion that two variables interact. The term $c_3X \cdot Z$ in the two-variable quadratic relationship is called the interaction term. For example, Bowen and Baumol[1] in their study of the performing arts found a strong relationship between theater attendance and both educational level and income level. The effect of educational level, however, may be reinforced by income. That is, suppose there are three individuals— one with a high level of education but low income, a second with a low level of education but high income, and a third with both a high level of education and high income. If we average the annual number of theater attendances by the first two individuals, this average might be less than the annual number of theater attendances by the third individual who has both high income and a high level of education. A regression plane fit to data on theater attendances, income level, and educational level must have an interaction term $c_3X \cdot Z$ to catch this possible reinforcing effect.

Relationships of the logarithmic form are used when one is concerned with the relationship between percentage changes in the variables rather

[1] W. J. Baumol and W. G. Bowen, *Performing Arts: The Economic Dilemma* (Cambridge, Mass.: M.I.T. Press, 1966).

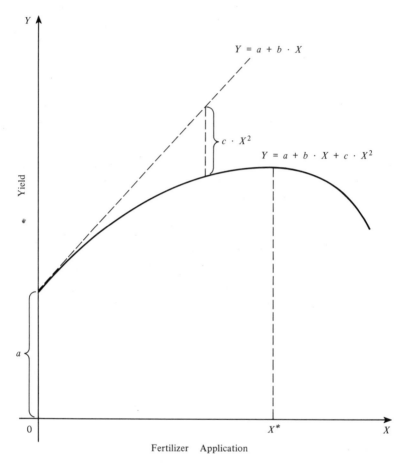

FIGURE 4.4   Quadratic Regression Curve

than absolute changes. The double logarithmic form, when graphed with $\log_e Y$ on the vertical axis and $\log_e X$ on the horizontal axis, is a straight line as in Fig. 4.5. The slope of this line measures the ratio of the percentage change in the X variable to the percentage change in the Y variable. In the terminology of economic theory, the slope is the elasticity of the Y variable in terms of the X variable.

A regression line that is logarithmic in the dependent variable is illustrated in Fig. 4.6. $\log_e Y$ is on the vertical axis and X is on the horizontal axis. The slope b is the percentage change in the Y variable *per unit* change in the X variable. For example, if the vertical axis measures the log of Gross National Product and the horizontal axis measures time, the slope is the growth rate (percentage change per unit of time) of Gross National Product.

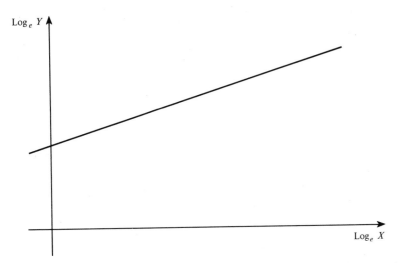

FIGURE 4.5   Logarithmic Regression in Dependent and Independent Variables

The slope of a regression line that is logarithmic in the independent variable is the ratio of the *absolute* change in the Y variable to the percentage change in the X variable. For example, if the Y variable is consumption expenditure and the X variable is income, then the slope indicates the ratio of the

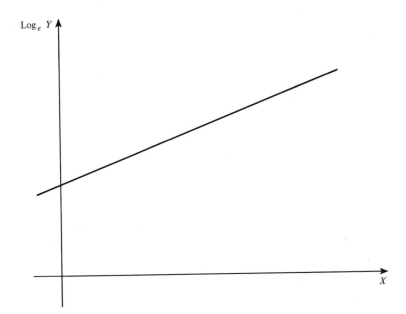

FIGURE 4.6   Logarithmic Regression in Dependent Variable

expected increase in consumption to a given percentage increase in income. For example, suppose the slope is 20. If consumption is \$80 and income is \$100, then a 10 percent increase in income (a \$10 increase) results in an increase in consumption (ΔY) which is calculated as follows:

$$\frac{\Delta Y}{.10} = 20, \qquad \Delta Y = (.10)20 = \$2.$$

In order to interpret a regression in terms of percentages, a regression line must be in terms of logarithms to the base e. The actual calculation of a regression line, however, is easier to perform in terms of logarithms to the base 10. Let us discuss the method of conversion for double logarithmic regression lines and regression lines that are logarithmic in either the dependent or independent variable.

By definition, the logarithm of a number Y to the base b ($\log_b Y$) is the number one gets by raising b to the power $\log_b Y$. That is,

$$b^{\log_b Y} = Y. \tag{4.25}$$

Thus

$$10^{\log_{10} Y} = Y \tag{4.26}$$

and

$$e^{\log_e Y} = Y. \tag{4.27}$$

Equating (4.26) and (4.27), we get

$$e^{\log_e Y} = 10^{\log_{10} Y}. \tag{4.28}$$

But from (4.25) also

$$10 = e^{\log_e 10}. \tag{4.29}$$

Substitute (4.29) into (4.28).

$$e^{\log_e Y} = \left(e^{\log_e 10}\right)^{\log_{10} Y} = e^{(\log_e 10)(\log_{10} Y)} \tag{4.30}$$

or

$$\log_e Y = (\log_e 10)(\log_{10} Y)$$

and

$$\log_e Y \cong (2.30259) \log_{10} Y. \tag{4.31}$$

Now let

$$\begin{aligned} Y_i' &= \log_e Y_i, & X_i' &= \log_e X_i, \\ Y_i'' &= \log_{10} Y_i, & X_i'' &= \log_{10} X_i, \end{aligned} \tag{4.32}$$

and from (4.31)

$$Y_i' \cong (2.30259) Y_i'', \qquad X_i' \cong (2.30259) X_i''. \tag{4.33}$$

Let a' and b' be the regression coefficients for a logarithmic regression line in which logarithms are to the base e and let a'' and b'' be the regression coefficients when logarithms are to the base 10.

First consider a double logarithmic regression line.

$$\begin{aligned} Y' &= a' + b'X' \quad \text{(to base } e\text{)}, \\ Y'' &= a'' + b''X'' \quad \text{(to base 10)}. \end{aligned} \tag{4.34}$$

From (3.13) we have

$$b' = \frac{\Sigma\, X_i' Y_i' - n\bar{X}' \cdot \bar{Y}'}{\Sigma\, X_i'^2 - n(\bar{X}')^2}. \tag{4.35}$$

From (4.34)

$$b = \frac{(\log_e 10)^2 [\Sigma\, X_i'' Y_i'' - n\bar{X}'' \cdot \bar{Y}'']}{(\log_e 10)^2 [\Sigma\, (X'')^2 - n(\bar{X}'')^2]} \tag{4.36}$$

or

$$b' = b''. \tag{4.37}$$

From (3.14)

$$\begin{aligned} a' &= Y' - b'X' \cong (2.30259)\,Y'' - b'(2.30259)X'' \\ &\cong (2.30259)[Y'' - b''X''] \end{aligned} \tag{4.38}$$

or

$$a' \cong (2.30259)a''. \tag{4.39}$$

Next consider a regression line that is logarithmic in the dependent variable.

$$\begin{aligned} Y' &= a' + b'X \quad \text{(to base } e\text{)}, \\ Y'' &= a'' + b''X \quad \text{(to base 10)}. \end{aligned} \tag{4.40}$$

From (3.13) we have

$$b' = \frac{\Sigma\, X_i Y_i' - n\bar{X} \cdot \bar{Y}'}{\Sigma\, X_i^2 - n(\bar{X})^2} \cong \frac{(2.30259)[\Sigma\, X_i Y_i'' - n\bar{X}\bar{Y}'']}{\Sigma\, X_i^2 - n(\bar{X})^2} \tag{4.41}$$

or

$$b' \cong (2.30259)b''. \tag{4.42}$$

From (3.14)

$$a' = \bar{Y}' - b'\bar{X} \cong (2.30259)\bar{Y}'' - (2.30259)b'' \cdot \bar{X} \tag{4.43}$$

or

$$a' \cong (2.30259) \cdot a''. \tag{4.44}$$

Finally, in the case of a regression line that is logarithmic in the independent variable, we have

$$Y = a' + b'X' \qquad \text{(to base } e\text{)},$$
$$Y = a'' + b''X'' \qquad \text{(to base 10)}. \tag{4.45}$$

It is easy to show that

$$b' \cong \frac{1}{(2.30259)} b'' \tag{4.46}$$

and

$$a' = a''. \tag{4.47}$$

These results may be summarized in the form of a table:

FORM OF REGRESSION	CONVERSION FROM $\log_e$ TO $\log_{10}$
Double logarithmic	$a' \cong (2.30259)a''$ $b' = b''$
Logarithmic in dependent variable	$a' \cong (2.30259)a''$ $b' \cong (2.30259)b''$
Logarithmic in independent variable	$a' = a''$ $b' \cong \dfrac{b''}{2.30259}$

Example 4.9. Let the Y variable represent Kenya exports and the X variable Gross Domestic Product. Let us use the data from Table 3.1 to fit a double logarithmic regression line.

Y_i	X_i	$Y_i'' = \log_{10} Y_i$	$X_i'' = \log_{10} X_i$
28.0	180.5	1.44716	2.25648
33.0	193.2	1.51851	2.28601
31.2	205.9	1.49415	2.31366
33.2	208.1	1.52114	2.31827
38.4	214.8	1.58433	2.33203
40.2	225.5	1.60423	2.35315
41.7	224.7	1.62014	2.35160
45.1	244.1	1.65418	2.38757
51.0	259.1	1.70757	2.41347
53.5	277.7	1.72835	2.44358

$(Y_i'')^2$	$(X_i'')^2$	$(X_i'') \cdot (Y_i'')$
2.09427	5.09170	3.26549
2.30587	5.22584	3.47133
2.23248	5.35302	3.45696
2.31387	5.37438	3.52641
2.51010	5.43836	3.69471
2.57355	5.53731	3.77499
2.62485	5.53002	3.80992
2.73631	5.70049	3.94947
2.91580	5.82484	4.12117
2.98719	5.97108	4.22336

$$\bar{Y}'' = 1.58798,$$
$$\bar{X}'' = 2.34558,$$
$$n(\bar{Y}'')^2 = 25.21680,$$
$$n(\bar{X}'')^2 = 55.01745,$$
$$n\bar{X}'' \cdot \bar{Y}'' = 37.24734.$$

$$\Sigma\,(Y_i'')^2 = 25.29429,$$
$$\Sigma\,(X_i'')^2 = 55.04704,$$
$$\Sigma\,(X_i'')(Y_i'') = 37.29381.$$

$$\Sigma\,(Y_i'')^2 - n(\bar{Y}'')^2 = .07749,$$
$$\Sigma\,(X_i'')^2 - n(\bar{X}'')^2 = .02959,$$
$$\Sigma\,(X_i'')(Y_i'') - n(\bar{X}'')(\bar{Y}'') = .04647.$$

$$b'' = \frac{.04647}{.02959} = 1.57046,$$
$$a'' = 1.58798 - 1.57046(2.34558),$$
$$a'' = -2.09566.$$

The regression line is

$$\log_{10} Y = -2.09566 + 1.57046 \log_{10} X.$$

The coefficient of determination is

$$R^2 = \frac{(.04647)^2}{(.07749)(.02959)} = .9420.$$

Let us convert the regression line into a line in terms of logarithms to the base e.

$$a' \cong (2.30259)(-2.09566) = -4.82545,$$
$$b' = 1.57046,$$

or

$$\log_e Y = -4.82545 + 1.57046 \log_e X. \tag{4.48}$$

The slope of this regression line is 1.57046. It may be interpreted as the elasticity of exports with respect to Gross Domestic Product—that is, the percentage change in exports divided by the percentage change in Gross Domestic Product.

Example 4.10.   Using the same data as in the previous example, let us fit two other regression lines—one which is logarithmic in the dependent variable only and the other which is logarithmic in the independent variable only. We get

$$\log_{10} Y = 0.91535 + (.003011)X$$

and

$$Y = -299.15 + (144.39) \log_{10} X.$$

These regression equations may be converted to logarithms to the base e as follows:

$a' \cong (2.30259)(0.91535) = 2.10768$ (logarithmic in
$b' \cong (2.30259)(.003011) = .006934$ dependent variable)

$a' = -299.15$ (logarithmic in
$b' \cong \dfrac{144.39}{2.30259} = 62.708$ independent variable)

The regression equations are

$$\log_e Y = 2.10768 + (.006934)X \qquad (R^2 = .9390) \tag{4.49}$$

and

$$Y = -299.15 + (62.708) \log_e X \qquad (R^2 = .9499). \tag{4.50}$$

Note that the slope (1.57046) of the double logarithmic regression line which was determined in Example 4.9 is the elasticity of exports with respect to Gross Domestic Product. The single logarithmic regression lines computed in Example 4.10 have an implied elasticity. The slope of the regression line logarithmic in the dependent variable may be interpreted as

$$b' = \frac{\text{percentage change in } Y}{\text{absolute change in } X}$$

or

$$b' \cdot X = \frac{\text{percentage change in } Y}{(\text{absolute change in } X)/X}$$

$$= \frac{\text{percentage change in } Y}{\text{percentage change in } X}.$$

Thus the elasticity $b' \cdot X$ changes with X. The term *average elasticity* is used often in this case and refers to the elasticity valued at $\bar{X}$.

Example 4.11. The slope of the regression line logarithmic in the dependent variable as computed in Example 4.10 was

$b' = .006934.$

The mean of the X variable (Gross Domestic Product) was computed to be

$\bar{X} = 223.36.$

Thus the average elasticity of exports (Y variable) with respect to Gross Domestic Product (X variable) is

$b' \cdot \bar{X} = (.006934)(223.36) = 1.54878.$

That is, a 1 percent change in exports accompanies, on the average, an increase of 1.5 percent in Gross Domestic Product.

Similarly, we may compute an implied elasticity for the regression line logarithmic in the independent variable. The slope b' may be interpreted as

$$b' = \frac{\text{absolute change in } Y}{\text{percentage change in } X}$$

or

$$b' \cdot \frac{1}{Y} = \frac{(\text{absolute change in } Y)/Y}{\text{percentage change in } X}$$

$$= \frac{\text{percentage change in } Y}{\text{percentage change in } X}.$$

Thus the elasticity $b' \cdot (1/Y)$ varies inversely with the value of Y. The average elasticity for the regression line logarithmic in the independent variable is $b' \cdot (1/\bar{Y})$.

Example 4.12. The slope of the regression line logarithmic in the independent variable was computed in Example 4.10 to be

$b' = 62.708.$

The average value of the Y variable (exports) was computed to be

$\bar{Y} = 39.53.$

Thus the average elasticity is

$$b' \cdot \frac{1}{\bar{Y}} = 62.708 \cdot \frac{1}{39.53} = 1.58634.$$

That is, a 1 percent rise in Gross Domestic Product is associated, on the average, with an increase of 1.6 percent in exports.

Of the three regression lines computed in Examples 4.9 and 4.10, the regression line logarithmic in the independent variable (4.50) computed in Example 4.10 gives the highest coefficient of determination [.9499 as compared to .9390 for the line logarithmic in the dependent variable (4.49) and .9420 for the double logarithmic regression line (4.48)]. In this case, one naturally asks whether it is possible to conclude that the line with the highest coefficient of determination gives a better "fit" than all others. The answer is no! The coefficient of determination gives the percentage of variation in the dependent variable that is "explained" by variation in the independent variable. The dependent variable, however, may be different. For example, in the double logarithmic line and the line logarithmic in the dependent variable, the coefficient of determination relates to variations in $\log_e Y$. In the regression line logarithmic in the independent variable, the coefficient relates to variations in Y. By transforming the variable from Y to $\log_e Y$ the variance is changed. Thus two different regression lines may explain different amounts of different variations, and straight comparisons of coefficients of determination are not legitimate. The only way to make comparisons is to convert all variables to a common basis and calculate coefficients of determination for the converted variables.

Example 4.13.   Since the dependent variables in (4.48) and (4.49) are the same—$\log_e Y$—the coefficients are directly comparable. That is, $R^2 = .9420$ for (4.48) and $R^2 = .9390$ for (4.49) so that (4.48), the double logarithmic equation, gives a better fit. In order to compare (4.50) with either (4.48) or (4.49), however, the predicted values of Y from equation (4.50) must be transformed into logarithms. This is done below.

$\hat{Y}_i = -299.15$ $+ (62.708) \log_e X_i$	$\log_{10} \hat{Y}_i$	$\log_{10} Y_i$
26.66	1.42586	1.44716
30.93	1.49038	1.51851
34.92	1.54307	1.49415
35.59	1.55133	1.52114
37.57	1.57484	1.58433
40.62	1.60874	1.60423
40.40	1.60638	1.62014
45.59	1.65887	1.65418
49.33	1.69311	1.70757
53.68	1.72981	1.72835
	15.88239	

$(\log_{10} Y_i)^2$	$\log_{10} \hat{Y}_i - \log_{10} Y_i$	$(\log_{10} \hat{Y}_i - \log_{10} Y_i)^2$
2.094	$-.02130$	.000454
2.306	$-.02813$	.000791
2.232	.04892	.002393
2.314	.03019	.000911
2.510	$-.00949$	.000090
2.574	.00451	.000020
2.625	$-.01376$	.000189
2.736	.00469	.000022
2.916	$-.01446$	.000209
2.987	.00146	.000002
25.294		.005081

The sum of the squared residuals in terms of logarithms is

$$\Sigma\,(u_i^*)^2 = \Sigma\,(\log_{10} \hat{Y}_i - \log_{10} Y_i)^2 = .005081.$$

Thus in terms of the logarithms we have a modified standard error of the estimate:

$$(s_u^*)^2 = \frac{\Sigma\,(u_i^*)^2}{n} = .0005081.$$

Similarly, the variance of the logarithm of the dependent variable is

$$(s_Y^*)^2 = \frac{\Sigma\,(\log_{10} Y_i)^2}{n} - (\overline{\log_{10} Y})^2$$

$$= \frac{25.294}{10} - (1.5882)^2$$

$$= .00700$$

The coefficient of determination in terms of logarithms is the explained variation divided by the total variation, or

$$(R^*)^2 = \frac{(s_Y^*)^2 - (s_u^*)^2}{(s_Y^*)^2}$$

$$= \frac{.00700 - .00051}{.00820}$$

$$= .9271.$$

We see that this adjusted R^2 for regression (4.50) is less than the original R^2 of .9499. Furthermore, the adjusted R^2 is less for equation (4.50) than (for 4.48) or (4.49). On a comparable basis, then, equation (4.48) gives the best fit of all three equations.

A regression equation logarithmic in the dependent variable is often used to compute the compound percentage rate of increase of a variable (Y) over time (t). The regression equation estimated is

$$\log_e Y = a + b \cdot t, \tag{4.51}$$

where b represents the compound rate of increase.

Example 4.14.   Let us calculate the rate of growth of exports for the Kenya economy, using the data in Table 3.1. The estimated value of b is the same if, instead of using the years (1956, 1957, ...) as representations of the time variable, one uses $t = 0, 1, 2, \ldots$, with $t = 0$ corresponding to 1956.

Log$_{10}$ of Kenya Exports

YEAR	$\log_{10} Y = Y''$	t
1956	1.519	0
1957	1.494	1
1958	1.521	2
1959	1.584	3
1960	1.604	4
1960	1.620	5
1961	1.654	6
1962	1.708	7
1963	1.728	8

We have

$$\bar{Y}'' = 1.604,$$
$$\bar{t} = 4,$$
$$\Sigma\, t^2 = 204,$$
$$\Sigma\, (\log_{10} Y_i) \cdot t = 59.508.$$

From (3.13), we obtain

$$b = \frac{\Sigma\, (\log_{10} Y_i) \cdot t - 9 \cdot \bar{t} \cdot \bar{Y}''}{\Sigma\, t^2 - 9(4)^2} = \frac{1.7}{60} = .0294.$$

Since we have done the computations in terms of logarithms to the base 10, we must use Equation (4.42) to convert to logarithms to the base e to determine the rate of growth.

$$(2.30259)b = (2.30259)(.0294) = .068.$$

Thus the percentage rate of growth is 6.8 percent.

APPENDIX Derivation of Formulae

Normal Equations for Multiple Regression

The normal equations for multiple regression analysis are derived by minimizing the sum of the squared vertical deviations from the regression line:

$$S = \sum_{i=1}^{n} [Y_i - a_0 - a_1X_{1i} - a_2X_{2i} - \cdots - a_kX_{ki}]^2 \qquad (4.52)$$

We may differentiate S partially with respect to each of the parameters $a_0, a_1, a_2, \ldots, a_k$ and set each derivative equal to zero.

$$\frac{\partial S}{\partial a_0} = 2 \sum_{i=1}^{n} [Y_i - a_0 - a_1X_{1i} - a_2X_{2i} - \cdots - a_kX_{ki}](-1) = 0$$

$$\frac{\partial S}{\partial a_1} = 2 \sum_{i=1}^{n} [Y_i - a_0 - a_1X_{1i} - a_2X_{2i} - \cdots - a_kX_{ki}](-X_{1i}) = 0$$

$$\frac{\partial S}{\partial a_2} = 2 \sum_{i=1}^{n} [Y_i - a_0 - a_1X_{1i} - a_2X_{2i} - \cdots - a_kX_{ki}](-X_{2i}) = 0 \qquad (4.53)$$

.
.
.

$$\frac{\partial S}{\partial a_k} = 2 \sum_{i=1}^{n} [Y_i - a_0 - a_1X_{1i} - a_2X_{2i} - \cdots - a_kX_{ki}](-X_{ki}) = 0$$

Divide both sides of (4.53) by 2 and multiply each term in the brackets by (-1), $(-X_{1i})$, $(-X_{2i})$, $\ldots$, $(-X_{ki})$, respectively:

$$\Sigma \, (-Y_i + a_0 + a_1X_{1i} + a_2X_{2i} + \cdots + a_kX_{ki}) = 0$$
$$\Sigma \, (-Y_iX_{1i} + a_0X_{1i} + a_1X_{1i}^2 + a_2X_{2i}X_{1i} + \cdots + a_kX_{ki}X_{1i}) = 0$$
$$\Sigma \, (-Y_iX_{2i} + a_0X_{2i} + a_1X_{1i}X_{2i} + a_2X_{2i}^2 + \cdots + a_kX_{ki}X_{2i}) = 0$$

.
.
.

$$\qquad (4.54)$$

$$\Sigma \, (-Y_iX_{ki} + a_0X_{ki} + a_1X_{.i}X_{ki} + a_2X_{2i}X_{ki} + \cdots + a_kX_{ki}^2) = 0$$

Using the rules for summation signs (see the Appendix to Chapter 2) and bringing the first term in each equation over to the right-hand side, we get

$$n \cdot a_0 + (\Sigma \, X_{1i})a_1 + (\Sigma \, X_{2i})a_2 + \cdots + (\Sigma \, X_{ki})a_k = (\Sigma \, Y_i)$$
$$(\Sigma \, X_{1i})a_0 + (\Sigma \, X_{1i}^2)a_1 + (\Sigma \, X_{2i}X_{1i})a_2 + \cdots + (\Sigma \, X_{ki}X_{1i})a_k = (\Sigma \, Y_iX_{1i})$$
$$(\Sigma \, X_{2i})a_0 + (\Sigma \, X_{1i}X_{2i})a_1 + (\Sigma \, X_{2i}^2)a_2 + \cdots + (\Sigma \, X_{ki}X_{2i})a_k = (\Sigma \, Y_iX_{2i})$$

$$(\Sigma \, X_{ki})a_0 + (\Sigma \, X_{1i}X_{ki})a_1 + (\Sigma \, X_{2i}X_{ki})a_2 + \cdots + (\Sigma \, X_{ki}^2)a_k = (\Sigma \, Y_iX_{ki})$$

$$(4.55)$$

These are the normal equations in the form given in (4.14). These are linear equations in the parameters $a_0, a_1, \ldots, a_k$. One solves the normal equations to obtain least-squares estimates of $a_0, a_1, \ldots, a_k$.

In order to simplify later algebraic manipulation let

$$y_i = Y_i - \bar{Y},$$
$$x_{1i} = X_{1i} - \bar{X}_1,$$
$$x_{2i} = X_{2i} - \bar{X}_2,$$

$$(4.56)$$

$$x_{ki} = X_{ki} - \bar{X}_k,$$

where $\bar{X}_1, \bar{X}_2, \ldots, \bar{X}_k$ represent the sample means for $X_1, X_2, \ldots, X_k$— that is,

$$\bar{X}_j = \frac{\displaystyle\sum_{i=1}^{n} X_{ji}}{n} \qquad \text{for } j = 1, 2, \ldots, k. \tag{4.57}$$

The first normal equation in (4.55) may be divided by n on both sides to give

$$a_0 = \bar{Y} - a_1\bar{X}_1 - a_2\bar{X}_2 - \cdots - a_k\bar{X}_k. \tag{4.58}$$

If this value of a_0 is substituted into S in (4.52), we get

$$S = \sum_{i=1}^{n} (y_i - a_1x_{1i} - \cdots - a_kx_{ki})^2. \tag{4.59}$$

Differentiate S and arrange terms to obtain the normal equations:

$$(\Sigma \, x_{1i}^2)a_1 + (\Sigma \, x_{2i}x_{1i})a_2 + \cdots + (\Sigma \, x_{ki}x_{1i})a_k = \Sigma \, y_ix_{1i}$$
$$(\Sigma \, x_{1i}x_{2i})a_1 + (\Sigma \, x_{2i}^2)a_2 + \cdots + (\Sigma \, x_{ki}x_{2i})a_k = \Sigma \, y_ix_{2i}$$

$$(4.60)$$

$$(\Sigma \, x_{1i}x_{ki})a_1 + (\Sigma \, x_{2i}x_{ki})a_2 + \cdots + (\Sigma \, x_{ki}^2)a_k = \Sigma \, y_ix_{ki}$$

The least-squares estimates may be determined by solving (4.60) for $a_1, a_2, \ldots, a_k$, and a_0 is then determined from (4.58).

Standard Error of the Estimate

The standard error of the estimate is

$$s_u^2 = \frac{\sum (y_i - a_1 x_{1i} - \cdots - a_k x_{ki})^2}{n},\tag{4.61}$$

where $a_1, a_2, \ldots, a_k$ are the least-squares estimates of the parameters $a_1, a_2, \ldots, a_k$. For computational purposes another form of s_u^2 is desirable. We shall derive this form for $k = 2$ and then state the general expression for arbitrary k. Let us rewrite the numerator of (4.61) as follows:

$$\sum (y_i - a_1 x_{1i} - a_2 x_{2i})^2 = \sum y_i^2 - 2 \sum y_i(a_1 x_{1i} + a_2 x_{2i}) + \sum (a_1 x_{1i} + a_2 x_{2i})^2.\tag{4.62}$$

Let us expand the last term on the right-hand side of (4.62):

$$\sum (a_1 x_{1i} + a_2 x_{2i})^2 = \sum (a_1 x_{1i})^2 + 2 \sum a_1 a_2 x_{1i} x_{2i} + \sum (a_2 x_{2i})^2$$

or

$$\sum (a_1 x_{1i} + a_2 x_{2i})^2 = [a_1^2 \sum x_{1i}^2 + a_1 a_2 \sum x_{1i} x_{2i}] + [a_2 a_1 \sum x_{1i} x_{2i} + a_2^2 \sum x_{2i}^2]$$

or

$$\sum (a_1 x_{1i} + a_2 x_{2i})^2 = a_1[a_1 \sum x_{1i}^2 + a_2 \sum x_{1i} x_{2i}] + a_2[a_1 \sum x_{1i} x_{2i} + a_2 \sum x_{2i}^2].\tag{4.63}$$

Now for $k = 2$, the normal equations are

$$a_1(\sum x_{1i}^2) + a_2(\sum x_{2i} x_{1i}) = \sum y_i x_{1i},$$
$$a_1(\sum x_{1i} x_{2i}) + a_2(\sum x_{2i}^2) = \sum y_i x_{2i}.\tag{4.64}$$

Substitute each of the these expressions into (4.61).

$$\sum (a_1 x_1 + a_2 x_{2i})^2 = a_1(\sum y_i x_{1i}) + a_2(\sum y_i x_{2i}).\tag{4.65}$$

Substitute (4.65) for the last term on the right-hand side of (4.62).

$$\sum (y_i - a_1 x_{1i} - a_2 x_{2i})^2 = \sum y_i^2 - 2[a_1 \sum y_i x_{1i} + a_2 \sum y_i x_{2i}] + [a_1 \sum y_i x_{1i} + a_2 \sum y_i x_{2i}].\tag{4.66}$$

Using (4.66) in (4.61), we derive the following expression for s_u^2.

$$s_u^2 = \frac{\sum y_i^2 - \hat{a}_1 \sum y_i x_{1i} - \hat{a}_2 \sum y_i x_{2i}}{n}.\tag{4.67}$$

This is a convenient expression for the calculation of s_u^2, where

$$\sum y_i^2 = \sum Y_i^2 - n(\bar{Y})^2,$$
$$\sum y_i x_{1i} = \sum Y_i X_{1i} - n \cdot \bar{Y} \cdot \bar{X}_1,\tag{4.68}$$
$$\sum y_1 x_{2i} = \sum Y_i X_{2i} - n \cdot \bar{Y} \cdot \bar{X}_2.$$

These expressions are derived in the same way as (3.60) and (3.61) of Appendix 2 to Chapter 3.

Coefficient of Determination

The definition of the coefficient of determination is

$$R^2 = \frac{s_y^2 - s_u^2}{s_y^2}. \tag{4.69}$$

$$s_Y^2 = \frac{1}{n} \sum (Y_i - \bar{Y})^2 = \frac{1}{n} \sum y_i^2 \tag{4.70}$$

and from (4.69)

$$R^2 = \frac{n \cdot s_Y^2 - n \cdot s_u^2}{n \cdot s_Y^2}$$

or

$$R^2 = \frac{\sum y_i^2 - \sum y_i^2 + \hat{a}_1 \sum y_i x_{1i} + \hat{a}_2 \sum y_i x_{2i}}{\sum y_i^2}$$

and

$$R^2 = \frac{\hat{a}_1 \sum y_i x_{1i} + \hat{a}_2 \sum y_i x_{2i}}{\sum y_i^2}. \tag{4.71}$$

The expression (4.71) combined with (4.68) is very efficient for purposes of calculation. If we use the notation in the chapter ($a_0 = a$; $a_1 = b$; and $a_2 = c$), the expressions (4.71) and (4.67) combined with (4.68) are equivalent to (4.8) and (4.9) in the main body of this chapter.

Generalization to k Variables

The expressions (4.67) and (4.71) may be generalized to the case of k variables $X_1, X_2, \ldots, X_k$.

$$s_u^2 = \frac{\sum y_i^2 - \hat{a}_1 \sum y_i x_{1i} - \cdots - \hat{a}_k \sum y_i x_{ki}}{n} \tag{4.72}$$

and

$$R^2 = \frac{\hat{a}_1 \sum y_i x_{1i} + \cdots + \hat{a}_k \sum y_i x_{ki}}{\sum y_i^2}, \tag{4.73}$$

where

$$\sum y_i^2 = \sum_{i=1}^{n} Y_i^2 - n \cdot (\bar{Y})^2,$$

$$\sum y_i x_{ji} = \sum_{i=1}^{n} Y_i X_{ji} - n \cdot \bar{Y} \cdot \bar{X}_j \quad \text{for } j = 1, 2, \ldots, k. \tag{4.74}$$

Nonlinear Regression

Consider the following regression equation, which is linear in the logarithms of all variables:

$$\log_e Y = a_0 + a_1 \log_e X_1 + \cdots + a_k \log_e X_k. \tag{4.75}$$

If we differentiate this equation partially and implicitly with respect to X_1, we obtain

$$\frac{1}{Y}\frac{\partial Y}{\partial X_1} = a_1 \frac{1}{X_1} \quad \text{or} \quad \frac{\partial Y}{\partial X_1}\cdot\frac{Y}{X_1} = a_1$$

and

$$\frac{\partial Y/Y}{\partial X_1/X_1} = a_1. \tag{4.76}$$

The numerator of this expression is the change in Y (that is, ∂Y) divided by Y, or the percentage change in Y. The denominator is the percentage change in X_1. Thus a_1 is the percentage change in Y divided by the percentage change in X_1, and it is called the elasticity of Y with respect to X_1. In general, the elasticity of Y with respect to X_j is

$$\frac{\partial Y/Y}{\partial X_j/X_j} = a_j. \tag{4.77}$$

The estimated value of a_j is equivalent to an estimate of this elasticity.

If the regression equation is logarithmic in the dependent variable, then

$$\log_e Y = a_0 + a_1 X_1 + \cdots + a_k X_k, \tag{4.78}$$

and

$$\frac{1}{Y}\frac{\partial Y}{\partial X_j} = a_j$$

or

$$\frac{\partial Y/Y}{\partial X_j/X_j} = a_j X_j \tag{4.79}$$

is the elasticity of Y with respect to X_j. An implicit, estimated average elasticity is obtained by multiplying the estimated value of the coefficient X_j by the mean value of X_j.

Finally, consider a regression equation that is logarithmic in the independent variables

$$Y = a_0 + a_1 \log_e X_1 + \cdots + a_k \log_e X_k. \tag{4.80}$$

Differentiating, we get

$$\frac{\partial Y}{\partial X_j} = a_j \frac{1}{X_j}$$

or

$$\frac{\partial Y/Y}{\partial X_j/X_j} = a_j \frac{1}{Y},$$

(4.81)

which is the elasticity of Y with respect to X_j. The implicit, estimated average elasticity is the estimated value of a_j divided by the mean value of the variable Y.

In the particular case where time is the only independent variable, a regression equation logarithmic in the dependent variable may be written

$$\log_e Y = a + b \cdot t,$$

(4.82)

where t is a time variable. Differentiating, we get

$$\frac{1}{Y}\frac{dY}{dt} = b$$

or

$$\frac{dY/Y}{dt} = b.$$

(4.83)

This expression may be interpreted as the percentage change in Y per unit of time or, as it is sometimes called, the rate of growth of the variable Y. The estimated value of the coefficient b is the estimated rate of growth of Y.

PROBLEMS

1. The following table contains data on the IQ's of a group of students, the number of hours they spent studying for a particular test, and their scores on this test.

(X) IQ	(Z) HOURS	(Y) SCORE
90	12	50
150	0	50
110	8	70
125	8	85
125	3	50
100	7	65
115	12	95
120	1	40
135	9	93

Using test score (Y) as the dependent variable and IQ (X) and hours of study (Z) as the independent variable, what is the linear regression line?

2. Given the following data on sales (Y), radio advertisement expenditures (X), and newspaper advertisement expenditures, (Z).

Y	X	Z
2	3	8
4	6	13
6	10	17
8	13	26

(a) Find the linear regression line. (b) Demonstrate that the regression line passes through the point of means ($Y = \bar{Y}$, $X = \bar{X}$, and $Z = \bar{Z}$).

3. (a) Using the data from problem 1, calculate the simple correlation coefficient between hours studied and test score. (b) What is the partial correlation coefficient where the effects of IQ are taken into account?

4. (a) Using the data from problem 2, calculate a regression equation that is logarithmic in all of the variables. (b) What is the elasticity of sales with respect to radio advertising expenditures? (c) Which provides a better fit—a linear regression line or a regression line in the logarithms of the variables?

5. (a) Using the data on sales (Y) and radio expenditures (X) from problem 2, calculate a regression line that is logarithmic in the Y variable only and one that is logarithmic in the X variable only. (b) Which line gives a higher estimate for the elasticity of sales with respect to radio advertising expenditures at the mean level of sales and advertising expenditures?

6. What is the rate of growth of exports in Table 3.1 of Chapter 3?

*7. Show that

$$\sum_{i=1}^{n} y_i x_{1i} = \sum_{i=1}^{n} Y_i X_{1i} - n \cdot \bar{Y} \cdot \bar{X}_1,$$

where $y_i = Y_i - \bar{Y}$, $x_{1i} = X_{1i} - \bar{X}_1$.

*8. Show that

$$R^2_{YX \cdot Z} = \hat{b}_{YXZ} \cdot \hat{b}_{XYZ},$$

where $\hat{b}_{YXZ}$ is the estimated regression coefficient in the regression equation (4.18) and $\hat{b}_{XYZ}$ is the estimated regression coefficient in

$$X = a_{XYZ} + b_{XYZ} X + c_{XYZ} Z.$$

Outline of Further Reading   for Part I

A. General Description
   1. Huff [1954], *passim*
   2. Levinson [1963], *passim*
   3. Wallis and Roberts [1962], *passim*

* Problems based on the appendix to this chapter.

The Huff book, *How to Lie with Statistics*, is an excellent primer on the misuse of statistics in a variety of circumstances. The Levinson, and Wallis and Roberts texts are both very well-written, nontechnical books which attempt to give an intuitive feel for the uses of statistics.

B. The Basic Elements of Descriptive Statistics—Charts, Graphs, Histograms or Frequency Distributions, Measures of Central Tendency and Dispersion
   1. Allen [1966], pp. 1–95
   2. Chou [1969], pp. 22–86
   3. Croxton and Cowden [1955], pp. 1–239
   4. Freund [1970], pp. 10–51
   5. Freund [1967], pp. 9–82
   6. Freund and Williams [1965], pp. 1–96
   7. Hoel [1966], pp. 7–44
   8. Mills [1955], pp. 1–133
   9. Neter and Wasserman [1966], pp. 1–186
   10. Richmond [1964], pp. 1–100
   11. Suits [1963], pp. 7–53
   12. Wallis and Roberts [1956], pp. 167–308
   13. Wonnacott and Wonnacott [1969], pp. 8–26
   14. Yamane [1967], pp. 1–77
   15. Yule and Kendall [1950], pp. 1–198

The Croxton and Cowden and Wallis and Roberts books are particularly elementary. The former has a very comprehensive coverage of descriptive techniques and is very useful as a reference work. The Allen book is a very handy short treatment of descriptive statistics particularly useful for economists.

C. Correlation and Regression—Heavy Emphasis on Descriptive Approach
   1. Allen [1966], pp. 115–152
   2. Croxton and Cowden [1955], pp. 261–319, 451–587
   3. Freund [1970], pp. 261–283
   4. Wonnacott and Wonnacott [1969], pp. 220–233
   5. Yule and Kendall [1950], pp. 199–365

These books introduce methods of correlation and regression primarily as techniques of descriptive statistics. They assume no knowledge of probability theory and have only passing references to the concept of statistical inference.

D. Correlation and Regression—Some Reference to Probability, Sampling, Statistical Inference
   1. Bryant [1966], pp. 123–146, 212–243
   2. Chou [1969], pp. 595–672

3. Freund [1967], pp. 331–378
4. Freund and Williams [1964], pp. 296–351
5. Hoel [1966], pp. 192–234
6. Mills [1955], pp. 246–318, 579–611
7. Neter and Wasserman [1966], pp. 512–567
8. Suits [1963], pp. 189–202
9. Wallis and Roberts [1956], pp. 524–558

These books introduce methods of regression and correlation analysis without reference to probability theory, but also cover tests of hypotheses in regression analysis which require some familiarity with the elements of probability theory.

E. Regression and Correlation—Heavy Emphasis on Probability and Inference
1. Hoel [1962], pp. 160–211
2. Kane [1968], pp. 217–282
3. Richmond [1964], pp. 424–474
4. Wonnacott and Wonnacott [1969], pp. 234–311
5. Wonnacott and Wonnacott [1970], pp. 1–131
6. Yamane [1967], pp. 368–472 and 752–844

These books rely quite heavily on probability theory and the theory of joint probability distributions in their development of regression and correlation methods.

F. Computation Techniques
1. Dwyer [1940–1941]
2. Ezekiel [1941], pp. 455–485

Some of the older textbooks, such as Ezekiel, describe in detail the Doolittle method for solving normal equations. Dwyer is a short theoretical piece on the Doolittle method which contains some illustrative examples. Since many packaged programs for solving the normal equations on modern high-speed electronic computers are available, the average user of regression and correlation techniques does not usually need to know the details of efficient computation techniques. The computer does the work for him.

Part II
Probability

Chapter 5

Foundations of Probability

At this point it is very useful to introduce the notion of probability, random variables, and probability distributions. The rest of this book depends a great deal on the foundation laid here and in the next chapter.

Sets

Fundamental to the theory of probability is the notion of a set. A *set* is a collection of objects called the *elements* of that set. The number of elements in a set may be finite or infinite.

Example 5.1. The books on a shelf constitute a set of books. Any particular book on that shelf is a member of that set.

Example 5.2. The numbers 1 through 10 constitute a set which we can denote by

$\{1, 2, 3, 4, 5, 6, 7, 8, 9, 10\}$

The number 5 is an element of this set. This set has a finite number (10) of elements.

Example 5.3. All positive even integers constitute a set which we can denote by

$\{2, 4, 6, 8, \ldots\}$

This set has an infinite number of elements.

In this volume we shall concern ourselves mostly with the notion of a *point set*. A set of points along a line or in a plane is a point set.

Example 5.4.   The set of points between 10 and 15 is a point set. (See Fig. 5.1.) This set has an infinite number of elements, since there are an infinite number of points between 10 and 15.

$$-5 \qquad 0 \qquad 5 \qquad 10 \qquad 15$$

FIGURE 5.1   Point Set on a Line

Example 5.5.   The set of *all* points along a line is also a point set with an infinite number of elements.

Example 5.6.   In Fig. 5.2, the set of points in the rectangle $ABCD$ is a point set in a plane.

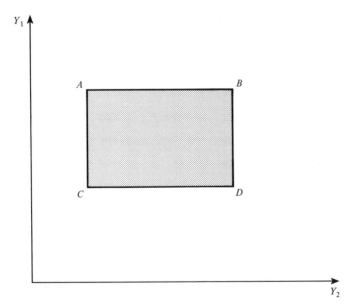

FIGURE 5.2   Point Set in a Plane

A member of the set of *all* points along a line may be represented by a variable Y. In such a case, the variable Y is said to be single-dimensional. A member of the set of *all* points in a plane can be represented by an ordered pair of variables $Y = (Y_1, Y_2)$. The ordered pair $Y = (Y_1, Y_2)$ is called a two-dimensional variable (or vector). The points in a three-dimensional space are represented by an ordered *triplet* $Y = (Y_1, Y_2, Y_3)$

of single-dimensional variables. The notion of point sets may be extended to n-dimensional spaces. The ordered n-tuplet $Y = (Y_1, Y_2, \ldots, Y_n)$ represents a member of the set of all points in an n-dimensional space.

Example 5.7. A member of the set of points along a line can be represented by the variable Y, as shown in Fig. 5.3.

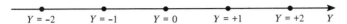

$Y = -2$ $Y = -1$ $Y = 0$ $Y = +1$ $Y = +2$ Y

FIGURE 5.3   Single Dimensional Variable

Example 5.8. In a two-dimensional space or plane we represent a point by a pair of numbers (Y_1, Y_2), as in Fig. 5.4. The number Y_1 is determined by reading off the vertical axis or ordinate and the number Y_2 by reading off the horizontal axis or abscissa. The points $(Y_1, Y_2) = (4, 3); (Y_1, Y_2) =$

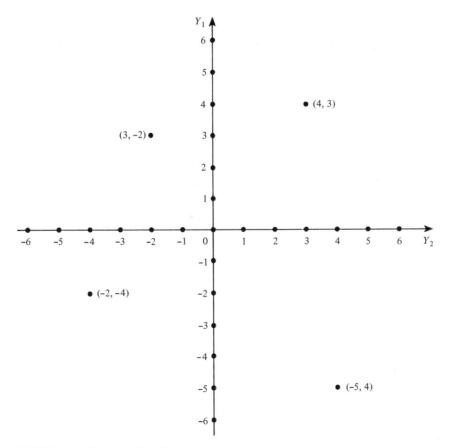

FIGURE 5.4   Points in Two Dimensions

$(-5, 4)$; $(Y_1, Y_2) = (-2, -4)$; and $(Y_1, Y_2) = (3, -2)$ are shown in Fig. 5.4.

Sets may be added. If A is a point set and B is a point set, their sum (or union) $A + B$ is the set of points each member of which is either an element of A *or* an element of B.

Example 5.9.   The set of points on a line between 5 and 10 and the set between 10 and 15 may be added to form a set of points between 5 and 15. (See Fig. 5.5.)

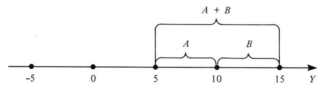

FIGURE 5.5   Addition of Sets on a Line

Example 5.10.   The set of points contained in the area marked A in Fig. 5.6 and those contained in B may be added to form a set of points, the shaded area in the diagram. This new set includes *both* the points in A and the points in B. Each member of this combined set is either a member of A *or* a member of B.

Sets may be multiplied. The product (or intersection) $A \cdot B$ is the set of points each element of which is *both* in A *and* in B.

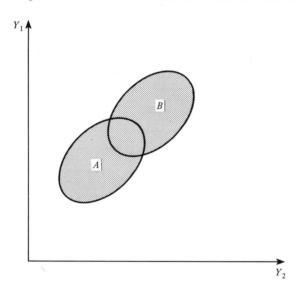

FIGURE 5.6   Addition of Sets in Two Dimensions

Example 5.11. The intersection or product of the set of points from 5 to 10 and the set of points from 10 to 15 contains one point, the point $Y = 10$. (See Fig. 5.7.)

FIGURE 5.7   Multiplication of Sets on a Line

Example 5.12. The intersection of the sets A and B is the set of points in the shaded area of Fig. 5.8. The points in the shaded area are members of *both A and B*.

If the product or intersection of two sets A and B contains no points (if A and B have no points in common), the sets A and B are said to be *disjoint*.

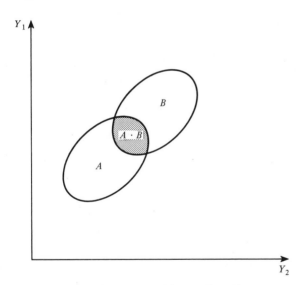

FIGURE 5.8   Multiplication of Sets in Two Dimensions

Example 5.13. The sets A and C in Fig. 5.9 are disjoint. They have no points in common and their intersection contains no points.

A *subset* of a set A is a set each element of which is also an element of A.

Example 5.14. The set of even integers $\{\ldots, -4, -2, 0, 2, 4, \ldots\}$ is a subset of the set of all integers $\{\ldots, -4, -3, -2, -1, 0, 1, 2, 3, 4, \ldots\}$.

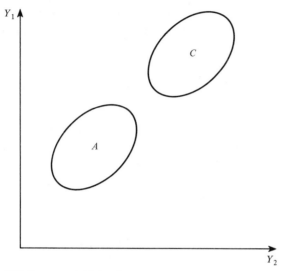

FIGURE 5.9 **Disjoint Sets**

Example 5.15. In Fig. 5.10, the set B is a subset of A since every point in B is also a point in A.

If B is a subset of A, then we may subtract B from A. The difference $A - B$ is the set of points belonging to A but not to B.

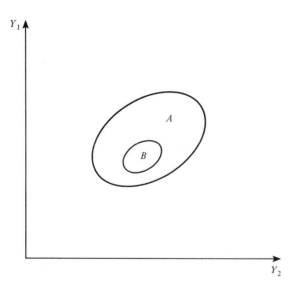

FIGURE 5.10 **Subset in Two Dimension**

Example 5.16. If A is the set of integers between 1 and 10 and B is the set of even integers between 1 and 10, then

$$A - B = \{1, 3, 5, 7, 9\}$$

is the set of odd integers between 1 and 10.

Example 5.17. In Fig. 5.11 the set B is a subset of the set A. The set $A - B$ is the shaded area.

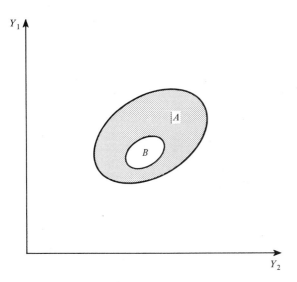

FIGURE 5.11   Subtraction of Sets

The operations of addition and multiplication follow certain laws.
The associative law:

$$(A + B) + C = A + (B + C),$$
$$(A \cdot B) \cdot C = A \cdot (B \cdot C).$$

The distributive law:

$$A \cdot (B + C) = A \cdot B + A \cdot C.$$

The commutative law:

$$A + B = B + A,$$
$$A \cdot B = B \cdot A.$$

Example 5.18. Suppose A is the set of all even *integers* between 1 and 10,

$$A = \{2, 4, 6, 8, 10\},$$

B is the set of integers between 2 and 5,

$B = \{2, 3, 4, 5\}$,

C is the set of integers between 4 and 7,

$C = \{4, 5, 6, 7\}$.

Now

$A \cdot B = \{2, 4\}$ and $(A \cdot B) \cdot C = \{4\}$.

On the other hand,

$B \cdot C = \{4, 5\}$ and $A \cdot (B \cdot C) = \{4\}$.

This demonstrates the associative law of multiplication. Now let us demonstrate the distributive law.

$(B + C) = \{2, 3, 4, 5, 6, 7\}$
$A \cdot (B + C) = \{2, 4, 6\}$

Likewise

$$A \cdot B = \{2, 4\},$$
$$A \cdot C = \{4, 6\},$$
$$A \cdot B + B \cdot C = \{2, 4, 6\}.$$

Intervals

An *interval* is a special type of point set. In one dimension an interval is a set of points Y on a line that lie between two numbers a and b. There are several types of one-dimensional intervals. A *closed* interval is a set of points Y where $\{a \leq Y \leq b\}$. The endpoints a and b of a closed interval are elements of the interval. With an *open* interval $\{a < Y < b\}$, however, the endpoints do not belong to the interval. In addition to open and closed intervals, there are partially open intervals—that is, $\{a < Y \leq b\}$ or $\{a \leq Y < b\}$.

Example 5.19.   In Fig. 5.12 the set of points lying between 3 and 7 and including the points $Y = 3$ and $Y = 7$ is a closed interval, denoted by $\{3 \leq Y \leq 7\}$. The set of points lying between 3 and 7 but not including the points $Y = 3$ and $Y = 7$ is an open interval denoted by $\{3 < Y < 7\}$. The set of points between 3 and 7 but not including the point 7 is a half-open interval denoted by $\{3 \leq Y < 7\}$.

Intervals may be *bounded* or *unbounded*. An interval that contains all points less than a certain number b is said to be unbounded from below

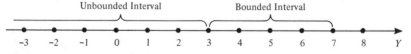

FIGURE 5.12   Intervals on a Line

and is denoted by $\{-\infty < Y \leq b\}$ (closed) or $\{-\infty < Y < b\}$ (open). An interval unbounded from above is denoted by $\{a \leq Y < +\infty\}$ (closed) or $\{a < Y < +\infty\}$ (open). The whole line of points $\{-\infty < Y < \infty\}$ is unbounded from below *and* from above. An interval $\{a \leq Y \leq b\}$, where a and b are given numbers, is said to be bounded.

Example 5.20.   In Fig. 5.12 the set of points to the left of $Y = 3$ is an unbounded interval from below.

Example 5.21.   The set of points Y that are positive is an open interval, unbounded from above.

The intersection or product of two intervals $I_1 = \{a \leq Y \leq b\}$ and $I_2 = \{c \leq Y \leq d\}$ is also an interval. The sum or union of the intervals I_1 and I_2 is not necessarily an interval.

Example 5.22.   The intervals $I_1 = \{5 \leq Y \leq 8\}$ and $I_2 = \{7 \leq Y \leq 10\}$ have as their intersection $I_1 \cdot I_2 = \{7 \leq Y \leq 8\}$, which is also an interval.

Example 5.23.   The intervals

$$I_1 = \{-2 \leq Y \leq 8\} \quad \text{and} \quad I_2 = \{10 \leq Y \leq 18\}$$

have as their sum or union the set of points that satisfy *either* of the two inequalities $-2 \leq Y \leq 8$ and $10 \leq Y \leq 18$, which is not an interval. The intersection of these two intervals is $I_1 \cdot I_2 = \{10 \leq Y \leq 8\}$, which is an interval containing no points, since Y cannot be simultaneously greater than 10 and less than 8.

A *degenerate* interval is an interval $\{a \leq Y \leq b\}$ where $a = b$. A degenerate interval reduces to a single point on a line. An interval $\{a \leq Y \leq b\}$ is called the *null* interval if $a > b$. The null interval contains no points.

Example 5.24.   The interval $\{5 \leq Y \leq 0\}$ is null, since we cannot have both inequalities $Y \geq 5$ and $Y \leq 0$ satisfied for any number Y.

Intervals may be defined in n dimensions. A closed interval for an n-dimensional point $Y = (Y_1, Y_2, \ldots, Y_n)$ is the set of points Y satisfying the inequalities

$$a_1 \leq Y_1 \leq b_1$$
$$a_2 \leq Y_2 \leq b_2$$

.

.

.

$$a_n \leq Y_n \leq b_n$$

A closed interval in n-dimensional space is denoted by

$$\{(a_1, a_2, \ldots, a_n) \leqq (Y_1, Y_2, \ldots, Y_n) \leqq (b_1, b_2, \ldots, b_n)\}$$

or for brevity $\{a \leqq Y \leqq b\}$, where it is understood that a, b, and Y stand for n-tuples of numbers.

Example 5.25.   In two-dimensional space, a closed interval $\{(5, 3) \leqq (Y_1, Y_2) \leqq (7, 4)\}$ is shown in Fig. 5.13. The interval is the set of points in the rectangle whose corners are the points $(5, 4)$, $(7, 4)$, $(7, 3)$, and $(5, 3)$.

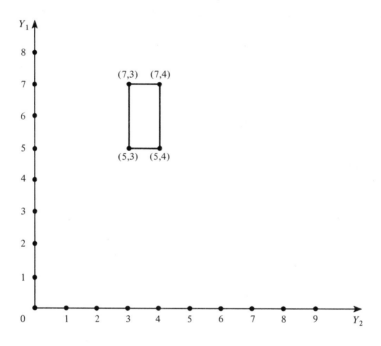

FIGURE 5.13   Interval in Two-dimensional Space

Intervals in n-dimensional space may be open or closed, bounded or unbounded, degenerate, and null in a manner analogous to intervals along a line (in one-dimensional space).

Events and Random Variables

Let Y represent the value of a variable. If the value of Y falls in a set A, then we say that the *event* associated with the set A occurs. For brevity, we sometimes say that the *event* A occurs. The variable Y is called a *random variable*.

Example 5.26. Suppose a die is tossed. The die can come up any number from 1 to 6. Let the random variable Y assume the number shown on the die. Consider the following six sets, each consisting of a single point.

$I_1 = \{Y = 1\}$,

$I_2 = \{Y = 2\}$,

$I_3 = \{Y = 3\}$,

$I_4 = \{Y = 4\}$,

$I_5 = \{Y = 5\}$,

$I_6 = \{Y = 6\}$.

Now, for example, if the number 3 comes up on the die, we say the event I_3 occurs. If the die is tossed again and the number 5 comes up, the event I_5 occurs.

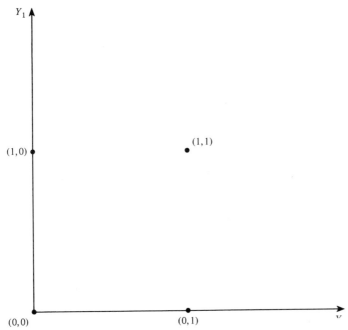

FIGURE 5.14   Representation of Events

Example 5.27. Suppose a coin is tossed twice. If the first toss comes up heads, let $Y_1 = 0$. If tails comes up on the first toss, let $Y_1 = 1$. If heads occurs on the second toss, let $Y_2 = 0$. If tails occurs, $Y_2 = 1$. Then (Y_1, Y_2) is an ordered pair (a point in two-dimensional space). The possible values are shown in Fig. 5.14. We may associate each of the four points with a degenerate interval

$I_1 = \{(0, 0) \leq (Y_1, Y_2) \leq (0, 0)\}$ $\{(Y_1, Y_2) = (0, 0)\}$,

$I_2 = \{(1, 0) \leq (Y_1, Y_2) \leq (1, 0)\}$ $\{(Y_1, Y_2) = (1, 0)\}$,

$I_3 = \{(0, 1) \leq (Y_1, Y_2) \leq (0, 1)\}$ $\{(Y_1, Y_2) = (0, 1)\}$,

$I_4 = \{(1, 1) \leq (Y_1, Y_2) \leq (1, 1)\}$ $\{(Y_1, Y_2) = (1, 1)\}$.

If the first toss is heads and the second toss is tails, then $(Y_1, Y_2) = (0, 1)$, and the event I_3 occurs.

Events may be added, multiplied, and subtracted in the same way that these operations may be applied to point sets. The result of any of these operations is to produce another event.

1. $A + B = C$ is an event indicating that either event A or event B occurs.
2. $A \cdot B = C$ is an event indicating that both events A and B occur.
3. $A - B = C$ is an event indicating that event A occurs but *not* event B.

The operations of addition, subtraction, and multiplication follow the associative, commutative, and distributive laws in the same ways as the corresponding operations on sets. In fact, associated with each event is a particular point set. The addition, subtraction, or multiplication of events produces another point set, which is determined by performing the corresponding operations on the original point sets.

Example 5.28.   Consider three intervals (point sets) A, B, and C and denote the events corresponding to those intervals by the same letters.

$A = \{2 \leq Y \leq 10\}$,
$B = \{4 \leq Y \leq 11\}$,
$C = \{11 \leq Y \leq 12\}$.

The event A, for example, means that the value of Y falls in the closed interval between 2 and 10. The event $A + C = D$ is the event that Y falls in either the interval A or the interval C—that is, in the point set

$$A + C = D = \{2 \leq Y \leq 10 \quad \text{or} \quad 11 \leq Y \leq 12\}.$$

The event $A + B = E$ is the event that Y falls in the interval $A + B = E = \{2 \leq Y \leq 11\}$. The event $F = D \cdot E$ is the event that Y falls in the point set

$$F = D \cdot E = \{2 \leq Y \leq 10 \quad \text{or} \quad Y = 11\}.$$

When the point sets A and B associated with events A and B are disjoint (that is, $A \cdot B$ contains no points), then the events are said to be *mutually exclusive*.

Example 5.29.   The events associated with the intervals

$A = \{8.5 \leq Y \leq 11.4\}$,
$B = \{12.1 \leq Y \leq 12.9\}$,

are mutually exclusive, since Y cannot lie in both intervals A and B at the same time.

Example 5.30.   Suppose a coin is tossed twice. Let the event A represent the occurrence of at least one head. Let the event B be the occurrence of two tails. These events are mutually exclusive. Suppose $Y_1 = 1$ when the first toss is heads and $Y_1 = 0$ when the first toss is tails. Similarly, $Y_2 = 1$ when the second toss is heads and $Y_2 = 0$ when the second toss is tails. Then the event A corresponds to the set of points $Y = (Y_1, Y_2)$ given by

$$A = \{(0, 1), (1, 0), (1, 1)\}.$$

The event B corresponds to the set

$$B = \{(0, 0)\}.$$

These sets are disjoint, since $A \cdot B$ contains no points. If the event C is the occurrence of two heads, then

$$C = \{(1, 1)\}.$$

The events A and C are not mutually exclusive, since

$$A \cdot C = \{(1, 1)\}$$

contains one point. The events B and C are mutually exclusive, since $B \cdot C$ contains no points.

The Axioms of Probability

The theoretical foundations of probability may be stated in the form of axioms. A probability is a number $P(A)$ associated with an event A. The number $P(A)$ is nonnegative and finite.

Axiom 1.   Associated with each possible event A is a probability $P(A)$.

Axiom 2.   If the events A and B are mutually exclusive, then

$$P(A + B) = P(A) + P(B).$$

Axiom 3.   If A is the event associated with the unbounded interval $\{-\infty < Y < +\infty\}$, then $P(A) = 1$.

These axioms form the basis for the whole theory of probability. In practical applications of the theory, however, we wish to know how to define events and how to assign probabilities to various events. There are three basic ways to do this:

1. The principle of insufficient reason.
2. The experimental approach.
3. The subjective approach.

The principle of insufficient reason may be applied whenever the value of the variable Y assumes only some finite number n of values. Each of these n values corresponds to a *basic* event. If each basic event is "equally likely" (on the grounds that we cannot argue that any event is more likely than any other), then we assign the probability $1/n$ to each one of these events. The probability assigned to any other (nonbasic) event is merely the number of basic events included in the (nonbasic) event divided by n, the total number of events.

Example 5.31. Let Y stand for the number of dots on the face of a die. Thus the random variable Y can assume any one of the values 1, 2, 3, 4, 5, 6 and no other value. The probability associated with $Y = 1$ is $1/6$. The same probability is assigned to $Y = 2$, $Y = 3$, and so on. We have no reason to believe that a die will come up with $Y = 1$ any more often than $Y = 2$, 3, 4, 5, or 6. The probability associated with the event that Y lies in the interval $\{2 \le Y \le 5\}$ is $4/6 = 2/3$, since the number of basic events included in this event is four ($Y = 2$, $Y = 3$, $Y = 4$, $Y = 5$) and there are a total of six basic events.

Example 5.32. Suppose a coin is tossed twice. The variable $Y_1 = 0$ or 1 depending on whether a head or tail occurs on the first toss, and similarly $Y_2 = 0$ or 1 depending on whether a head or tail appears on the second toss. The composite variable $Y = (Y_1, Y_2)$ can assume one of four values

$$(Y_1, Y_2) = (0, 0), \qquad\qquad (Y_1, Y_2) = (1, 0),$$
$$(Y_1, Y_2) = (0, 1), \qquad\qquad (Y_1, Y_2) = (1, 1).$$

These points are shown in Fig. 5.14. We have no reason to believe that any one of these events is more likely than another, so we assign the probability of $1/4$ to each. Then the probability assigned to the interval

$$\{(0, 0) \le Y \le (1/2, 1)\}$$

is $2/4$, since this interval contains two of the basic events.

The experimental or frequency approach is one in which probabilities are assigned to any event by performing a series of n experiments. Suppose the number of times the event A occurs is k. Then the probability associated with the event A is k/n.

Example 5.33. Suppose we toss two coins. This experiment is performed a large number of times, say 1,000 times. Suppose two heads come up 252 times. Then the probability assigned to the event that two heads come up ($Y_1 = 0$, $Y_2 = 0$ in the notation of Example 5.32) is $252/1000 = .252$.

The rationale behind the frequency or experimental approach is that as the number of experiments is made larger and larger, the frequency of an

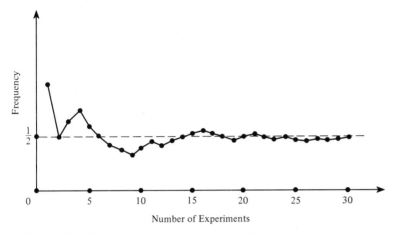

FIGURE 5.15   **Frequency Approach to Probability**

event relative to the total number of experiments tends to fluctuate less and less around some stable value. This is illustrated in Fig. 5.15.

The subjective approach to probability does not require any experimentation nor any assumption that certain basic events are equally likely. Probabilities are assigned on the basis of an individual's "degree of belief" in the possibility of an event.

Example 5.34.   Suppose a weatherman on the basis of his observations feels there is a 50-50 chance of rain on a certain day. The probability associated with the event of rain then is .50.

Regardless of how one assigns probabilities, the theory assumes that the axioms are satisfied and the basic rules of probability remain the same.

Rules of Probability

Let us derive some probability rules that are useful in terms of practical applications of probability theory.

Rule 1   (rule of subtraction)

$$P(A - B) = P(A) - P(B). \tag{5.1}$$

This rule may be derived from the three axioms above. Now B is a subset of itself. Therefore we may subtract B from B to get $B - B = \{0\}$, the null set. Now $A + (B - B) = A + \{0\} = A$. Thus

$$P(A) = P(A + (B - B)),$$

or, by the associative and commutative laws,

$P(A) = P((A - B) + B).$

Now obviously the sets $A - B$ and B are mutually exclusive (contain no points in common). Thus, according to Axiom 2,

$P(A) = P(A - B) + P(B).$

If we subtract $P(B)$ from both sides of this equality, we arrive at Rule 1.

Rule 2   (rule of addition for non-mutually exclusive events)

$$P(A + B) = P(A) + P(B) - P(A \cdot B). \tag{5.2}$$

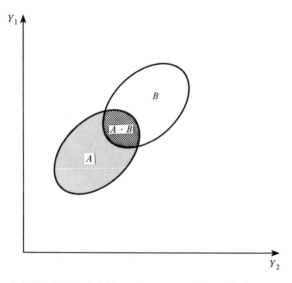

FIGURE 5.16   **Addition of Non-mutually Exclusive Events**

It is obvious (see Fig. 5.16) that $A + B = (A - A \cdot B) + B$, where $(A - A \cdot B)$ and B are mutually exclusive. Thus from Axiom 2 we have

$P(A + B) = P((A - A \cdot B) + B) = P(A - A \cdot B) + P(B).$

Using Rule 1, the term $P(A - A \cdot B)$ can be written $P(A) - P(A \cdot B)$. Thus

$P(A + B) = P(A) - P(A \cdot B) + P(B)$

and we have proved the validity of Rule 2.

Example 5.35.   Suppose a ball is drawn from an urn containing 4 white balls numbered 1 through 4 and 4 black balls numbered 3 through 6. Let A stand for the event that a white ball is drawn from the urn. Let B stand for the event that the number 3 occurs on the ball drawn from the urn. We assume that each ball is equally likely to be drawn. Thus $P(A = \text{white}) = 4/8 = 1/2$, since there are 4 out of 8 white balls, and $P(B = 3) = 2/8 = 1/4$, since there are two balls, a black one and white one, numbered 3. The event $A \cdot B$ stands for the event of drawing a white ball numbered 3. Since there is only one such ball, we have $P(A \cdot B) = 1/8$. Thus $P(A + B)$, the probability of *either* drawing a white ball *or* a ball with the number 3, is

$$P(A + B) = 1/2 + 1/4 - 1/8 = 5/8.$$

We will describe several other rules of probability. First, however, we must define two terms: conditional probabilities, and independent events. The *conditional probability* $P(A \mid B)$ is the probability of A given that the event B has occurred. The definition is

$$P(A \mid B) = \frac{P(A \cdot B)}{P(B)}. \qquad (5.3)$$

If we multiply both sides of (5.3) by $P(B)$, we have

Rule 3   (multiplication)

$$P(A \cdot B) = P(A \mid B) \cdot P(B). \qquad (5.4)$$

Example 5.36.   In the previous example, what is the conditional probability of a ball being drawn with the number 3, given that it is white? We calculated $P(A \cdot B) = 1/8$, the probability of drawing a white ball numbered 3. $P(B)$, the probability of a ball numbered 3, is 1/4. Thus the conditional probability is

$$P(A \mid B) = \frac{P(A \cdot B)}{P(B)} = \frac{1/8}{1/4} = 1/2.$$

That is, the probability of drawing a white ball given that it is numbered 3 is 1/4.

An event A is said to be independent of the event B if

$$P(A \mid B) = P(A). \qquad (5.5)$$

Rule 4   If the event A is independent of the event B, then the event B is independent of the event A—that is,

$$P(B \mid A) = P(B). \qquad (5.6)$$

The proof of this rule is as follows. From (5.3) and (5.5) we have

$$P(A \mid B) = \frac{P(A \cdot B)}{P(B)} = P(A)$$

or

$$P(A \cdot B) = P(A) \cdot P(B). \tag{5.7}$$

Reversing the roles of B and A in (5.3), we get

$$P(B \mid A) = \frac{P(A \cdot B)}{P(A)}.$$

From (5.7), it follows that

$$P(B \mid A) = \frac{P(A) \cdot P(B)}{P(A)} = P(B),$$

which verifies Rule 4.

   Thus if A is independent of B, then B is independent of A, and we say merely that events A and B are independent. From the proof of Rule 4, we have proved another rule.

Rule 5   (rule of multiplication for independent events) If A and B are independent, then the probability of $A \cdot B$ is given by (5.7).

Example 5.37.   From Example 5.35 let the event A be a ball drawn numbered 3 and the event B be a black ball drawn. The probability of drawing a black ball numbered 3 (the event $A \cdot B$) is

$$P(A \cdot B) = 1/8,$$

since there is only one black ball numbered 3. The probability of drawing a black ball is

$$P(B) = 4/8 = 1/2,$$

since there are 4 black balls out of 8 balls. Now

$$P(A \mid B) = \frac{P(A \cdot B)}{P(B)} = \frac{1/8}{1/2} = 1/4.$$

However, there are two balls labeled 3, so that

$$P(A) = 1/4.$$

Thus we see that the event A is independent of the event B. Similarly, the event B is independent of the event A—that is,

$$P(B \mid A) = \frac{P(A \cdot B)}{P(A)} = \frac{1/8}{1/4} = 1/2 = P(B).$$

Rule 4 applies, so that

$$P(A \cdot B) = P(A) \cdot P(B) = 1/4 \cdot 1/2 = 1/8.$$

Example 5.38. If a coin is tossed twice, the probability of a head on the first toss (event A) is 1/2. The probability of a head on the second toss (event B) is also 1/2. The events A and B are independent; a head on the first toss would not affect the probability of a head on the second toss and vice versa. Rule 5 applies, and the probability of obtaining two heads in succession (the event $A \cdot B$) is

$$P(A \cdot B) = P(A) \cdot P(B) = 1/2 \cdot 1/2 = 1/4$$

A series of events $A_1, A_2, \ldots, A_k$ are said to be *exhaustive* if the probability of their sum is unity—that is,

$$P(A_1 + A_2 + \cdots + A_k) = 1. \tag{5.8}$$

Example 5.39. Suppose a television manufacturing firm has five factories. In 1967, the five factories manufactured 1,000 television sets. The number of sets manufactured by each factory is as follows:

FACTORY	SETS MANUFACTURED
1	100
2	200
3	300
4	100
5	300

Suppose some particular television set is selected at random from those manufactured by the firm. Consider the following list of events

A_1: Set manufactured by factory 1 or 2.
A_2: Set manufactured by factory 2 or 3.
A_3: Set manufactured by factory 4 or 5.

The probability of any events occurring is the number of sets falling into the category described by the event divided by the total number of sets.

$$P(A_1) = \frac{100 + 200}{1000} = 3/10,$$

$$P(A_2) = \frac{200 + 300}{1000} = 1/2,$$

$$P(A_3) = \frac{100 + 300}{1000} = 2/5.$$

Now $P(A_2 + A_3) = P(A_2) + P(A_3) = 9/10$ since A_2 and A_3 are mutually exclusive events. The probability

$$P[A_1 + (A_2 + A_3)] = P(A_1) + P(A_2 + A_3) - P[A_1 \cdot (A_2 + A_3)],$$

because A_1 and the event $(A_2 + A_3)$ are not mutually exclusive. The event $A_1 \cdot (A_2 + A_3)$ is the event that the set was manufactured by factory 2, since this is the only factory that falls into event A_1 and either event A_2 or A_3. Thus

$$P[A_1 \cdot (A_2 + A_3)] = 200/1000 = 1/5$$

and

$$P[A_1 + (A_2 + A_3)] = 3/10 + 9/10 - 1/5$$

or

$$P(A_1 + A_2 + A_3) = 12/10 - 2/10 = 1.$$

Thus A_1, A_2, and A_3 is an exhaustive series of events.

If two events A and B are exhaustive and mutually exclusive (that is, the intersection of the sets associated with events A and B is the null set), then Axiom 2 holds and

$$P(A + B) = P(A) + P(B) = 1. \tag{5.9}$$

In this case, the event B is said to be the complement of the event A and, conversely, the event A is the complement of the event B. If A is an event, we denote its complement by $\bar{A}$, and

$$P(A) + P(\bar{A}) = 1. \tag{5.10}$$

Suppose A_1, A_2, ... , A_k is a series of events that are exhaustive and mutually exclusive. Again applying Axiom 2, we have

$$P(A_1 + A_2 + \cdots + A_k) = P(A_1) + P(A_2) + \cdots + P(A_k) = 1. \tag{5.11}$$

Suppose B is any other event.

Rule 6   (rule of total probability). If A_1, A_2, ... , A_k are mutually exclusive and exhaustive events,

$$P[B \cdot (A_1 + A_2 + \cdots + A_k)] = P(B \cdot A_1) + P(B \cdot A_2) \\ + \cdots + P(B \cdot A_k) = P(B). \tag{5.12}$$

This rule is a little more difficult to prove than some of the others. From the distribution law

$$P[B \cdot (A_1 + A_2 + \cdots + A_k)] = P(B \cdot A_1 + B \cdot A_2 + \cdots + B \cdot A_k).$$

Now $B \cdot A_i$ is a subset of the set A_i. Since A_1, A_2, ... , A_k are mutually exclusive, certainly $B \cdot A_1$, $B \cdot A_2$, ... , $B \cdot A_k$ are mutually exclusive. Thus from Axiom 2

$$P(B \cdot A_1 + B \cdot A_2 + \cdots + B \cdot A_k) = P(B \cdot A_1) + P(B \cdot A_2)$$
$$+ \cdots + P(B \cdot A_k).$$

This demonstrates the validity of the first equality of (5.12).

Now $B \cdot \sum_{i=1}^{k} A_i$ is a subset of B. Let $\bar{A}$ be the difference between B and $\sum_{i=1}^{k} A_i$.

$$\bar{A} = B - B \cdot \sum_{i=1}^{k} A_i \tag{5.13}$$

or

$$B \cdot \left(\sum_{i=1}^{k} A_i \right) = B - \bar{A}. \tag{5.14}$$

Now, using Rule 1 for subtraction,

$$P \left[B \cdot \left(\sum_{i=1}^{k} A_i \right) \right] = P(B - \bar{A}) = P(B) - P(\bar{A}). \tag{5.15}$$

Now $\bar{A}$ and $\sum_{i=1}^{k} A_i$ are mutually exclusive by definition. Applying Axiom 2,
we have

$$P \left(\sum_{i=1}^{k} A_i + \bar{A} \right) = P \left(\sum_{i=1}^{k} A_i \right) + P(\bar{A}). \tag{5.16}$$

But since $A_1, A_2, \ldots, A_k$ are mutually exclusive and exhaustive,

$$P \left(\sum_{i=1}^{k} A_i \right) = 1. \tag{5.17}$$

According to Axiom 3,

$$P \left(\sum_{i=1}^{k} A_i \right) + P(\bar{A}) = 1. \tag{5.18}$$

Combining (5.17) and (5.18), we have

$$P(\bar{A}) = 0. \tag{5.19}$$

Substituting this value in (5.15), we have

$$P\left[B\cdot\left(\sum_{i=1}^{k} A_i\right)\right] = P(B),\tag{5.20}$$

which is the same as the second equality in (5.12), and Rule 6 is shown to be true.

Example 5.40.   Let us return to the problem discussed in Example 5.39. We shall define a new set of events:

   A_1: Set manufactured by factory 1
   A_2: Set manufactured by factory 2
   A_3: Set manufactured by factory 3
   A_4: Set manufactured by factory 4
   A_5: Set manufactured by factory 5

Since there are 1,000 sets manufactured, of which 100, 200, 300, 100, and 300 were manufactured by factories 1 through 5, respectively,

$$P(A_1) = 1/10, \quad P(A_2) = 1/5, \quad P(A_3) = 3/10,$$
$$P(A_4) = 1/10, \quad P(A_5) = 3/10.$$

These events are mutually exclusive and exhaustive, and

$$P(A_1 + A_2 + \cdots + A_5) = P(A_1) + P(A_2) + \cdots + P(A_5) = 1.$$

Suppose each factory turns out a certain proportion of defective sets:

FACTORY	PERCENT DEFECTIVE
1	3
2	2
3	4
4	1
5	8

Let B be the event of a defective set. Then the probability of choosing a defective set manufactured by the ith factory $(B \cdot A_i)$ is the number of defective sets produced by the ith factory divided by the total number of sets produced:

$$P(B \cdot A_1) = 3/1{,}000, \quad P(B \cdot A_2) = 4/1{,}000, \quad P(B \cdot A_3) = 12/1{,}000,$$
$$P(B \cdot A_4) = 1/1{,}000, \quad P(B \cdot A_5) = 24/1{,}000.$$

The probability of choosing a defective set regardless of manufacturer is obtained by applying Rule 6.

$$P(B) = P(B \cdot A_1) + \cdots + P(B \cdot A_5) = 44/1{,}000.$$

The final rule that we will discuss is called Bayes' theorem. Let A and B be two events.

Rule 7   (Bayes' theorem)

$$P(A \mid B) = \frac{P(B \mid A) \cdot P(A)}{P(B)}.$$ (5.21)

This rule follows immediately from Rule 3 for multiplication:

$$P(B \cdot A) = P(B \mid A) \cdot P(A)$$ (5.22)

or alternatively

$$P(A \cdot B) = P(A \mid B) \cdot P(B).$$ (5.23)

Setting the right-hand sides of (5.22) and (5.23) equal to each other,

$$P(B \mid A) \cdot P(A) = P(A \mid B) \cdot P(B).$$ (5.24)

Divide both sides of (5.24) by $P(B)$ to obtain (5.21) and prove Rule 7.

Bayes' theorem is often combined with Rule 6 to produce another form of the theory. If $A_1, A_2, \ldots, A_k$ are mutually exclusive and exhaustive events, then

$$P(A_i \mid B) = \frac{P(B \mid A_i) \cdot P(A_i)}{\sum\limits_{i=1}^{k} P(B \cdot A_i)}.$$ (5.25)

This rule is often used in practical problems to give answers that fly in the face of (some people's) intuition.

Example 5.41.   Using the same problem as in Example 5.40, let us ask the question: Given that a defective set is chosen, what is the probability that it was manufactured by factory 5? This is the conditional probability

$$P(A_5 \mid B),$$

where the event B is that a defective set is chosen. One may be tempted to argue that in order to know whether a set is defective or not, a set first has to be chosen. Once a set is chosen, it makes no difference as far as the probability that it was manufactured by one factory or another, since the information that it was defective could not be known prior to the choosing of the set. Thus the probability that a set was manufactured by factory 5 does not depend on any such prior information, and $P(A_5 \mid B) = P(A_5)$. This argument is false, however, if we are going to calculate probabilities according to our three axioms and our definition of conditional probability. Let us use (5.25) to calculate $P(A_5 \mid B)$. The probability of a defective set regardless of source of manufacture (called the total probability) is

$$P(B) = P(B \cdot A_1) + \cdots + P(B \cdot A_5) = 44/1,000,$$

as we calculated in the previous example. The probability of choosing a set manufactured by factory 5 is

$$P(A_5) = 3/10.$$

The probability of choosing a defective set given that it comes from factory 5 is

$$P(B \mid A_5) = \frac{P(B \cdot A_5)}{P(A_5)} = \frac{24/1,000}{3/10} = 8/100,$$

where $P(B \cdot A_5)$ was calculated in the previous example. Thus from (5.25)

$$P(A_5 \mid B) = \frac{P(B \mid A_5) \cdot P(A_5)}{P(B \cdot A_1) + \cdots + P(B \cdot A_5)} = \frac{8/100 \cdot 3/10}{44/1,000} = 6/11.$$

Thus we see that although the probability of choosing a set manufactured by factory 5 is only 3/10, if we know that it is defective, the probability rises to 6/11. The reason for this, of course, is that factory 5 manufactures a high proportion of defective sets (8 percent as opposed to no more than 4 percent for other factories).

APPENDIX Inequalities

Given any two numbers a and b, one of the following three statements must be true: (1) a is greater than b, (2) a is less than b, or (3) a is equal to b. If a is greater than b [statement (1)], then we write

$$a > b. \tag{5.26}$$

For example, $5 > 4$, $10.37 > 2.1$, and $|\sqrt{2}| > 0$. If a is less than b, [statement (3)], we write

$$a < b. \tag{5.27}$$

For example, $-8 < -2.9$, $-4 < 3$, and $10^2 < 10^3$. If *either* statement (1) or statement (3) is true, we write

$$a \geqq b, \tag{5.28}$$

which reads: a is greater than or equal to b. Thus $2 \geqq 2$ and $5 \geqq -1$. Similarly if a is less than or equal to b [statement (2) or statement (3)], we write

$$a \leqq b. \tag{5.29}$$

The expressions (5.26), (5.27), (5.28), and (5.29) are called *inequalities*. Expressions (5.26) and (5.27) are called *strict* inequalities.

Inequalities follow certain rules, which we will illustrate by examples:

Rule 1 If and only if $a \geqq b$ and $b \geqq a$, then $a = b$.

For example, $2.7 \geqq 2.7$ and $2.7 \leqq 2.7$; thus $2.7 = 2.7$. For any two numbers that are not equal, say 5 and 4, the inequality can be written in one direction only—that is, $4 \leqq 5$ but not $4 \geqq 5$.

Rule 2 If $a > b$ $(a \geqq b)$, then $-a < -b$ $(-a \leqq -b)$.

Thus the sense of an inequality is reversed whenever multiplication of both sides by -1 is performed. For example, since $3 > 1$, then $-3 < -1$, or since $7 \geqq -4$, then $-7 \leqq 4$.

Rule 3 If $a > b$ $(a \geqq b)$ and $c > 0$, then $c \cdot a > c \cdot b$ $(c \cdot a \geqq c \cdot b)$.

Thus $4 > 1$ implies that $3 \times 4 > 3 \times 1$ or $12 > 3$.

Rule 4 If $a > b$ $(a \geqq b)$ and $c < 0$, then $c \cdot a < c \cdot b$ $(c \cdot a \leqq c \cdot b)$.

This is a combination of Rules 1 and 2. If we multiply both sides of an inequality by a negative number, we reverse the sense of the inequality. Thus since $4 > -3$ and $-2 < 0$, we have $(-2) \cdot 4 < (-2) \cdot (-3)$ or $-8 < 6$. Similarly $(-1/3) \cdot 4 < (-1/3)(-3)$ or $-4/3 < 1$.

Rule 5   If $a > b$ $(a \geq b)$, then $a + c > b + c$ $(a + c \geq b + c)$.

For example, since $8 \geq -3$, we have $8 + 4 \geq (-3) + 4$ or $12 \geq 1$. Similarly $9 \geq 2$ implies $9 - 7 \geq 2 - 7$ or $-2 \geq -5$.

To sum up, the rules of inequalities allow us to add any number, subtract any number, multiply positive numbers, and divide positive numbers on both sides of an inequality and it remains the same inequality. If we multiply or divide by a negative number, we reverse the sense of the inequality.

PROBLEMS

1. Given the two point sets $A = \{3, 2, 1\}$ and $B = \{4, 5, 6\}$, what is the union $A + B$? What is $A \cdot B$? What does one call such an intersection?
2. Using the three sets $A = \{2, 4\}$, $B = \{1, 5\}$, and $C = \{3, 7\}$. Demonstrate the associative and commutative laws for addition of sets and the distributive governing addition and multiplication of sets.
3. What is the intersection of the intervals $A = \{2 \leq X \leq 10\}$ and $B = \{8 \leq X \leq 10\}$?
4. Suppose a card is drawn from a deck. Apply the principle of insufficient reason to determine the probability that it will be a spade.
5. In the following group of scores, apply the principle of insufficient reason to determine the probability that a score is below 90. Between 70 and 78.

99	92	87	83	81	80	79	79	78	73
71	71	70	69	69	68	68	63	58	50

6. An urn contains eight red balls and four black balls. The red balls are numbered in two sets from 1–4 and the black balls are numbered from 1–4. One ball is drawn. What is the probability of drawing a red ball, but not one with the number 3? What is the probability of drawing either a red ball numbered 2 or a black ball? A ball numbered 2 or a red ball?
7. What is the conditional probability that a drawn ball will have the number 4, given that it is red?
8. If I draw two cards from a deck in succession (without replacement), what is the probability that the first will be a spade and the second a heart? That both will be hearts? What is the probability of drawing three queens in succession?
9. The following represent highway deaths over a nine-month period in each of five regions into which the United States is divided (measured in deaths per 10,000 passenger miles).

A	20
B	30
C	10
D	40
E	20

There are 30,000,000 drivers in the country. The following represent the number of 10,000 units driven in each region in nine months.

A	300
B	200
C	600
D	100
E	500

Suppose that an army enlisted man confronts a probability of 1/2 that he will be sent to a combat zone for nine months with a .98 chance of survival. Is it safer to drive or get drafted?

10. Compute the following probabilities:
    (a) If four coins are tossed, they will not all turn up tails.
    (b) If five coins are tossed, at least one will turn up heads and one tails.
    (c) In a poker game a man picks up four of five cards and they are all spades. The fifth card is a spade.

11. Three cards are drawn from a pack and the first two replaced; the deck is shuffled; three are drawn, two replaced, and the deck shuffled; three are drawn and two replaced. Three cards are now held outside the deck. What is the probability that at least one of the cards is a king?

12. What is the probability of finding two defectives when a random sample of four is drawn from a process producing 30 percent defectives? What is the probability of all defectives? No defectives?

13. Suppose there are two urns. Urn I has three white balls and seven green balls; urn II has eight white balls and two green balls. An urn is selected at random and a ball is drawn. The ball turns out to be white. What is the probability that the ball came from urn I?

14. With the same two urns as in problem 13, an urn is selected at random and two balls are drawn (without replacement) from the urn. One of the balls drawn is white and the other green. What is the probability that the balls came from urn II?

15. Two machines are used in a factory to manufacture light bulbs. Machine I produces 2 percent defectives and machine II produces 1 percent defectives. Machine I produces 30 percent of the factory's output and machine II 70 percent. Three bulbs are selected at random from the factory's output at the end of the day. Two are defective and one is not defective. What is the probability that the nondefective bulb comes from machine I?

Chapter 6

Probability Distributions and Expectations

In the last chapter we introduced the notion of probability in a very special (and somewhat roundabout) way. An elementary event was said to occur if the value of a random variable fell within a point set. The summation, subtraction, and multiplication of events produced other events. To each possible event we assigned a nonnegative probability and specified certain axioms concerning probabilities and events.

Probability Distributions in One Dimension

The set of probabilities associated with each and every possible event is called a probability distribution. Two simple kinds of probability distributions are applied to most practical problems: those which are *discrete* and those which are *continuous*. First, let us consider the former kind. A discrete probability distribution for a random variable with one dimension can be represented by a diagram in which the vertical axis gives the probability and the horizontal axis the value of the random variable, as in Fig. 6.1.

DISCRETE PROBABILITY DISTRIBUTIONS

Consider a finite number of specific values of Y, say $Y_1, Y_2, \ldots, Y_n$. The probability that Y equals any one of these specific values is positive. The probability that Y equals any other specific value is 0. The positive probabilities are written

$$P(Y_1) = p_1$$
$$P(Y_2) = p_2$$
$$\cdot \qquad \cdot$$
$$\cdot \qquad \cdot \qquad (6.1)$$
$$\cdot \qquad \cdot$$
$$P(Y_n) = p_n$$

The sum of the probabilities $p_1 + p_2 + \cdots + p_n = 1$ (Axiom 3 of Chapter 5).

Example 6.1. Suppose a die is tossed. Let the random variable Y represent the number of dots that come up. The probability of any number from 1 through 6 is $1/6$. The probability that Y is anything but a number from 1 through 6 is zero. This probability distribution is represented in Fig. 6.1.

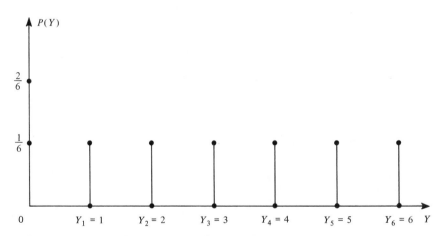

FIGURE 6.1   Probability Distribution for Toss of a Die

Here we have

$$Y_1 = 1, \qquad P(Y_1) = p_1 = 1/6,$$
$$Y_2 = 2, \qquad P(Y_2) = p_2 = 1/6,$$
$$Y_3 = 3, \qquad P(Y_3) = p_3 = 1/6,$$
$$Y_4 = 4, \qquad P(Y_4) = p_4 = 1/6,$$
$$Y_5 = 5, \qquad P(Y_5) = p_5 = 1/6,$$
$$Y_6 = 6, \qquad P(Y_6) = p_6 = 1/6.$$

The probability for any other event—say, for example, the event that Y lies in the sum of the two intervals $\{-3 \leqq Y \leqq 2\}$ and $\{6 \leqq Y < 10\}$—is determined by adding the probabilities for any of the points Y_1, Y_2, $\ldots$, Y_6 contained in the sum of the two intervals. The points $Y_1 = 1$,

$Y_2 = 2$, and $Y_6 = 6$ lie in the sum of the two intervals, and the probability is $p_1 + p_2 + p_6 = 3/6 = 1/2$.

Example 6.2.   Suppose two dice are tossed. Let the random variable Y represent the sum of the dots on the two dice. Now there are 36 possible combinations of the numbers on the two dice. These are listed below:

FIRST DIE	SECOND DIE	SUM (Y)
1	1	2
2	1	3
1	2	3
3	1	4
2	2	4
1	3	4
4	1	5
3	2	5
2	3	5
1	4	5
5	1	6
4	2	6
3	3	6
2	4	6
1	5	6
6	1	7
5	2	7
4	3	7
3	4	7
2	5	7
1	6	7
6	2	8
5	3	8
4	4	8
3	5	8
2	6	8
6	3	9
5	4	9
4	5	9
3	6	9
6	4	10
5	5	10
4	6	10
6	5	11
5	6	11
6	6	12

Assuming that each combination is equally likely, the probability that

Y equals some value from 2 **through** 12 is the number of combinations for which the sum Y is that value divided by 36.

$P(Y = 2) = 1/36,$	$Y_1 = 2,$	$p_1 = 1/36,$
$P(Y = 3) = 2/36,$	$Y_2 = 3,$	$p_2 = 2/36,$
$P(Y = 4) = 3/36,$	$Y_3 = 4,$	$p_3 = 3/36,$
$P(Y = 5) = 4/36,$	$Y_4 = 5,$	$p_4 = 4/36,$
$P(Y = 6) = 5/36,$	$Y_5 = 6,$	$p_5 = 5/36,$
$P(Y = 7) = 6/36,$	$Y_6 = 7,$	$p_6 = 6/36,$
$P(Y = 8) = 5/36,$	$Y_7 = 8,$	$p_7 = 5/36,$
$P(Y = 9) = 4/36,$	$Y_8 = 9,$	$p_8 = 4/36,$
$P(Y = 10) = 3/36,$	$Y_9 = 10,$	$p_9 = 3/36,$
$P(Y = 11) = 2/36,$	$Y_{10} = 11,$	$p_{10} = 2/36,$
$P(Y = 12) = 1/36,$	$Y_{11} = 12,$	$p_{11} = 1/36.$

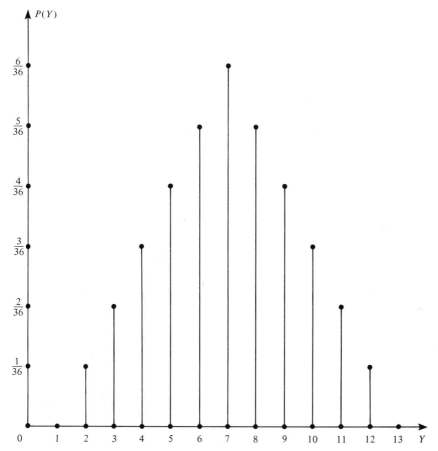

FIGURE 6.2   Probability Distribution of Sum of Pair of Dice

The probability that Y equals anything but a number from 2 through 12 is zero. The probability distribution is shown in Fig. 6.2. Let us determine the probability that Y lies in the intersection of the two intervals $\{4 < Y \leq 7.2\}$ and $\{2 \leq Y < 7\}$. The intersection is the interval $\{4 < Y < 7\}$. The points $Y_4 = 5$ and $Y_5 = 6$ belong to this intersection. Thus the probability is

$$p_4 + p_5 = 4/36 + 5/36 = 9/36 = 1/4.$$

Example 6.3. Suppose a coin is tossed four times. Let Y, the random variable, be the number of heads that come up on those four tosses. The possible outcomes of the four tosses are listed below (H stands for heads and T for tails):

TTTT:	$Y = 0$	HHTT:	$Y = 2$	HHHT:	$Y = 3$
		HTHT:	$Y = 2$	HHTH:	$Y = 3$
HTTT:	$Y = 1$	HTTH:	$Y = 2$	HTHH:	$Y = 3$
THTT:	$Y = 1$	THHT:	$Y = 2$	THHH:	$Y = 3$
TTHT:	$Y = 1$	THTH:	$Y = 2$		
TTTH:	$Y = 1$	TTHH:	$Y = 2$	HHHH:	$Y = 4$

There are a total of 16 different outcomes. The probability distribution is given by

$$P(Y = 0) = 1/16 \qquad\quad = p_1,$$
$$P(Y = 1) = 4/16 = 1/4 = p_2,$$
$$P(Y = 2) = 6/16 = 3/8 = p_3,$$
$$P(Y = 3) = 4/16 = 1/4 = p_4,$$
$$P(Y = 4) = 1/16 \qquad\quad = p_5.$$

This is shown in Fig. 6.3. This distribution is a special case of what is called the binomial distribution.

CONTINUOUS PROBABILITY DISTRIBUTIONS

A continuous probability distribution in one dimension is one in which the probability of a random variable Y assuming any specific value is zero. The probability that Y lies in any nondegenerate interval, $P(a \leq Y \leq b)$, where $a \neq b$, is given by the area underneath some curve. The equation for the curve is denoted by $f(Y)$, which is called the *density function*.

Example 6.4. Suppose we have a calibrated dial with a spinner attached to the middle (see Fig. 6.4). The dial has numbers on it from 1 through 10. The spinner is spun and we set the random variable Y equal to the closest

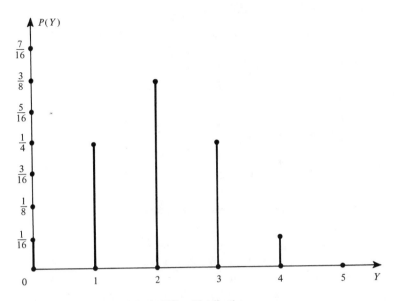

FIGURE 6.3 **Binomial Probability Distribution**

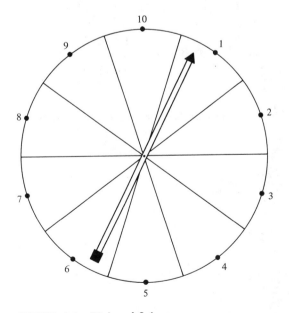

FIGURE 6.4 **Dial and Spinner**

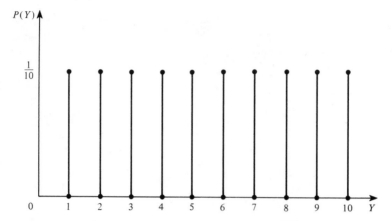

FIGURE 6.5   Probability Distribution for Numbers on Dial

number to which the arrow is pointing. Assuming that each number is equally likely, the probability of any particular number is 1/10. We have the probability distribution shown in Fig. 6.5. This is a discrete probability distribution. Let us calibrate the dial, however, so that we can read up to one decimal place between 0.1 and 10.0. Thus if the arrow is closest to the mark on the dial indicating 6.4, we say that the random variable $Y = 6.4$. If the dial is spun, the probability of any particular value coming up, assuming that each number is equally likely, is equal to 1/100, since there are 100 different values of Y between 0.1 and 10.0. The probability distribution—still discrete—is shown in Fig. 6.6. The calibration of the dial may be made finer and finer. The probability that any particular value of Y, say $Y = 5$, occurs becomes smaller and smaller, from 1/10 to 1/100 to 1/1,000 to 1/10,000 and so on. The probability distribution remains discrete.

If the calibration is very fine, however, the discrete probability distribu-

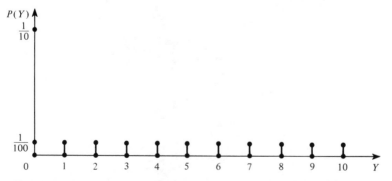

FIGURE 6.6   Partial Probability Distribution for Numbers on Calibrated Dial

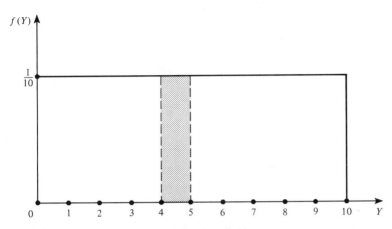

FIGURE 6.7   Rectangular Probability Distribution

tion can be approximated by a continuous distribution called a rectangular distribution. This distribution is shown in Fig. 6.7. The horizontal axis no longer represents the probability that a random variable Y obtains any particular value, but is the value of the density function $f(Y)$. The density function $f(Y)$ is given by

$$f(Y) = 1/10 \qquad \text{for } 0 \leq Y \leq 10,$$
$$f(Y) = 0 \qquad \text{for } Y < 0 \text{ or } Y > 10. \tag{6.2}$$

The probability that Y falls in any interval, say $4 < Y < 5$, is given by the area under the density function $f(Y)$ in Fig. 6.7. This area has been shaded and is a rectangle. The base of the rectangle is one unit in length and the height is 1/10. The area, therefore, is $1 \cdot 1/10 = 1/10$. Note that the probability that Y is exactly equal to 4 is zero, since the rectangle that denotes the probability shrinks to a straight line, which has no area.

Example 6.5.   Suppose we wish to measure the weight of all male students attending a university lecture. Suppose the scale is calibrated roughly to weigh only to the nearest 10 pounds. The weight of the students ranges from 110 to 220 pounds, which gives 11 intervals of 10 pounds. Suppose there are 120 students and their weights are given in the table on p. 134.

If we were to choose any student at random from this lecture and weigh him, the probability that he would weigh between 150 and 160 pounds ($Y = 155$) is the number of students whose weight is in this range divided by the total number of students in the lecture, or $40/120 = 1/3$. The discrete probability distribution is shown in Fig. 6.8. Each point denotes the probability that the random variable Y equals any value—115, 125,

Y	FROM	TO	NUMBER OF STUDENTS
115	110 lb	120 lb	2
125	120	130	3
135	130	140	8
145	140	150	26
155	150	160	40
165	160	170	22
175	170	180	11
185	180	190	5
195	190	200	2
205	200	210	0
210	210	220	1

... , 215. If this excercise is performed (a) on a larger group of students and (b) on a more finely calibrated scale, a probability distribution will be defined for a larger number of values on the random variable Y, and the probability that Y will assume any particular value generally tends to become smaller and smaller. Experience on measurement of weights indicates that the random variable Y has a probability distribution that

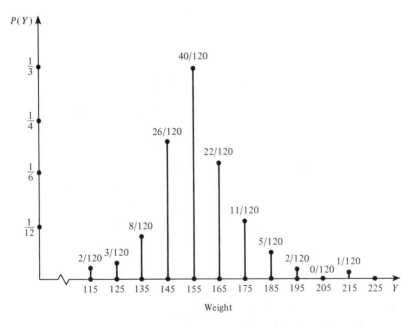

FIGURE 6.8   Probability Distribution for Weights of Students in a Class

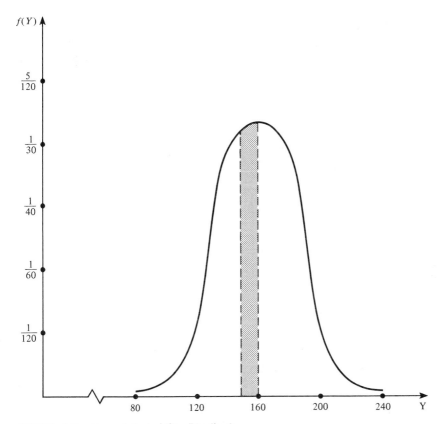

FIGURE 6.9   Normal Probability Distribution

closely approximates a continuous distribution called a *normal* distribution for large groups of people. This distribution has a density function $f(Y)$ like that shown in Fig. 6.9. The density function is "bell-shaped." The area under this bell-shaped curve for any nondegenerate interval $\{a \leqq Y \leqq b\}$ closely approximates the probability that any individual selected at random from the group has a weight lying in the interval. For example, the area under the curve between 150 and 160 pounds in Fig. 6.9 would be approximately 1/3.

The normal distribution is useful not only in the case of weight measurement of groups of individuals but describes fairly accurately the distribution of many natural phenomena, such as performance of individuals on examinations, height of individuals, errors in rifle marksmanship, average thickness of steel sheets produced in a steel rolling mill, and many others.

The third axiom of probability states that the probability for the event that the random variable Y lies in the unbounded interval $\{-\infty < Y < \infty\}$

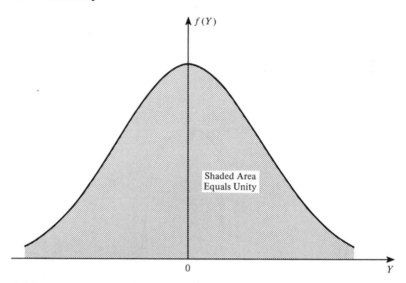

FIGURE 6.10 **Area under Density Function**

is unity. For a discrete probability distribution this unbounded interval contains all the points Y_1, Y_2, ... , Y_n, so that

$$P(-\infty < Y < \infty) = \sum_{i=1}^{n} p_i = 1. \tag{6.3}$$

For a continuous probability distribution the total area under the density function is equal to unity (see Fig. 6.10).

Example 6.6. In Example 6.2 we discussed the probability distribution of the random variable Y, representing the sum of dots obtained by throwing a pair of dice. If we add the probabilities $p_1 + p_2 + \cdots + p_{12}$, we get

$$1/36 + 2/36 + 3/36 + 4/36 + 5/36 + 6/36 + 5/36 + 4/36$$
$$+ 3/36 + 2/36 + 1/36 = 1.$$

Example 6.7. For the normal distribution pictured in Fig. 6.10, the total shaded area under the curve is equal to unity.

Distribution Functions in One Dimension

One of the important concepts concerning probability distributions is the notion of a distribution function. Consider any interval $\{-\infty < Y \leq X\}$ that is unbounded from below. The probability that the random variable Y lies in this interval is written as

$$F(X) = P(-\infty < Y \leq X). \tag{6.4}$$

The endpoint X of the interval may be viewed as a variable, and the probability $F(X)$ is called a distribution function of the variable X.

For a discrete probability distribution, positive probabilities are assigned to a number of specific values Y_1, Y_2, $\ldots$, Y_n. If $Y_k \leq X \leq Y_{k+1}$, then the probability that Y lies in the interval $\{-\infty < Y \leq X\}$ is the sum of the first k probabilities, or

$$F(X) = p_1 + p_2 + \cdots + p_k = \sum_{i=1}^{k} p_i. \tag{6.5}$$

Example 6.8.   Let us use the example of the discrete probability distribution in Example 6.3. The probability that the variable Y equals any of the specific values $Y_1 = 0$, $Y_2 = 1$, $Y_3 = 2$, $Y_4 = 3$, $Y_5 = 4$ is given by

$$P(Y = 0) = 1/16 = p_1,$$
$$P(Y = 1) = 1/4 \ = p_2,$$
$$P(Y = 2) = 3/8 \ = p_3,$$
$$P(Y = 3) = 1/4 \ = p_4,$$
$$P(Y = 4) = 1/16 = p_5.$$

Suppose $X = 2.7$. Then $2 \leq X < 3$ or $Y_3 \leq X < Y_4$. The probability $F(X) = F(2.7) = P(-\infty < Y \leq 2.7)$. Since the interval $\{-\infty < Y < 2.7\}$ contains the points $Y_1 = 0$, $Y_2 = 1$, and $Y_3 = 2$, we have

$$F(2.7) = \sum_{i=1}^{k} p_i = \sum_{i=1}^{3} p_i = p_1 + p_2 + p_3$$
$$= 1/16 + 1/4 + 3/8 = 11/16.$$

If we let X trace out all possible values, we can draw a graph of $F(X)$. The graph is shown in Fig. 6.11. As long as $X < 0$, the interval $\{-\infty < Y < X\}$ contains none of the points that have been assigned positive probabilities. Thus $F(X)$ is zero. When $X = 0$ exactly, one point $Y_1 = 0$ is contained in the interval and $F(0) = p_1 = 1/16$. As long as $0 \leq X < 1$, the interval still only contains one point, $Y_1 = 0$, so that $F(X) = p_1 = 1/16$. Whenever $X = 1$ exactly, however, two points $Y_1 = 0$ and $Y_2 = 1$ are contained in the interval and

$$F(X) = p_1 + p_2 = 1/16 + 1/4 = 5/16.$$

Continuing in this way, one can trace out a series of horizontal line segments as in Fig. 6.11. These segments trace out the function $F(X)$. If the horizontal line segments are connected by vertical (dotted) lines, as in Fig. 6.11, one gets what looks like a series of steps. For this reason $F(X)$ is called a "step" function for discrete probability distributions.

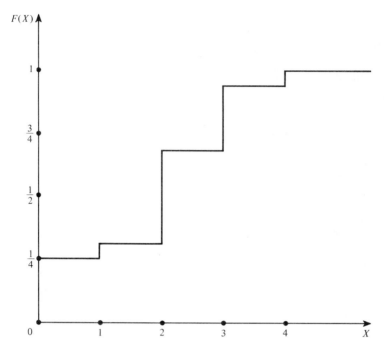

FIGURE 6.11   Distribution Function for Discrete Probability Distribution

For continuous probability distributions, the distribution function $F(X)$ may be viewed as the area under the density function curve $f(Y)$ to the left of the value $Y = X$.

Example 6.9.   Figure 6.12 shows the density function $f(Y)$ for a normal distribution. The shaded area to the left of $Y = X$ is the value of $F(X)$.

The distribution function for all values of X is shown in Fig. 6.13. Note that as X becomes larger and larger $F(X)$ approaches the value of unity.

The distribution function $F(X)$ for any type of probability distribution has the important property that it can be used to measure the probability associated with any half-open interval (or any point in the case of discrete distributions). Let $P(a < Y \leqq b)$ be the probability that Y lies in the interval $\{a < Y \leqq b\}$. Then

$$P(a < Y \leqq b) = F(b) - F(a). \tag{6.6}$$

Example 6.10.   Let us consider the distribution function derived in Example 6.8. The distribution function is shown in Fig. 5.11. The probability that Y equals 2, 3, or 4 is

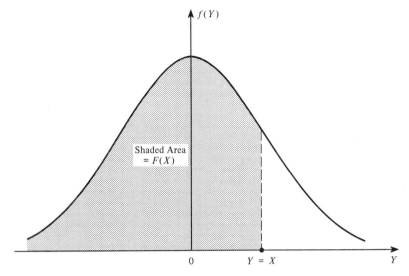

FIGURE 6.12   **Relation between Area and Distribution Function for Normal Probability Distribution**

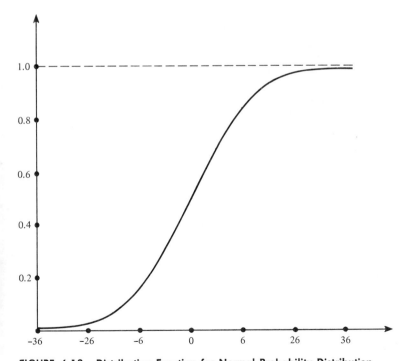

FIGURE 6.13   **Distribution Function for Normal Probability Distribution**

$P(1 < Y \leq 4) = F(4) - F(1) = 1 - 1/16 = 15/16.$

The probability that Y equals 3 is $P(Y = 3)$ or

$P(2 < Y \leq 3) = F(3) - F(2) = 15/16 - 11/16 = 1/4.$

This may be checked with the value of $P(Y = 3)$ obtained by other means in Example 6.3.

Example 6.11.   Figure 6.14 shows a graph of the normal distribution. The horizontally shaded area is $F(b)$. The vertically shaded area is $F(a)$. The difference $F(b) - F(a)$ is the area in between a and b or the probability $P(a < Y \leq b)$.

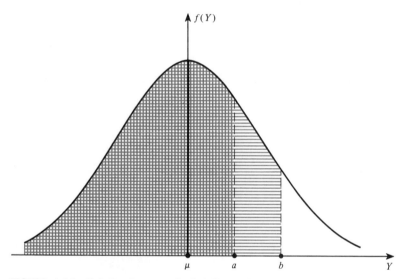

FIGURE 6.14   Relation between Probability and Density Functions

Expectations in One Dimension

The concept of *expected value* of a random variable is best understood in terms of discrete probability distributions. If p_1, p_2, $\ldots$, p_n are the n probabilities associated with the values Y_1, Y_2, $\ldots$, Y_n of the random variable Y, then the expected value of Y is

$$E(Y) = p_1 Y_1 + p_2 Y_2 + \cdots + p_n Y_n. \tag{6.7}$$

Example 6.12.   If a die is tossed, the probability of any particular number of dots from 1 through 6 is $1/6$. Let Y represent the number of dots. The expected value of Y is

$$E(Y) = (1/6)1 + (1/6)2 + (1/6)3 + (1/6)4 + (1/6)5 + (1/6)6 \qquad (6.8)$$
$$= 21/6 = 7/2 = 3\tfrac{1}{2}.$$

Let $h(Y)$ be any function of the random variable Y. The expected value of $h(Y)$ is

$$E(h(Y)) = p_1 h(Y_1) + \cdots + p_n h(Y_n). \qquad (6.9)$$

Example 6.13.   Let a coin be tossed. If heads occurs, the random variable $Y = 1$, and $Y = 0$ if tails occurs. Suppose an individual A is promised $1 for a head and $3 for a tail. The payoff then for a head is

$$h(1) = \$1 \qquad (6.10)$$

and for a tail is

$$h(2) = \$3. \qquad (6.11)$$

Since the probability of either a head or a tail is 1/2, the expected payoff is

$$E(h(Y)) = \frac{1}{2}(\$1) + \frac{1}{2}(\$3) = \$2. \qquad (6.12)$$

Example 6.14.   Suppose an individual A proposes to an individual B the following bet: A will throw a pair of dice. If the sum of the dots is a multiple of 3, then B pays A $2. If any other number comes up, A pays B $1. Let Y be the number of dots appearing on the dice and let $h(Y)$ be the amount received by individual A. Then from the table of probabilities in Example 6.13, we have:

Y_i	$h(Y_i)$	$p_i = p(Y_i)$
$Y_1 = 2$	-1	$1/36$
$Y_2 = 3$	$+2$	$2/36$
$Y_3 = 4$	-1	$3/36$
$Y_4 = 5$	-1	$4/36$
$Y_5 = 6$	$+2$	$5/36$
$Y_6 = 7$	-1	$6/36$
$Y_7 = 8$	-1	$5/36$
$Y_8 = 9$	$+2$	$4/36$
$Y_9 = 10$	-1	$3/36$
$Y_{10} = 11$	-1	$2/36$
$Y_{11} = 12$	$+2$	$1/36$

The expected amount received by A—that is, the expected payoff to A—is

$$E(h(Y)) = \frac{1}{36}(-1) + \frac{2}{36}(2) + \frac{3}{36}(-1) + \cdots + \frac{2}{36}(-1) + \frac{1}{36}(2) = 0. \qquad (6.13)$$

If instead A received \$2.50 for a multiple of 3 while giving up \$1 for a nonmultiple of 3, then the expected payoff to A would be \$ .17.

Expected values for continuous probability distributions are most easily defined in terms of integrals. This is done in the appendix to this chapter.

Algebra of Expectations

Expectations follow a certain set of rules:

Rule 1   The expected value of a constant c is equal to the constant c.

We shall prove the validity of each rule for the discrete case. The proofs for the continuous case are very similar but more lengthy, but all the algebraic rules stated below apply to the continuous case.

The expected value of a constant c is the special case in which $h(Y) = c$. From (6.7)

$$
\begin{aligned}
E(c) &= \sum_{i=1}^{n} c \cdot P(Y_i) \\
&= c \sum_{i=1}^{n} P(Y_i) \\
&= cP(-\infty < Y < +\infty) \\
&= c.
\end{aligned}
$$

(6.14)

Rule 2   The expected value of

$$
E(c \cdot h(Y)) = c \cdot E(h(Y)).
$$

(6.15)

From (6.7), we have

$$
\begin{aligned}
E(c \cdot h(Y)) &= \sum_{i=1}^{n} c \cdot h(Y_i) \cdot P(Y_i) \\
&= c \cdot \sum_{i=1}^{n} h(Y_i) \cdot P(Y_i) \\
&= c \cdot E(h(Y)).
\end{aligned}
$$

(6.16)

Rule 3   The expected value of the sum of two different functions of Y, $g(Y)$ and $h(Y)$, is

$$
E[g(Y) + h(Y)] = E(g(Y)) + E(h(Y)).
$$

(6.17)

This follows also from (6.7) in a very straightforward manner.

$$E[g(Y) + h(Y)] = \sum_{i=1}^{n} [g(Y_i) + h(Y_i)] \cdot P(Y_i)$$

$$= \sum_{i=1}^{n} [g(Y_i) \cdot P(Y_i) + h(Y_i) \cdot P(Y_i)] \qquad (6.18)$$

$$= \sum_{i=1}^{n} [g(Y_i) \cdot P(Y_i) + \sum_{i=1}^{n} h(Y_i) \cdot P(Y_i)]$$

$$= E[g(Y)] + E[h(Y)].$$

Theoretical Mean and Variance

A probability distribution may be characterized by a number of parameters. The two most important are the theoretical mean μ and the theoretical variance σ^2. The theoretical mean of the probability distribution for a random variable Y is the expected value of the random variable Y.

$$\mu = E(Y). \qquad (6.19)$$

The theoretical variance σ^2 of a probability distribution is the expected value of $h(Y) = (Y - \mu)^2$.

$$\sigma^2 = E(Y - \mu)^2. \qquad (6.20)$$

Thus for a discrete probability distribution

$$\sigma^2 = (Y_1 - \mu)^2 p_1 + (Y_2 - \mu)^2 p_2 + \cdots + (Y_n - \mu)^2 p_n. \qquad (6.21)$$

Example 6.15.   If a coin is tossed, the probability of a head is $p_1 = 1/2(Y = Y_1 = 1)$ and a tail is $p_2 = 1/2(Y = Y_2 = 0)$. The theoretical mean of this probability distribution is

$$\mu = Y_1 p_1 + Y_2 p_2 = 1 \cdot 1/2 + 0 \cdot 1/2 = 1/2. \qquad (6.22)$$

The theoretical variance is

$$\sigma^2 = (Y_1 - \mu)^2 p_1 + (Y_2 - \mu)^2 p_2$$
$$= (1 - 1/2)^2 \cdot 1/2 + (0 - 1/2)^2 \cdot 1/2$$
$$= (1/2)^2 \cdot 1/2 + (-1/2)^2 \cdot 1/2$$
$$= 1/8 + 1/8 = 1/4.$$

Example 6.16.   The discrete probability distribution in Example 6.2 referred to the probability that the sum Y of the number of dots obtained by throwing a pair of dice was equal to any specific value. The distribution is as follows:

$$P(Y = Y_1 = 2) = p_1 = 1/36,$$
$$P(Y = Y_2 = 3) = p_2 = 2/36,$$
$$P(Y = Y_3 = 4) = p_3 = 3/36,$$
$$P(Y = Y_4 = 5) = p_4 = 4/36,$$
$$P(Y = Y_5 = 6) = p_5 = 5/36,$$
$$P(Y = Y_6 = 7) = p_6 = 6/36,$$
$$P(Y = Y_7 = 8) = p_7 = 5/36,$$
$$P(Y = Y_8 = 9) = p_8 = 4/36,$$
$$P(Y = Y_9 = 10) = p_9 = 3/36,$$
$$P(Y = Y_{10} = 11) = p_{10} = 2/36,$$
$$P(Y = Y_{11} = 12) = p_{11} = 1/36.$$

The expected value of Y is

$$\mu = \sum_{i=1}^{11} p_i Y_i = 2 \cdot (1/36) + 3 \cdot (2/36) + \cdots + 11 \cdot (2/36) + 12 \cdot (1/36)$$

or

$$\mu = 7.$$

The mean or expected value of the sum of dots obtained by throwing a pair of dice is 7.

Example 6.17. For a normal distribution, the theoretical variance and theoretical standard deviation have a particular significance. In Fig. 6.15,

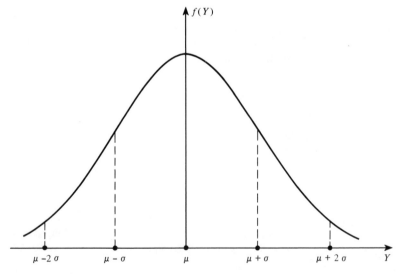

FIGURE 6.15   Normal Distribution and Its Variance

a normal distribution is drawn. The mean μ occurs where the normal distribution reaches its maximum value. The probability that the random variable Y falls within two standard deviations on either side of the mean is approximately .95. The probability that Y falls within one standard deviation on either side of the mean is approximately .67. For example, if $\mu = 5$ and $\sigma = 2$, then

$$P(1 \leq Y \leq 9) \cong .95, \qquad P(3 \leq Y \leq 7) \cong .67.$$

Probability Distributions in More Than One Dimension—Joint Distributions

Suppose $Y_1, Y_2, \ldots, Y_k$ are each random variables. These may be combined into a k-dimensional random variable $Y = (Y_1, Y_2, \ldots, Y_k)$. The k-dimensional random variable Y has a k-dimensional probability distribution. As in the single-dimensional case, a k-dimensional probability distribution may be discrete or continuous. For a discrete probability distribution, each of the k random variables $Y_1, Y_2, \ldots, Y_k$ may take on a finite number of different values and the k-dimensional variable $Y = (Y_1, Y_2, \ldots, Y_k)$ takes on n different values,

$$Y^1 = (Y_1^1, Y_2^1, \ldots, Y_k^1); \quad Y^2 = (Y_1^2, Y_2^2, \ldots, Y_k^2); \quad \ldots;$$
$$Y^n = (Y_1^n, Y_2^n, \ldots, Y_k^n).$$

Each one of these n different values has a probability

$$P(Y^1) = p_1, \quad P(Y^2) = p_2, \quad \ldots, \quad P(Y^n) = p_n.$$

The sum of the probabilities, of course, equals one;

$$\sum_{i=1}^{n} p_i = 1.$$

Example 6.18.   Suppose a coin is tossed twice. $Y_1 = 0$ or 1 depending on whether the first toss is heads or tails, and $Y_2 = 0$ or 1 depending on whether the second toss is heads or tails. There are four $(n = 4)$ possible values of the random variable $Y = (Y_1, Y_2)$.

$$Y = (Y_1^1, Y_2^1) = (0, 0),$$
$$Y = (Y_1^2, Y_2^2) = (1, 0),$$
$$Y = (Y_1^3, Y_2^3) = (0, 1),$$
$$Y = (Y_1^4, Y_2^4) = (1, 1).$$

Since each of these values is equally likely, the probability of each is $1/4$. The points are represented in the three-dimensional diagram in Fig. 6.16 by the dots in the (Y_1, Y_2) plane. The probability $P(Y_1, Y_2)$ associated with

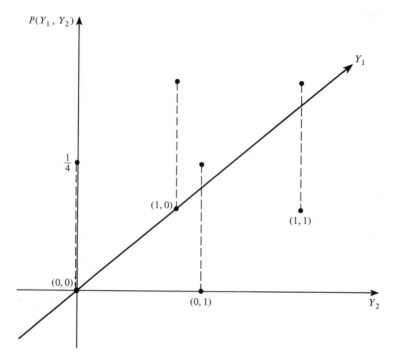

FIGURE 6.16 **Discrete Probability Distribution in Two Dimensions**

each point is 1/4 and is represented by the vertical distance above the (Y_1, Y_2) plane.

For a continuous probability distribution in k dimensions, there is a density function $f(Y_1, Y_2, \ldots, Y_k)$. The probability that the variable $Y = (Y_1, Y_2, \ldots, Y_k)$ lies in any nondegenerate interval $\{a \leq Y \leq b\}$ depends on the density function. In particular, if $k = 2$, the probability that Y lies in the rectangle, $\{(a_1, a_2) \leq (Y_1, Y_2) \leq (b_1, b_2)\}$ is given by the volume beneath the surface of the density function $f(Y_1, Y_2)$.

Example 6.19. Figure 6.17 depicts a joint normal distribution for the two-dimensional random variable $Y = (Y_1, Y_2)$. A rectangular interval $\{(a_1, a_2) \leq (Y_1, Y_2) \leq (b_1, b_2)\}$ is shown. The volume above the rectangle and below the surface of the density function is the probability that the random variable Y lies in that interval.

The expected value $E(Y)$ of a k-dimensional random variable Y is defined in a fashion similar to that of a single-dimensional variable.

$$E(Y) = (Y_1^1 \cdot p_1, Y_2^1 \cdot p_1, \ldots, Y_k^1 \cdot p_1) + \cdots + (Y_1^n \cdot p_n, Y_2^n \cdot p_n, \cdots, Y_k^n \cdot p_n)$$

$$= \left(\sum_{i=1}^{n} p_i Y_1^i, \sum_{i=1}^{n} p_i Y_2^i, \ldots, \sum_{i=1}^{n} p_i Y_k^i \right). \tag{6.23}$$

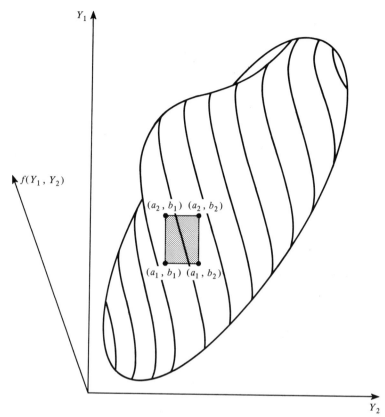

FIGURE 6.17   Joint Normal Distribution in Two Dimensions

Each of the k components of Y^j is multiplied by the appropriate probability, p_j.

Example 6.20. Suppose an urn contains three balls labeled with the number one and four balls labeled with the number zero. A ball is drawn, the number recorded as Y_1, and the ball replaced. Another ball is drawn and the number recorded as Y_2. The probability distribution may be characterized as follows:

$$P(Y_1 = 0,\ Y_2 = 0) = (4/7)(4/7) = 16/49,$$
$$P(Y_1 = 1,\ Y_2 = 0) = (3/7)(4/7) = 12/49,$$
$$P(Y_1 = 0,\ Y_2 = 1) = (4/7)(3/7) = 12/49,$$
$$P(Y_1 = 1,\ Y_2 = 1) = (3/7)(3/7) = 9/49.$$

The expected value of $Y = (Y_1,\ Y_2)$ is

$$E(Y_1, Y_2) = (0 \cdot 16/49, 0 \cdot 16/49) + (1 \cdot 12/49, 0 \cdot 12/49)$$
$$+ (0 \cdot 12/49, 1 \cdot 12/49) + (1 \cdot 9/49, 1 \cdot 9/49)$$
$$= (3/7, 3/7).$$

We may also define the expected value $E[h(Y)]$ of a function $h(Y)$ of the k-dimensional vector Y.

$$E[h(Y)] = h(Y^1)p_1 + h(Y^2)p_2 + \cdots + h(Y^n)p_n. \tag{6.24}$$

Example 6.21.   Using the probability distribution of Example 6.20, let us calculate the expected value of the sum $h(Y) = Y_1 + Y_2$. From the definition (6.24), we have

$$E(Y_1 + Y_2) = (0 + 0) \cdot \frac{16}{49} + (1 + 0) \cdot \frac{12}{49} + (0 + 1) \cdot \frac{12}{49}$$
$$+ (1 + 1) \cdot \frac{9}{49} = \frac{6}{7}.$$

Marginal Distributions and Independence

A probability distribution for a k-dimensional random variable $Y = (Y_1, Y_2, \ldots, Y_k)$ can be converted into a probability distribution for any one of the k components of Y separately. For example, the marginal distribution of the first component Y_1 is defined as follows:

1. The probability that the first component Y_1 lies in the interval $\{a_1 \leqq Y_1 \leqq b_1\}$ is the probability that Y lies in the k-dimensional interval $\{(a_1, -\infty, -\infty, \ldots, -\infty) \leqq Y \leqq (b_1, +\infty, +\infty, \ldots, +\infty)\}$.
2. In particular, for the case of a discrete probability distribution, the probability that the first component Y_1 equals some specific value a_1 is the sum of the probabilities for all vectors $Y = (a_1, Y_2, \ldots, Y_k)$ that have a_1 as a first component.
3. For continuous distributions, item 1 above enables us to convert the k-dimensional density function $f(Y_1, Y_2, \ldots, Y_k)$ into a single-dimensional marginal density function $f_1(Y_1)$.

Example 6.22.   In the coin-tossing case described in Example 6.16, the marginal distribution $P(Y_1)$ is the probability of obtaining a head ($Y_1 = 0$) or tail ($Y_1 = 1$) on the first toss of two tosses of a coin. The marginal distribution can be obtained by adding the probabilities for all values of Y_2 for which $Y_1 = 0$ and then repeating the process for $Y_1 = 1$. That is,

$$P(0, 0) = 1/4$$
$$\underline{P(0, 1) = 1/4}$$
$$P(Y_1 = 0) = 1/2$$

Similarly,

$$P(1, 0) = 1/4$$
$$\underline{P(1, 1) = 1/4}$$
$$P(Y_1 = 1) = 1/2$$

The single-dimensional random variables Y_1, Y_1, ... , and Y_n are said to be *independently distributed* if their joint distribution can be compiled by multiplying probabilities from the marginal distributions. That is,

$$P[(a_1, a_2, \ldots, a_n) < Y < (b_1, b_2, \ldots, b_n)]$$
$$= P(a_1 < Y < b_1) \cdot P(a_2 < Y_2 < b_2) \cdot \ldots \cdot P(a_n < Y_n < b_n). \qquad (6.25)$$

In terms of density functions, the variables Y_1, Y_2, ... , Y_n are independently distributed if

$$f(Y_1, Y_2, \ldots, Y_n) = f_1(Y_1) \cdot f_1(Y_2) \cdot \ldots \cdot f_n(Y_n). \qquad (6.26)$$

Example 6.23.   The probability of obtaining a head on the first toss of a coin is $P(Y_1 = 0) = 1/2$, and the probability of a tail is $P(Y_1 = 1) = 1/2$. Similarly, for a second toss $P(Y_2 = 0) = 1/2$ and $P(Y_2 = 1) = 1/2$. The variables Y_1 and Y_2 are said to be independently distributed, since the joint probability distribution $P(Y_1, Y_2)$ can be constructed as follows:

$$P(0, 0) = P(Y_1 = 0) \cdot P(Y_2 = 0) = (1/2)(1/2) = 1/4,$$
$$P(0, 1) = P(Y_1 = 0) \cdot P(Y_2 = 1) = (1/2)(1/2) = 1/4,$$
$$P(1, 0) = P(Y_1 = 1) \cdot P(Y_2 = 0) = (1/2)(1/2) = 1/4,$$
$$P(1, 1) = P(Y_1 = 1) \cdot P(Y_2 = 1) = (1/2)(1/2) = 1/4.$$

Let $Y = (Y_1, Y_2)$ be a two-dimensional random variable and let μ_1 and μ_2 be the theoretical means of the marginal distributions of Y_1 and Y_2, respectively. $E[(Y_1 - \mu_1)(Y_2 - \mu_2)]$ is called the theoretical covariance of the random variables Y_1 and Y_2. If the two variables Y_1 and Y_2 are independently distributed, then

$$E[(Y_1 - \mu_1)(Y_2 - \mu_2)] = 0. \qquad (6.27)$$

Let us prove this statement for the case of a two-dimensional, discrete probability distribution.

Consider the marginal distributions for Y_1 and Y_2:

$$P(Y_1^i) = p_i \qquad \text{for } i = 1, \ldots, n_1,$$
$$P(Y_1^j) = p_j \qquad \text{for } j = 1, \ldots, n_2,$$

respectively. That is, Y_1 assumes n_1 different values, each with a positive probability. Y_2 assumes one of n_2 different values. The two-dimensional variable $Y = (Y_1, Y_2)$ then assumes one of $n = n_1 \cdot n_2$ different values. The probability of any particular value $Y^{i,j} = (Y_1^i, Y_2^j)$ is $p_{i,j} = p_i \cdot p_j$. Then

$$E[(Y_1 - \mu_1) \cdot (Y_2 - \mu_2)] = \sum_{i=1}^{n_1} \sum_{j=1}^{n_2} (Y_1^i - \mu_1)(Y_2^j - \mu_2)p_i p_j$$

$$= \sum_{i=1}^{n_1} (Y_1^i - \mu_1)p_i \cdot \sum_{j=1}^{n_2} (Y_2^j - \mu_2)p_j \qquad (6.28)$$

$$= E(Y_1 - \mu_1) \cdot E(Y_2 - \mu_2).$$

From Rule 3 of the previous section

$$E(Y_1 - \mu_1) = E(Y_1) - E(\mu_1), \qquad E(Y_2 - \mu_2) = E(Y_2) - E(\mu_2).$$

Using Rule 1 and the definition of μ_1 and μ_2, we get

$$E(Y_1) - E(\mu_1) = \mu_1 - \mu_1 = 0, \qquad E(Y_2) - E(\mu_2) = \mu_2 - \mu_2 = 0.$$

Thus (6.27) must be true.

APPENDIX   Expectations for Continuous
Random Variables

With a continuous distribution function $f(Y)$, the expected value of Y is the indefinite integral

$$E(Y) = \int_{-\infty}^{\infty} Yf(Y)\, dY = \mu. \tag{6.29}$$

The expected value of Y is also called the theoretical mean of the distribution function $f(Y)$.

The expected value of $g(Y)$, a function of Y, is the indefinite integral

$$E[g(Y)] = \int_{-\infty}^{\infty} g(Y) \cdot f(Y)\, dY. \tag{6.30}$$

In particular if $g(Y) = (Y - \mu)^2$, the expected value is the theoretical variance σ^2.

$$E[(Y - \mu)^2] = \int_{-\infty}^{\infty} (Y - \mu)^2 f(Y)\, dY = \sigma^2. \tag{6.31}$$

One may define the expected value of a function of a multidimensional random variable $(Y_1, \ldots, Y_n)$ as the multiple integral

$$E[g(Y_1, \ldots, Y_n)]$$
$$= \int_{-\infty}^{\infty} \cdots \int_{-\infty}^{\infty} g(Y_1, \ldots, Y_n) f(Y_1, \ldots, Y_n)\, dY_1 \cdots dY_n. \tag{6.32}$$

If Y_1 and Y_2 are independently distributed—that is, if $f(Y_1, Y_2) = f_1(Y_1) \cdot f_2(Y_2)$—then the covariance equals zero.

$$E[(Y_1 - \mu_1)(Y_2 - \mu_2)] = 0, \tag{6.33}$$

where μ_1 is the theoretical mean of $f_1(Y_1)$, the marginal distribution of Y_1, and μ_2 is the theoretical mean of $f_2(Y_2)$, the marginal distribution of Y_2.

All the rules concerning the algebraic manipulations of expectations for discrete random variables are valid for continuous random variables. They are easily proved given the various properties of indefinite integrals.

PROBLEMS

1. A shoe factory contains three machines. Machine I manufactures 5 percent defectives, machine II, 4 percent defectives and machine III, 3 percent defectives. The three machines manufacture an equal number of shoes. Suppose a sample of two shoes is taken at random. Let the random variable Y represent the number of defectives. (a) What is the probability distribution of the random variable Y? (b) What is the expected number of defectives? (c) What is the variance in the number of defectives?

2. Suppose a spinner is attached to a dial with the numbers 1, 2, and 3. Any of the numbers is equally likely to turn up if the spinner is spun. Let the multidimensional random variable $Y = (Y_1, Y_2)$ represent the result of two spins on the dial. That is, Y_1 represents the number obtained by spinning on the dial once and Y_2 represents the number obtained by spinning on the dial a second time.
   (a) What is the probability distribution of the random variable Y?
   (b) What is the expected value of the random variable Y?
   (c) What is the expected value of the sum $Y_1 + Y_2$?
3. Average cost per freight-mile on a railroad line is 50 cents/ton, and standard deviation is 15 cents/ton. Assuming costs are approximately normal, what percentage of costs lies between 35 cents/ton and 65 cents/ton?
4. A proposes to B the following bet. A will throw a pair of dice. If both dice register even numbers, B pays A \$3. If one die registers even and the other odd, A pays B \$1. What is the expected payoff to A?
5. Given the following demand for shoes:

SHOE SIZE	RELATIVE FREQUENCY
9	.20
10	.40
11	.30
12	.10

(a) What is the expected value of shoe size? (b) What is the variance in shoe size? (c) What is the standard deviation?
6. (Use the data from Example 5.40 of Chapter 5.) A standard television set costs \$200, of which \$50 is profit for a local dealer. One in every ten defective sets has a defect that will cause a serious shock to the owner. In each shock case, the dealer can expect to be sued for an amount of \$8,000.00 by the injured party. Can the dealer order from any of the five manufacturers and expect to make a profit?
7. Assume total allegiance to ethnic and religious considerations on the part of a voting group. The breakdown of the adult Democratic population is as follows:

Ethnic background:	Afro-American	1,000
	Anglo-Saxon	7,000
	Puerto Rican	500
Religion:	Protestant	5,000
	Catholic	2,800
	Buddhist	700

In past elections, interested voters have abided by religious considerations 85 percent of the time and ethnic considerations 15 percent of the time. An election is so exciting that everyone plans to vote in the Democratic primary. One candidate is Protestant with Afro-American background. The other is a Catholic with Anglo-Saxon background. Assuming an even split between voters with no religious or ethnic interest, who can expect to win?

8. Given the following heights (in inches) of a sample population:

HEIGHT	NO. OF PEOPLE
64	15
65	18
66	22
67	26
68	26
69	21
70	16
71	13

(a) Draw a histogram. (b) What is the expected or mean height?

9. Given the data in Problem 8, what is the probability that a selected person will be 5'7" or below?

10. If a given subject is known with certainty to be above 5'7", what is the probability that he is 5'10"?

11. A continuous random variable can take values only between 0 and 4 (inclusive). Its density function is $Y = f(X) = X/8$. What is the probability that X is greater than 1?

12. In the last stages of ennui, 3 young nobles decide to play Russian roulette one evening. Each pulls the trigger of his six-shot revolver containing one live bullet. Each noble contributes $10,000, with the agreement that the survivors will split the pot. What is the expected take of each survivor?

13. My assets will be $10,000 at the end of this year. The probabilities of my expected debts by the end of the year are: $11,000—1/2; $12,000—1/5; $13,000—1/4; $14,000—1/20. What is the probability that the coming year is going to bankrupt me?

14. A real estate broker buys ten swamp lots for $2,000 each. He attempts to sell these lots at $3,000 each. The lots are worth only $1,000. The probability that the broker can find X unwary buyers for the lots is $(10 - X)/55$. Those lots for which he cannot find unwary buyers must be sold at the true worth or $1,000. What is the broker's expected profit or loss?

Outline of Further Reading for Part II

A. Elementary Introductions to Theory of Probability and Probability Distributions
   1. Bryant [1966], pp. 13–33
   2. Freund [1962], pp. 7–163
   3. Freund [1967], pp. 83–188
   4. Freund [1970], pp. 84–193
   5. Freund and Williams [1964], pp. 97–178
   6. Goldberg [1960], pp. 1–294
   7. Hoel [1966], pp. 45–116
   8. Wallis and Roberts [1956], pp. 309–344

B. Intermediate Treatment of Theory of Probability and Probability Distributions (some calculus used)
   1. Anderson and Bancroft [1952], pp. 9–57
   2. Chou [1969], pp. 114–236
   3. Hoel [1962], pp. 4–130
   4. Hogg and Craig [1970], pp. 1–115
   5. Kane [1968], pp. 98–171
   6. Mood and Graybill [1963], pp. 6–138
   7. Neter and Wasserman, pp. 187–243
   8. Neyman, pp. 15–95, 164–249
   9. Richmond [1964], pp. 101–157
   10. Wonnacott and Wonnacott [1969], pp. 27–101
   11. Yamane [1967], pp. 78–112, 129–167

C. Advanced Treatment of Probability Theory
   1. Cramér [1955], pp. 11–154
   2. Kendall and Stuart [1969], pp. 55–197
   3. Kolmogorov [1950], pp. 2–68
   4. Parzen [1960], *passim*

D. Advanced Theory and Applications of Discrete Probability Distributions
   1. Feller [1957], *passim*
   2. Feller [1966], *passim*

These two volumes by Feller provide an exceptionally comprehensive coverage of probability and discrete probability distributions. Many of the sections on applications of probability and some of the theoretical topics covered require relatively little mathematical sophistication, but other topics can be read with full appreciation only by expert mathematicians.

E. Advanced Theory of Continuous Probability Distributions
1. Cramér [1946], pp. 137–322
2. Wilks [1962], pp. 1–194

F. Special Coverage of the Multivariate Normal Distribution
1. Anderson [1958], pp. 5–43
2. Graybill [1961], pp. 48–73
3. Hogg and Craig [1970], pp. 379–384

The multivariate normal distribution is a generalization of the normal distribution of a single random variable to the case of an n-dimensional random variable. The multivariate normal distribution is extremely important to the theory of analysis of variance and analysis of covariance, topics covered in Chapter 10 of this book. The above three references are quite advanced.

Part III

Statistical Inference

Chapter 7

Sampling, Point Estimation, and Sampling Distributions

Let Y be a single-dimensional random variable with theoretical mean μ and theoretical variance σ^2. Sampling is a procedure by which we take n observations of this random variable. Let Y_1, Y_2, ... , and Y_n stand for the values of these n observations. Each of the observed values Y_i can be regarded as a random variable itself, and the multidimensional variable $(Y_1, Y_2, \ldots , Y_n)$ is also a random variable. The number n is called the *sample size*.

Example 7.1. Suppose an economist wishes to know the annual expenditure on food for a number of different households in the United States for 1968. He wishes to restrict himself to families with incomes in the $10,000 to $11,000 range. He may assume that within this group expenditure on food, denoted by the variable Y, is random and has a normal distribution about some mean μ.

The economist decides that he will choose ten households and determine their annual food expenditures for 1968. He will thus obtain $n = 10$ observations of the random variable Y. He denotes the expenditures for the ten households by Y_1, Y_2, ... , Y_{10}. Before he does the sampling, he does not know the particular values of these variables. He therefore may assume beforehand that each observation is random and distributed normally with the same mean μ as the general variable Y. If he were to draw several samples of 10 households, he could expect each sample observation Y_i and the total sample $(Y_1, Y_2, \ldots , Y_n)$ to vary in a random fashion.

Random Sampling

Generally, we shall assume that sampling is done on a *random* basis. By this we mean that the ith observation Y_i is independently distributed with respect to the jth observation Y_j. That is, the joint distribution of the sample $Y_1, Y_2, \ldots, Y_n$ is given by multiplying probabilities of the marginal distributions. In terms of density functions,

$$f(Y_1, Y_2, \ldots, Y_n) = f_1(Y_1) \cdot f_2(Y_2) \cdot \ldots \cdot f_n(Y_n), \tag{7.1}$$

where $f_i(Y_i)$ is the density function for Y_i.

Furthermore, *random sampling* requires that each of the marginal distributions be the same. In particular, the probability distribution or density function for each observation in the sample must be the same.

$$f_1(Y) = f_2(Y) = \cdots = f_n(Y) = f(Y), \tag{7.2}$$

where $f(Y)$ is the density function for the random variable Y being sampled.

Thus random sampling requires that density functions for each sample variable are the same and have the same theoretical mean and standard deviation.

The density function of the sample $f(Y_1, Y_2, \ldots, Y_n)$ is also called the *likelihood function*. In the case of random samples, (7.1) and (7.2) may be combined to give the following likelihood function:

$$f(Y_1, Y_2, \ldots, Y_n) = f(Y_1) \cdot f(Y_2) \cdot \ldots \cdot f(Y_n). \tag{7.3}$$

Example 7.2.   Suppose the economist making the study of food expenditure (Example 7.1) has reason to believe that food-expenditure patterns differ between urban and rural areas of the United States. If his sample is constructed by choosing names in *urban* telephone directories, he is not justified in calling his sample random. His sample values will have a different distribution than the distribution of food expenditure for the United States as a whole—that is, $f_i(Y_i) \neq f(Y)$.

His sample also will not be random if he chooses it in the following way. He goes to a group of telephone directories and chooses one at random. Then from that directory (say the one for Montgomery, Ala.) he chooses the first ten names on a page selected at random. This sampling procedure violates the conditions for a random sample in at least two ways. First, the observations are confined to one city, which is not necessarily representative of the nation as a whole. Second, since the first ten names in the book are chosen, the observations are not independent. If Smith, Jay, is chosen in the sample, then this means that Smith, John, who is next in the book, is inevitably chosen, and the probability that John Smith is chosen is not independent of the probability that Jay Smith is chosen.

There exists a fairly extensive literature concerning methods of choosing random samples and modifications of theory to handle certain types of nonrandom samples. In this volume, however, we shall generally assume that all samples are somehow chosen in a random fashion.

Statistics

Suppose a sample is drawn and n specific sample values $Y_1, Y_2, \ldots, Y_n$ are obtained. From these sample values, a single, real number Z may be obtained by means of a formula. For example, if

$$Z = \frac{Y_1 + Y_2 + \cdots + Y_n}{n} = \bar{Y}, \tag{7.4}$$

then the number Z is a statistic called the *sample mean*. Alternatively, if

$$Z = Y_n - Y_1, \tag{7.5}$$

where Y_n is the largest sample value and Y_1 is the smallest sample value, the statistic Z is called the *sample range*. Also if

$$Z = \frac{(Y_1 - \bar{Y})^2 + (Y_2 - \bar{Y})^2 + \cdots + (Y_n - \bar{Y})^2}{n} = s^2, \tag{7.6}$$

then Z is a statistic called the *sample variance*, denoted by s^2.

There are many other types of conceivable sample statistics: the sample median, the sample mode, the sample geometric mean, and so on. In general we write

$$Z = Z(Y_1, Y_2, \ldots, Y_n). \tag{7.7}$$

This indicates that a statistic Z is a function of the sample values $Y_1, Y_2, \ldots, Y_n$.

Remember that a random variable Y may be distributed with a theoretical mean of μ and a theoretical variance σ^2. These should be kept distinct from the statistics called the sample mean $\bar{Y}$ and the sample variance s^2. The theoretical μ and σ^2 are called *parameters* of the probability distribution of Y. These are not the only possible parameters of a distribution; there may be many others.

Example 7.3. Suppose a die is tossed and the random variable Y denotes the number of dots. The probability distribution may be represented by the following table:

Y_i	$P(Y_i)$
1	$P(1) = 1/6$
2	$P(2) = 1/6$
3	$P(3) = 1/6$
4	$P(4) = 1/6$
5	$P(5) = 1/6$
6	$P(6) = 1/6$

The theoretical mean (expected value of Y) is

$$\mu = E(Y) = \sum_{i=1}^{6} P(Y_i) \cdot Y_i$$
$$= (1/6) \cdot 1 + (1/6) \cdot 2 + \cdots + (1/6) \cdot 6 \tag{7.8}$$
$$= \frac{21}{6} = 3 \cdot 5.$$

The theoretical variance is

$$\sigma^2 = E(Y - \mu)^2 = \sum_{i=1}^{6} P(Y_i) \cdot (Y_i - 3.5)^2$$
$$= (1/6) \cdot (1 - 3.5)^2 + (1/6) \cdot (2 - 3.5)^2 + \cdots + (1/6) \cdot (6 - 3.5)^2 \tag{7.9}$$
$$= 2.9167.$$

Suppose the die is tossed twice and the numbers 3 and 5 come up. This constitutes a sample (Y_1, Y_2), where $Y_1 = 3$ and $Y_2 = 5$. The sample mean $\bar{Y}$ is

$$\bar{Y} = \frac{Y_1 + Y_2}{2} = \frac{3 + 5}{2} = 4. \tag{7.10}$$

The sample variance is

$$s^2 = \frac{(Y_1 - 4)^2 + (Y_2 - 4)^2}{2} = \frac{(3 - 4)^2 + (5 - 4)^2}{2} = 1. \tag{7.11}$$

The sample mean and variance are not the same as the theoretical mean and variance.

Estimation

The probability distribution for a random variable may have parameters that associate that distribution with a particular member of a class of distributions. The theoretical mean and theoretical standard deviation are specific examples. In Fig. 7.1 there are three different normal distributions, 7.1(a), 7.1(b), and 7.1(c), each with a different mean; and three normal distributions, 7.1(d), 7.1(e), and 7.1(f), each with a different theoretical

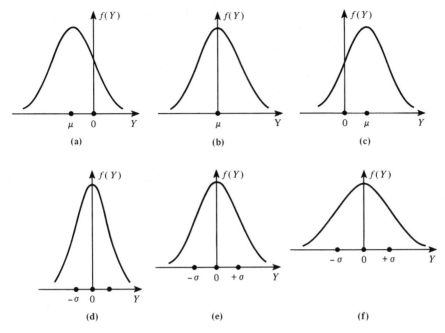

FIGURE 7.1   Normal Distributions with Different Means and Different Variances

variance σ^2. Each of these distributions can be regarded as a member of a family of distributions, the family of normal distributions, with different means and different variances.

Given a random variable Y for which a random sample $(Y_1, Y_2, \ldots, Y_n)$ is obtained, a statistic $Z(Y_1, Y_2, \ldots, Y_n)$ may be regarded as an *estimate* (or *estimator*) of a parameter of the probability distribution of Y. For example, the sample mean $\bar{Y}$ may be regarded as an estimate of the theoretical mean μ. The sample variance s^2 may be regarded as an estimate of the theoretical variance σ^2.

Sampling Distributions

A sample statistic Z is a function of the sample variables $Y_1, Y_2, \ldots, Y_n$, which are random variables. The statistic Z itself, as a function of random variables, is also a random variable. It is often possible to determine the distribution $f(Z)$ of the random variable Z, given information concerning the distribution of the random variable Y from which the sample is drawn. The distribution of the statistic Z is called the sampling distribution.

Example 7.4.   Suppose an urn contains five balls, two of which are labeled 3, one of which is labeled 2, and two of which are labeled 1. The probability of obtaining a 3 is 2/5, a 2 is 1/5, and a 1 is 2/5. If Y is the random variable denoting the number of balls, the theoretical mean of the probability distribution is

$$\mu = E(Y) = 3 \cdot (2/5) + 2 \cdot (1/5) + 1 \cdot (2/5) = 2.$$

The theoretical variance σ^2 is

$$
\begin{aligned}
\sigma^2 &= E[(Y - \mu)^2] \\
&= E[Y^2 - 2\mu Y + \mu^2] \\
&= E(Y^2) - 2\mu E(Y) + \mu^2 \\
&= E(Y^2) - 2\mu^2 + \mu^2 \\
&= E(Y^2) - \mu^2 \\
&= 3^2 \cdot (2/5) + 2^2 \cdot (1/5) + 1^2 \cdot (2/5) - 2^2 \\
&= 18/5 + 4/5 + 2/5 - 4 \\
&= 24/5 - 4 \\
&= 4/5.
\end{aligned}
$$

Now suppose we conduct a series of experiments in which two balls are drawn at random from the urn. After each ball is drawn, it is replaced in the urn. This experiment is repeated four times with the following results:

Y_1	Y_2	MEAN $= (Y_1 + Y_2)/2 = \bar{Y}$
3	2	2.5
1	3	2.0
2	1	1.5
3	2	2.5
		8.5

where Y_1 is the number on the first ball drawn at random and Y_2 is the number on the second ball. The sample mean of the four sample means is obtained by adding the sample means and dividing by 4.

Sample mean of sample means $= 8.5/4 = 2.125$. The sample variance of the sample means is

$$
\begin{aligned}
s^2 &= (2.5 - 2.125)^2 + (2.0 - 2.125)^2 + (1.5 - 2.125)^2 + (2.5 - 2.125)^2 \\
&= .7525.
\end{aligned}
$$

The sample mean of the sample means is an estimate of the theoretical mean of the random variable $\bar{Y}$, the sample mean. Similarly, the sample variance of the sample means is an estimate of the theoretical variance of the random variable $\bar{Y}$.

As we shall demonstrate in the next section, the theoretical mean of the sample mean $\bar{Y}$ is the theoretical mean of Y itself. The theoretical variance of the sample mean $\bar{Y}$ is the theoretical variance of Y divided by the size of the sample, or σ^2 divided by n. In this particular case,

$$\frac{\sigma^2}{n} = \frac{4/5}{2} = 2/5 = .400.$$

If we repeat the experiment above, this time, however, drawing three balls instead of two, the theoretical variance of the sample mean is reduced. That is, there is less variability of means drawn from large samples. In particular, for a sample of size 3,

$$\frac{\sigma^2}{n} = \frac{4/5}{3} = 4/15 = .267.$$

Example 7.4 is a special case of a more general proposition, which we state in the form of a theorem.

Theorem 7.1   If the random variable Y has a distribution with mean μ and variance σ^2, the sample mean from a random sample of size n has a distribution with mean $\mu_{\bar{Y}} = \mu$ and a variance $\sigma_{\bar{Y}}^2 = \sigma^2/n$. If the distribution of Y is normal, then the distribution of $\bar{Y}$ is normal.

We may use the rules of expectations in the third section of Chapter 6 and the definition of a random sample to prove Theorem 7.1. First, let us prove that $\mu_{\bar{Y}} = \mu$.

$$E(Y) = E\left(\sum_{i=1}^{n} \frac{Y_i}{n}\right). \tag{7.12}$$

From Rule 3 of Chapter 6, we may write

$$E(Y) = \sum_{i=1}^{n} E\left(\frac{Y_i}{n}\right). \tag{7.13}$$

From Rule 2 of the same chapter, we obtain

$$E(Y) = \frac{1}{n} \sum_{i=1}^{n} E(Y_i). \tag{7.14}$$

Since, however, in a random sample the distribution of Y_i is the same for all i and equal to the distribution of Y, we have

$$E(Y_i) = \mu \qquad \text{for } i = 1, \ldots, n \tag{7.15}$$

or, from (7.14),

$$E(Y) = \frac{1}{n} \sum_{i=1}^{n} \mu = \frac{n \cdot \mu}{n} = \mu. \tag{7.16}$$

Next let us show that the variance $\sigma_{\bar{Y}}^2$ of $\bar{Y}$ is equal to σ^2/n.

$$\sigma_{\bar{Y}}^2 = E(\bar{Y} - \mu)^2 = E\left(\frac{\Sigma Y_i - n\mu}{n}\right)^2$$

$$= \frac{1}{n^2} E[\Sigma (Y_i - \mu)]^2. \tag{7.17}$$

If we expand the right-hand side of (7.17), we obtain

$$\frac{1}{n^2} E[\Sigma (Y_i - \mu)]^2 = \frac{1}{n^2} E[\Sigma (Y_i - \mu)^2 + \text{cross products}], \tag{7.18}$$

where the cross products are terms that contain $(Y_i - \mu) \cdot (Y_j - \mu)$ for $i \neq j$. In a random sample, however, the probability distributions of Y_i and Y_j are independent. That is, the covariance

$$E(Y_i - \mu)(Y_j - \mu) = 0. \tag{7.19}$$

Thus the cross-product terms in (7.18) all drop out, and (7.18) becomes

$$\sigma_{\bar{Y}}^2 = \frac{1}{n^2} E[\Sigma (Y_i - \mu)^2]$$

$$= \frac{1}{n^2} \sum_{i=1}^{n} E(Y_i - \mu)^2. \tag{7.20}$$

Since the probability distributions are the same for all Y_i, we have

$$E(Y_i - \mu)^2 = \sigma^2 \qquad \text{for } i = 1, \ldots, n, \tag{7.21}$$

and (7.20) becomes

$$\sigma_{\bar{Y}}^2 = \frac{n \cdot \sigma^2}{n^2} = \frac{\sigma^2}{n}. \tag{7.22}$$

This proves the second part of the theorem.

Theorem 7.1 is illustrated with the use of Fig. 7.2. This shows a normal distribution of the variable Y and the distribution of the sample mean $\bar{Y}$. Note that the sample mean has a smaller variance than the variable Y. The larger the sample size n, the smaller is the variance of the sample mean.

Properties of Estimators

There is usually some rationale in choosing a particular test statistic Z as an estimator of a parameter θ. For example, your intuition may tell you

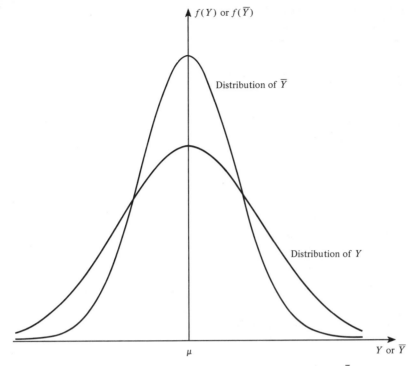

FIGURE 7.2   Distribution of a Random Variable Y and Its Mean Ȳ

that the sample range is not really an estimate of the theoretical mean μ or that sample mean $\bar{Y}$ is probably a good estimator of the theoretical mean μ. But is the median or the mode a good estimator of μ? In order to assess the "goodness" of estimators, we may set up certain standards or criteria by which to judge them. We shall discuss five criteria:

Criterion 1   *A statistic Z is an unbiased estimator of the parameter θ if $E(Z) = \theta$. That is, the statistic Z has a probability distribution whose expected value is equal to the parameter being estimated.*

The sample mean $\bar{Y}$ from a random sample is always an unbiased estimator of the theoretical mean μ of the distribution of $\bar{Y}$. This comes from Theorem 7.1, which states that the theoretical mean of the sample mean is the same as the theoretical mean μ—that is, $E(\bar{Y}) = \mu$. Note that Y may have any arbitrary distribution, not just a normal distribution, and the unbiasedness of $\bar{Y}$ still holds. We may remark also that the sample median and sample mode are unbiased estimates of the mean. They are not efficient estimators, however. The criterion of efficiency is defined and discussed below.

The property of unbiasedness is useful, since it means that when repeated samples are drawn and sample means calculated, we may expect "on the average" to obtain the value of the true parameter.

Criterion 2   *A statistic Z is an asymptotically unbiased estimator of the parameter θ if $E(\bar{Y})$ approaches the parameter θ as the sample size n approaches infinity.*

Thus for large samples (*n* large), we can say that an asymptotically unbiased estimate is approximately unbiased.

The sample variance s^2 from a random sample is not an unbiased estimate of the theoretical variance σ^2 of the distribution of a random variable. The sample variance s^2 is an asymptotically unbiased estimate of the theoretical variance σ^2, however, when the random variable Y has any arbitrary distribution. Let us demonstrate these propositions. The expected value of the sample variance is

$$E(s^2) = E\left[\frac{\sum_{i=1}^{n} (Y_i - \bar{Y})^2}{n}\right] = \left(\frac{1}{n}\right) \cdot E\left[\sum_{i=1}^{n} (Y_i - \bar{Y})^2\right], \tag{7.23}$$

where we have applied Rule 3 for expected values from Chapter 6. We may rewrite (7.16) as follows:

$$\begin{aligned}
E(s^2) &= \left(\frac{1}{n}\right) \cdot E\left[\sum_{i=1}^{n} (Y_i - \mu + \mu - \bar{Y})^2\right] \\
&= \left(\frac{1}{n}\right) \cdot E\left\{\sum_{i=1}^{n} [(Y_i - \mu) - (\bar{Y} - \mu)]^2\right\} \\
&= \left(\frac{1}{n}\right) \cdot E\left\{\sum_{i=1}^{n} [(Y_i - \mu)^2 - 2(Y_i - \mu) \cdot (\bar{Y} - \mu) + (\bar{Y} - \mu)^2]\right\}.
\end{aligned} \tag{7.24}$$

Applying Rules 2 and 3 from Chapter 6, we have

$$\begin{aligned}
E(s^2) &= \left(\frac{1}{n}\right) \cdot \sum_{i=1}^{n} E[(Y_i - \mu)^2] \\
&\quad - \left(\frac{2}{n}\right) \cdot E\left[(\bar{Y} - \mu)\left(\sum_{i=1}^{n} Y_i - n\mu\right)\right] + \frac{n}{n} E[(\bar{Y} - \mu)^2]. \tag{7.25}
\end{aligned}$$

Since

$$\bar{Y} = \frac{\sum_{i=1}^{n} Y_i}{n}, \tag{7.26}$$

we have

$$\sum_{i=1}^{n} Y_i = n \cdot \bar{Y}. \tag{7.27}$$

Thus

$$E\left\{(\bar{Y} - \mu)\left[\sum_{i=1}^{n} (Y_i) - n\mu\right]\right\} = E[(\bar{Y} - \mu)(n\bar{Y} - n\mu)] \tag{7.28}$$

$$= nE[(\bar{Y} - \mu)^2].$$

Thus (7.25) may be written

$$E(s^2) = \left(\frac{1}{n}\right) \cdot \sum_{i=1}^{n} E(Y_i - \mu)^2 - \left(\frac{2n}{n}\right) \cdot E[(\bar{Y} - \mu)^2]$$

$$+ \left(\frac{n}{n}\right) \cdot E[(\bar{Y} - \mu)^2]. \tag{7.29}$$

Now each Y_i has the same distribution and the same variance. Thus

$$E[(Y_1 - \mu)^2] = E[(Y_2 - \mu)^2] = \cdots = E[(Y - \mu)^2] = \sigma^2. \tag{7.30}$$

Furthermore, from Theorem 7.1, we know that the sample mean $\bar{Y}$ has a variance of σ^2/n. Thus

$$E[(\bar{Y} - \mu)^2] = \frac{\sigma^2}{n}. \tag{7.31}$$

From (7.30) and (7.31) we may write

$$E(s^2) = \frac{n}{n} \sigma^2 - \frac{2n}{n} \frac{\sigma^2}{n} + \frac{n}{n} \frac{\sigma^2}{n}$$

$$= \sigma^2 - \frac{\sigma^2}{n} \tag{7.32}$$

$$= \frac{n-1}{n} \sigma^2.$$

The expected value of s^2 is an underestimate of the theoretical variance by a factor of $(n - 1)/n$. It is easy to see, however, that this factor approaches unity as n becomes larger and larger. That is, s^2 is an asymptotically unbiased estimate of σ^2; that is, for large values of n, s^2 is approximately unbiased.

Although s^2 is not an unbiased estimate of σ^2, the statistic

$$\frac{n}{n-1} s^2 = \frac{n}{n-1} \frac{\sum_{i=1}^{n} (Y_i - \bar{Y})^2}{n} = \frac{\sum (Y_i - \bar{Y})^2}{n-1} \tag{7.33}$$

is an unbiased estimator of σ^2. To show this, apply Rule 3 to (7.33) as follows:

$$E\left(\frac{n}{n-1} s^2\right) = \frac{n}{n-1} E(s^2) = \frac{n}{n-1} \cdot \frac{n-1}{n} \sigma^2 = \sigma^2. \tag{7.34}$$

Criterion 3   *A statistic Z is a consistent estimator of the parameter θ if the probability that Z is different from θ approaches zero as the sample size n approaches infinity.*

This is a rather technical property which is closely related to the property of asymptotic unbiasedness. In particular, a consistent estimate is asymptotically unbiased. However, an unbiased estimate or an asymptotically unbiased estimate is not necessarily consistent.

Although an estimator may be biased, it may be asymptotically unbiased or consistent. A biased estimator that is also inconsistent may be a very misleading statistic. No matter how large a sample is drawn, we cannot expect a biased and inconsistent estimator to be equal to the parameter being estimated.

Criterion 4   *A statistic Z is a minimum-variance estimator (sometimes called an efficient or best estimator) of the parameter θ if the sampling distribution of any other estimator based on a sample of the same size has a variance at least as great as or greater than Z.*

In general, the variance of an estimator will vary inversely with the sample size, so that a minimum-variance estimator must be compared with other estimators based on the same sample size. Sometimes a biased estimator will have a small variance, while an unbiased estimator has a large variance. In such cases, the minimum-variance property may be more important than the unbiased property.

As stated above, the sample mean is an unbiased estimator of theoretical mean. The sample median is also an unbiased estimator of the theoretical mean, but it is not efficient. Likewise, in a sample of size 20, just looking at the first observation gives an unbiased estimate of the mean, but it is hardly efficient. The sample mean has a smaller variance than any other estimator of the theoretical mean; that is, it is a more efficient estimator. Therefore the sample mean is a minimum-variance, unbiased estimator of the theoretical mean.

Criterion 5   *A statistic Z is a maximum-likelihood estimator if, for any particular sample, the statistic maximizes the likelihood function when substituted for the parameter θ. The likelihood function is the density function for the sample as given by (7.1).*

Very roughly, we can say that the maximum-likelihood estimator for a particular sample maximizes the probability that the particular sample was chosen. This is again a technical property, used to derive methods of estimating parameters. Under certain conditions maximum-likelihood estimators possess desirable properties. In particular, they are consistent and asymptotically efficient.

The sample mean is a maximum-likelihood estimator of the mean μ of a random variable Y that has a normal distribution. The sample variance is a maximum-likelihood estimator of σ^2 if Y has a normal distribution. (For a proof of these properties see the appendix to this chapter.)

Some Important Sampling Distributions

In this section we discuss some of the properties of some important sampling distributions. All of the distributions discussed here are continuous. They are extremely important in the discussion of interval estimation and hypothesis testing in the next chapter.

THE NORMAL DISTRIBUTION

By now the reader should be familiar with the bell shape of the normal distribution. This distribution is important, because many events of nature seem to follow a normal distribution—the brightness of stars, errors of measurement, and weights of individuals to name a few. In addition, the sample mean of a normally distributed random variable is also normally distributed, as was stated in Theorem 7.1.

An even more important property of the normal distribution is that it approximately characterizes the distribution of the sample mean $\bar{Y}$, even when the original variable Y is not normally distributed.

Theorem 7.2 (Central Limit Theorem)   If the random variable Y (not necessarily normally distributed) has a theoretical mean μ and variance σ^2, the sample mean $\bar{Y}$ has a distribution that approaches a normal distribution with mean μ and variance σ^2/n as the sample size n becomes larger and larger.

This theorem enables us to use the properties of normal distributions whenever the size of the sample is large (a rule of thumb frequently offered is whenever n is greater than 30).

If a random variable Y is normally distributed with a theoretical mean μ and variance σ^2, the probability that Y falls in some interval $\{a \leq Y \leq b\}$ may be calculated by converting the variable Y to a standard normal variate z. The conversion formula is

$$z = \frac{Y - \mu}{\sigma}, \tag{7.35}$$

where σ is the theoretical standard deviation.

Theorem 7.3 If Y is normally distributed with theoretical mean μ and variance σ^2, the standard normal variate z in (7.35) has a normal distribution with a theoretical mean of 0 and a theoretical variance of 1.

Example 7.5. Suppose Y is normally distributed with a theoretical mean of 5 and a theoretical standard deviation of 10. What is the probability that Y lies between 2 and 7? When $Y = 2$, the value of the standard variate is

$$z = \frac{Y - \mu}{\sigma} = \frac{2 - 5}{10} = \frac{-3}{10} = -.300.$$

When $Y = 7$, the value of the standard normal variate is

$$z = \frac{Y - \mu}{\sigma} = \frac{7 - 5}{10} = \frac{2}{10} = .200.$$

The probability that Y lies between 2 and 7 is the same as the probability that z lies between $-.300$ and $+.200$. This probability is represented by the shaded area in Fig. 7.3. Note that the shaded area can be divided into two parts, the area between $z = -.300$ and $z = 0$ and the area between $z = 0$ and $z = +.200$. Since the normal distribution is symmetric (one half is a mirror reflection of the other), the area between $z = -.300$ and $z = 0$ is the same as the area between $z = 0$ and $z = +.300$. Thus the total shaded area is the sum of (i) the area between $z = 0$ and $z = +.300$, and (ii) the area between $z = 0$ and $z = +.200$.

Table A of the Appendix of Statistical Tables gives the area between $z = 0$ and $z =$ any positive number. The two areas above may be determined by finding the two values of z, $+.300$ and $+.200$, and noting the two areas. They are .11791 and .07926. The sum of these numbers is .19717, the probability that z lies between $-.300$ and $+.200$ or that Y lies between 2 and 7.

If a random variable Y has a normal distribution with mean μ and variance σ^2, from Theorem 7.1 we know that the sample mean $\bar{Y}$ has a normal distribution with mean μ and variance $\sigma_{\bar{Y}}^2 = \sigma^2/n$. From Theorem 7.3, however, we may infer that the standard normal variate

$$z = \frac{\bar{Y} - \mu}{\sigma_{\bar{Y}}^2} = \frac{\bar{Y} - \mu}{\sqrt{\sigma^2/n}} = \frac{(\bar{Y} - \mu)\sqrt{n}}{\sigma} \tag{7.36}$$

has a *standard* normal distribution—that is, a mean 0 and variance 1.

Example 7.6. Suppose the random variable Y is normally distributed with mean μ of 8 and a theoretical variance σ^2 of 25. What is the probability that the sample mean $\bar{Y}$ calculated from a sample of 9 lies between 7.0

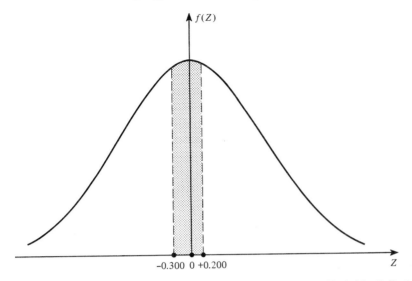

FIGURE 7.3   Probability that Normally Distributed Random Variable Falls in an Interval

and 8.5? The standard normal variate in this case is obtained from (7.36). Since $\sigma^2 = 25$, we have $\sigma = 5$, or from (7.36) we obtain

$$z = \frac{(\bar{Y} - \mu) \sqrt{n}}{\sigma} = \frac{(\bar{Y} - 8) \sqrt{9}}{5} = \frac{(\bar{Y} - 8) \cdot 3}{5}.$$

When $\bar{Y} = 7.0$, the standard normal variate is

$$z = \frac{(7.0 - 8) \cdot 3}{5} = -.6.$$

When $\bar{Y} = 8.5$, the standard normal variate is

$$z = \frac{(8.5 - 8) \cdot 3}{5} = .3.$$

From Table A of the Appendix of Statistical Tables, the area between $z = 0$ and $z = .6$ is .22575. Between $z = 0$ and $z = .3$ the area is .11791. The sum of these two areas is .34366. Thus the probability that $\bar{Y}$ lies between 7.0 and 8.5 is .34366.

THE CHI-SQUARE DISTRIBUTION

A second probability distribution of considerable importance is the chi-square distribution. One reason for its importance is its close relationship to the normal distribution. In general, if $y_1, y_2, \ldots, y_n$ are n random variables, each independently and normally distributed with a mean of zero and a theoretical variance σ^2, the statistic

$$\chi^2 = \frac{\displaystyle\sum_{i=1}^{n} y_i^2}{\sigma^2} \qquad (7.37)$$

has a chi-square distribution with n degrees of freedom. The number of degrees of freedom (d.f.) is an important parameter of the chi-square distribution. Figure 7.4 shows three chi-square distributions, each with different number of degrees of freedom. As the number of degrees of freedom becomes very large, the chi-square distribution begins to look like a normal distribution. Note that the chi-square variate has a zero density function to the left of the origin.

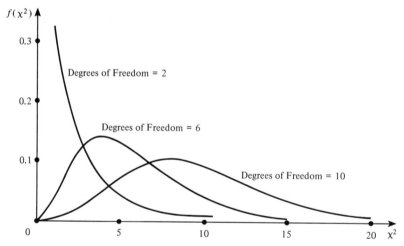

FIGURE 7.4   Chi-square Distribution

Since the sum of squares of a normally distributed variable divided by its variance has a chi-square distribution, the sample variance of a normal variable is also closely related to the chi-square distribution. In the form of a theorem, we have:

Theorem 7.4   If the random variable Y has a normal distribution, the sample standard deviation s^2 multiplied by the sample size n and divided by the theoretical variance σ^2 of the random variable Y,

$$\chi^2 = \frac{ns^2}{\sigma^2} = n \cdot \sum_{i=1}^{n} \frac{(Y_i - \bar{Y})^2}{n\sigma^2} = \sum_{i=1}^{n} \frac{(Y_i - \bar{Y})^2}{\sigma^2}, \qquad (7.38)$$

has a chi-square distribution with $n - 1$ degrees of freedom.

Although the expression (7.37) has a chi-square distribution with n degrees of freedom, note that the statistic ns^2/σ^2 as given in (7.38) has a chi-square distribution with $n - 1$ degrees of freedom, even though the squared quantities $y_i = Y_i - \bar{Y}$ are normally distributed with mean zero and variance σ^2. The difference arises because the individual y_i terms are not independent. In fact, the sum

$$y_1 + y_2 + \cdots + y_n = 0. \qquad (7.39)$$

To show this, note that

$$\sum_{i=1}^{n} y_i = \sum_{i=1}^{n} (Y_i - \bar{Y})$$

$$= \sum_{i=1}^{n} Y_i - n \cdot \bar{Y} \qquad (7.40)$$

$$= \sum_{i=1}^{n} Y_i - n \frac{\sum_{i=1}^{n} Y_i}{n}$$

$$= 0.$$

Since equation (7.39) must be satisfied, we may always determine one of the y_i's by specifying all the others. For example, if we know $y_1, y_2, \ldots, y_{n-1}$, then y_n is given by

$$y_n = -(y_1 + y_2 + \cdots + y_{n-1}). \qquad (7.41)$$

Usually, if the n random variables must satisfy p different, independent linear equations:

$$a_{11}y_1 + \cdots + a_{1n}y_n = r_1$$
$$a_{21}y_1 + \cdots + a_{2n}y_n = r_2$$
$$. \qquad\qquad\qquad\qquad\qquad (7.42)$$
$$.$$
$$.$$
$$a_{p1}y_1 + \cdots + a_{pn}y_n = r_p$$

where the a's and r's are constants and where p is less than n, then one may always determine p of the y_i variables by specifying the values of the other $n - p$ random variables. This reduces the degrees of freedom.

Theorem 7.5   If $y_1, y_2, \ldots, y_n$ are normally distributed random variables with mean zero and independent except for p linear restrictions of the form (7.42), then the statistic χ^2 in (7.38) has chi-square distribution with $n - p$ degrees of freedom.

Example 7.7. Suppose Y is a normally distributed variable with a variance of 10. What is the probability that the sample variance from a sample of size 20 lies between 5 and 15? If the sample variance is 5, then from (7.38), we have

$$\chi^2 = \frac{ns^2}{\sigma^2} = \frac{20 \cdot 5}{10} = 10.$$

If the sample standard deviation is 15, then

$$\chi^2 = \frac{ns^2}{\sigma^2} = \frac{20 \cdot 15}{10} = 30.$$

Thus the probability that s^2 lies between 5 and 15 is equivalent to the probability that the statistic χ^2 lies between 10 and 30. The shaded area between $\chi^2 = 10$ and $\chi^2 = 30$ is the probability. This area may be thought of as the area between $\chi^2 = 0$ and $\chi^2 = 30$ less than the area between $\chi^2 = 0$ and $\chi^2 = 10$. These areas may be determined from Table B in the Appendix of Statistical Tables.

Since the sample size is 20, the degrees of freedom are 19. Thus the degrees of freedom are given in the first column of the table. The top row of the table gives the shaded area to the right of a given value of χ^2—that is, the probability of obtaining a χ^2 value greater than the given value. For example, looking across row 19 corresponding to 19 degrees of freedom, we find that when the given value of χ^2 is 10.117, the area to the right of this value is .95. The area between 0 and 10.117 (to the left of 10.117) is $1 - .95 = .05$. Thus the area between $\chi^2 = 0$ and $\chi^2 = 10$ is approximately .95. The difference, $.95 - .05 = .90$ is the probability that χ^2 lies between 10 and 30 or that the sample standard deviation s^2 lies between 5 and 15.

THE STUDENT'S t DISTRIBUTION

Another important distribution is the Student's t distribution. Like the chi-square distribution, the Student's t distribution depends on a parameter called the degrees of freedom. Figure 7.5 shows a number of different Student's t distributions with differing degrees of freedom. The Student's t distribution approximates very closely the *standard* normal distribution when the degrees of freedom are very large.

Theorem 7.6   If u is a normally distributed random variable with mean μ and variance σ_u^2, and v^2 is a random variable with a chi-square distribution

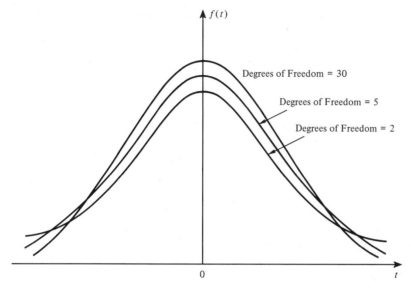

FIGURE 7.5   Student's _t_ Distribution

with f degrees of freedom, then the statistic,

$$t = \frac{(u - \mu)\sqrt{f}}{v \cdot \sigma_u} \tag{7.43}$$

has a Student's t distribution with $n - 1$ degrees of freedom. The statistic t is called the t ratio.

Example 7.8.   Suppose the variable Y has a normal distribution with theoretical mean μ and variance σ^2. If s^2 is the sample variance, then ns^2/σ^2 has a chi-square distribution with $n - 1$ degrees of freedom. Thus from Theorem 7.6

$$t = \frac{(u - \mu)\sqrt{f}}{v \cdot \sigma_u} = \frac{(Y - \mu)\sqrt{n - 1}}{\dfrac{\sqrt{n}\,s}{\sigma} \cdot \sigma} = \frac{(Y - \mu)\sqrt{(n - 1)}}{\sqrt{n}\,s}$$

has a Student's t distribution with $n - 1$ degrees of freedom, where $u = Y$ and $v^2 = n \cdot s^2/\sigma^2$ and $\sigma_u = \sigma$. Now from (7.27) we know that $ns^2/(n - 1)$ is an unbiased estimator of the theoretical variance σ^2. Thus $\sigma = \sqrt{n}\,s/\sqrt{(n - 1)}$ is an unbiased estimator of the theoretical standard deviation σ. Therefore t may be written

$$t = \frac{Y - \mu}{\hat{\sigma}},$$

which has a Student's t distribution with $n - 1$ degrees of freedom.

Example 7.8 proves a special case of the following corollary to Theorem 7.6.

Corollary 7.6　If Y is normally distributed with theoretical mean μ and variance σ^2, the ratio

$$t = \frac{Y - \mu}{\hat{\sigma}} \tag{7.44}$$

has a Student's t distribution, where $\hat{\sigma}$ is an unbiased estimator of the theoretical standard deviation σ.

THE F DISTRIBUTION

The final sampling distribution discussed in this chapter is the F distribution. The F distribution is closely related to the Student's t and the chi-square distributions. If two random variables u and v have independent chi-square distributions, we often wish to know how the ratio u/v is distributed. In particular, we have:

Theorem 7.7　If u and v have independent chi-square distributions with k_1 and k_2 degrees of freedom, respectively, the ratio

$$F = \frac{u}{v} \cdot \frac{k_2}{k_1} \tag{7.45}$$

has an F distribution with k_1 and k_2 degrees of freedom. The statistic F in (7.45) is called the F ratio.

The F distribution depends on two different degrees of freedom, k_1 and k_2. The parameter k_1 is called the degrees of freedom of the numerator and k_2 the degrees of freedom of the denominator. The F distribution for $k_1 = 2$ and $k_2 = 3$ is shown in Fig. 7.6. This diagram also includes two other F distributions, one for $k_1 = 5$ and $k_2 = 3$ and another for $k_1 = 2$ and $k_2 = 10$.

Example 7.9.　Suppose Y_1 and Y_2 are two normally distributed variables with the same variance σ^2. A sample of size n_1 is drawn for Y_1 and of size n_2 for Y_2. Let s_1^2 be the sample variance of Y_1 and s_2^2 be the sample variance for Y_2. Then $u = n_1 s_1^2 / \sigma^2$ has a chi-square distribution with $k_1 = n_1 - 1$ degrees of freedom, and $v = n_2 s_2^2 / \sigma^2$ has a chi-square distribution with $k_2 = n_2 - 1$ degrees of freedom. The F ratio is

$$\begin{aligned} F &= \frac{n_1 s_1^2}{(n_1 - 1)\sigma^2} \cdot \frac{(n_2 - 1)\sigma^2}{n_2 s_2^2} \\ &= \frac{n_1 s_1^2 (n_2 - 1)}{n_2 s_2^2 (n_1 - 1)} = \frac{[n_1/(n_1 - 1)]s_1^2}{[n_2/(n_2 - 1)]s_2^2}, \end{aligned} \tag{7.46}$$

which has an F distribution with $n_1 - 1$ and $n_2 - 2$ degrees of freedom. Note that the F ratio in (7.46) may be thought of as the ratio of two

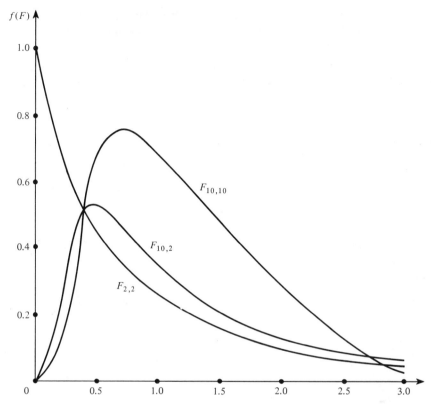

FIGURE 7.6   *F* Distribution

unbiased estimates, $[n_1/(n_1 - 1)]s_1^2$ and $[n_2/(n_2 - 1)]s_2^2$, of the theoretical variance σ^2 [see (7.34)].

Example 7.9 is an illustration of a more general proposition concerning the ratio of unbiased estimates of variance.

Corollary 7.7   If $\hat{\sigma}_1^2$ and $\hat{\sigma}_2^2$ are two independently distributed unbiased estimators of the variance of a normally distributed random variable, the ratio

$$F = \frac{\hat{\sigma}_1^2}{\hat{\sigma}_2^2} \tag{7.47}$$

has an F distribution.

Another characteristic of the F distribution is that if the degrees of freedom in the numerator $k_1 = 1$, the t and F distributions are quite similar.

Theorem 7.8   If the statistic F has an F distribution with $k_1 = 1$ and k_2 degrees of freedom, the statistic $t = \sqrt{F}$ has the Student's t distribution with k_2 degrees of freedom.

Table C in the Appendix of Statistical Tables allows the calculation of probabilities for the F distribution.

The Student's t and F distributions have important applications in the testing of hypotheses and in the estimation of confidence intervals. These are discussed in the next chapter.

APPENDIX Maximum-Likelihood Estimation of the Mean and Variance of a Normal Distribution

In this appendix we shall demonstrate that the sample mean Y is a maximum-likelihood estimator of the theoretical mean μ, and the sample variance s^2 is a maximum-likelihood estimator of σ^2, if the variable Y is normally distributed. The formula for the normal density function is the following:

$$f(Y) = \frac{1}{\sqrt{2\pi}\,\sigma} \exp\left[-\frac{(Y-\mu)^2}{2\sigma^2}\right].^\dagger \tag{7.48}$$

The likelihood function for a sample of size n is

$$
\begin{aligned}
L &= f(Y_1, Y_2, \ldots, Y_n) \\
&= \frac{1}{\sqrt{2\pi}\,\sigma} \exp\left[-\frac{(Y_1-\mu)^2}{2\sigma^2}\right] \cdot \ldots \cdot \frac{1}{\sqrt{2\pi}\,\sigma} \exp\left[-\frac{(Y_n-\mu)^2}{2\sigma^2}\right]. \quad (7.49)
\end{aligned}
$$

This may be rewritten as

$$L = \left(\frac{1}{\sqrt{2\pi}\,\sigma}\right)^n \cdot \exp\left[-\frac{(Y_1-\mu)^2 + \cdots + (Y_n-\mu)^2}{2\sigma^2}\right]. \tag{7.50}$$

The maximization of L may be undertaken more easily if we maximize the natural logarithm of L, which attains its maximum value whenever L attains a maximum, since $\log L$ is a monotonic transformation of L.

$$\log L = n \cdot \log\left(\frac{1}{\sqrt{2\pi}\,\sigma}\right) - \frac{(Y_1-\mu)^2 + \cdots + (Y_n-\mu)^2}{2\sigma^2}. \tag{7.51}$$

To maximize $\log L$ with respect to μ and σ^2, we differentiate partially with respect to μ and σ^2 and set each derivative equal to zero.

$$\frac{\partial(\log L)}{\partial\mu} = \frac{(-1)}{2\sigma^2}[2(Y_1-\mu) + \cdots + 2(Y_n-\mu)] = 0. \tag{7.52}$$

$$\frac{\partial(\log L)}{\partial\sigma} = \frac{n(\sqrt{2\pi})(-1)}{\sqrt{2\pi}\,\sigma} - \frac{(-2)[(Y_1-\mu)^2 + \cdots + (Y_n-\mu)^2]}{2\sigma^3} = 0. \tag{7.53}$$

Equation (7.52) may be multiplied on both sides by $(-\sigma^2)$ to give

$$(Y_1-\mu) + \cdots + (Y_n-\mu) = 0 \tag{7.54}$$

or

$$Y_1 + \cdots + Y_n - n\mu = 0 \tag{7.55}$$

---

† The notation exp (x) represents e^x, where e is the base of the natural (Naperian) logarithms.

or

$$\mu = \frac{Y_1 + \cdots + Y_n}{n} = \bar{Y}. \tag{7.56}$$

Thus the maximum-likelihood estimator of the theoretical mean μ is the sample mean $\bar{Y}$.

Substitute this value for μ into (7.53) and multiply both sides of this equation by $(-\sigma^3)$ to obtain

$$n\sigma^2 - [(Y_1 - \bar{Y})^2 + \cdots + (Y_n - \bar{Y})^2] = 0 \tag{7.57}$$

or

$$\sigma^2 = \frac{(Y_1 - \bar{Y})^2 + \cdots + (Y_n - \bar{Y})^2}{n} = s^2. \tag{7.58}$$

The maximum-likelihood estimator of the theoretical variance σ^2 is the sample variance s^2.

PROBLEMS

1. Suppose the country of Uganda wishes to conduct a survey of farms in a certain district to determine acreages under various crops. The village chiefs are asked to submit lists of farmers paying personal taxes to the district head. Not all village chiefs comply with this request. Of the 56 lists submitted, 10 are chosen at random. From the chosen lists, the first ten farmers are placed in the sample so that there are 100 farms in the sample. Is this sample of farms likely to be random? Why or why not?

2. A political scientist wishes to conduct a survey of voting habits in a big city. Voter registration lists are available that list voters by street address. Every twentieth name on each registration list is chosen. Is the sample of voters likely to be random? Why or why not?

3. Suppose a fair coin is tossed. Let the random variable $Y = 1$ if the coin turns up heads and $Y = 0$ if the coin turns up tails. (a) What is the probability distribution of the random variable Y? (b) What is the mean, $E(Y) = \mu$, of the probability distribution? (c) What is the variance, $E(Y - \mu)^2 = \sigma^2$? (d) Suppose the coin is tossed five times to obtain five sample values of Y—Y_1, Y_2, Y_3, Y_4, and Y_5. What is the expected value of the mean $\bar{Y} = \Sigma Y_i/5$? (e) What is the variance of the sample mean $\bar{Y}$? (f) Suppose a sample of five tosses turns up two heads and three tails. What is the sample mean? (g) Calculate an unbiased estimate of the variance σ^2 of Y from this sample.

4. Suppose the height of American men aged 21 has a mean of 70 inches and a standard deviation of 3 inches. The requirement for admission to a state police training institute is that the applicant be at least 6 feet or 72 inches. What percentage of males aged 21 would qualify for admission?

5. Suppose a random variable Y has a mean of 20 and a standard deviation of 10. (a) For samples of 400, what is the approximate probability (use the normal

distribution) that the sample mean $\bar{Y}$ lies between 19 and 22? (b) What is the probability that $\bar{Y}$ is less than 18?

6. Show that the mean of the odd observations of a sample of size n of the random variable Y is an unbiased estimate of the theoretical mean μ of Y.

7. Suppose the mean life of light bulbs manufactured by a particular firm is 720 hours. The standard deviation is 60 hours. A sample of 20 light bulbs is tested and the sample standard deviation s of these light bulbs is 80. What is the probability that a sample standard deviation of 80 or greater will be obtained from a sample of size 21?

8. What is the theoretical mean of the chi-square variate $n \cdot s^2 / \sigma^2$?

Chapter 8

Hypothesis Testing and Internal Estimation

One may wish to test hypotheses for a number of reasons. For example, a business firm is faced with a decision as to whether to install new machinery. Since the machinery is expensive, the management does not want to install the new machinery unless there is very strong evidence that the new equipment improves efficiency. In other instances, statistical tests of hypotheses are used for quality control—for example, to test the strength of construction materials, the purity of drugs, or the life of electrical equipment. Tests of hypotheses have a bearing on social questions and on government decisions concerning social programs. For example, one may wish to test whether wage levels between whites or nonwhites are equivalent in a certain area, or whether reading skills are a function of expenditures per pupil by a school system.

Formally, we define a *hypothesis* as an assumption about the value of a parameter or parameters of a probability distribution. In order to test a hypothesis, a random sample of values of the variable Y is used to compute an estimate of the value of the parameter about which the hypothesis has been made. A test is based on a comparison of the hypothesized value of the parameter and the value of the statistic.

Example 8.1. One might wish to establish the fact that the average income of New York City households is about $7,000. This is equivalent to the hypothesis that the mean value (μ) of the random variable Y, the income of a New York City household, is about 7,000. This hypothesis may be tested by choosing a number of households at random and determining the average income of these households for a particular year. The test could be based on the divergence of the average income of the sample households from the hypothesized income level of $7,000.

Example 8.2.   Suppose one wants to determine whether the average income of New York City households differs from that of Chicago households. Let the random variable Y_1 denote the income of a New York City household, and the random variable Y_2 denote the income of a Chicago household. We may construct another random variable, namely the difference between Y_2 and Y_1. This difference $(Y_2 - Y_1)$ also has a probability distribution, and the hypothesis that average Chicago and New York household incomes do not differ is equivalent to the hypothesis that the mean value μ of the random variable $(Y_2 - Y_1)$ is zero. A way to test this hypothesis is to draw a random sample of households in Chicago and one in New York, calculate the average for each sample, and determine the difference between these averages. The test of the hypothesis may then be based on the magnitude of this difference.

The Null Hypothesis

The hypothesis being tested is called the *null hypothesis*. In order to test a null hypothesis, we need an *alternative hypothesis* which is true if the null hypothesis is not true. The null hypothesis is denoted by $H0$. The alternative hypothesis is denoted by $H1$.

Example 8.3.   In Example 8.1 the null hypothesis would be that $\mu = 7,000$. If one suspects that income may be larger than 7,000, the alternative hypothesis would be that $\mu \geq 7,000$. This may be written:

$$H0: \mu = 7,000 \quad \text{and} \quad H1: \mu \geq 7,000.$$

Example 8.4.   In Example 8.2 the null hypothesis is that the theoretical mean of the random variable $Y_2 - Y_1$ is equal to zero. If we have no reason to believe that household incomes in New York are greater than those in Chicago or vice versa, then the alternative hypothesis would be that μ is any possible value, positive or negative or zero.

$$H0: \mu = 0 \quad \text{and} \quad H1: -\infty < \mu < +\infty.$$

If, on the other hand, we would expect household incomes to be higher in New York, the hypotheses could be stated as:

$$H0: \mu = 0 \quad \text{and} \quad H1: \mu \leq 0.$$

In terms of the theory of sets, the null hypothesis is equivalent to saying that the parameter θ lies in a set Ω_0. The alternative hypothesis is that θ lies in a set Ω_1. We assume that θ lies in either Ω_0 or Ω_1 or both. That is, either $H0$ is true or $H1$ is true. The set corresponding to the alternative hypothesis Ω_1 usually properly contains the set Ω_0 corresponding to the null hypothesis. That is, if the null hypothesis is true, then the alternative hypothesis is true, but not the other way around. The alternative hypothesis may be true without the null hypothesis being true. We may say that the null hypothesis is more *restrictive* than the alternative hypothesis. The alternative hypothesis is more *general* or *less restrictive* than the null hypothesis.

Example 8.5.   In Example 8.3 we had

$H0: \mu = 7,000$ and $H1: \mu \geq 7,000$.

If $H0$ is true ($\mu = 7,000$), then $H1$ is true ($\mu \geq 7,000$), but $H1$ may be true (for example, $\mu = 8,000$) and $H0$ not true ($\mu = 7,000$).

Hypothesis Tests and Errors in Testing

A *test* of a hypothesis is a method by which one either accepts or rejects the null hypothesis. Rejection of the null hypothesis is equivalent to acceptance of the alternative (less restrictive) hypothesis.

In performing a test, two types of errors can be made.

Type I *error:* One rejects the null hypothesis when it is true.
Type II *error:* One accepts the null hypothesis when it is false.

Suppose the null hypothesis is true and consists of hypothesizing that the parameter θ equals some specific value θ^*. Suppose also that a test consists of rejecting the null hypothesis whenever the statistic Z differs from θ^*, either above or below, by some specified amount X. The null hypothesis is accepted whenever Z falls within this range. That is, the null hypothesis is accepted if Z falls in the interval $\{\theta^* - X \leq Z \leq \theta^* + X\}$ and rejected if Z does not lie in this interval. If we know the sampling distribution of Z, it is possible to determine the probability α of a type I error by calculating the probability that Z falls outside the interval $\{\theta^* - X \leq Z \leq \theta^* + X\}$ whenever the hypothesis is true.

Example 8.6.   Suppose the statistic Z is an unbiased estimator of θ. Then if the null hypothesis is true, the mean value of Z is θ^*. If Z has a normal distribution, then the probability that Z lies outside the interval of acceptance $\{\theta^* - X \leq Z \leq \theta^* + X\}$ is the shaded area under the two tails of the normal distribution curve shown in Fig. 8.1. This area is the probability of a type I error.

It is often not possible to define a probability for a type II error. The alternative hypothesis involves an assumption that does not specify the exact value of the parameter θ. If the distribution of the statistic Z depends

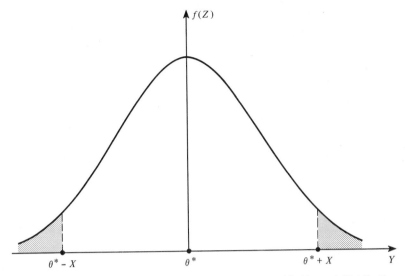

FIGURE 8.1   **Regions of Acceptance and Rejection with Normal Distribution**

on the parameter θ, then the alternative hypothesis is consistent with a family of distributions. In such a case one can, however, determine the probability of a type II error as a function of the true value θ. The type II error generally decreases as the true value of θ diverges more and more from the hypothesized value.

Example 8.7.   As in Example 8.6, suppose the statistic Z is an unbiased estimator of θ and Z has a normal distribution. The null hypothesis that $\theta = \theta^*$ is false. Suppose that the true value of θ is θ^0. Then Z will have a normal distribution with mean θ^0, as shown in Fig. 8.2(a). The interval of acceptance is $\{\theta^* - X \leq Z \leq \theta^* + X\}$. The probability of acceptance (type II error since $H0$ is false) is the shaded area in Fig. 8.2(a). It is obvious that the type II error can be quite large, especially if the true value is close to the hypothesized value. As the true value diverges more and more from the hypothesized value, the probability of a type II error declines, as shown in Figs. 8.2(b) and 8.2(c).

The Critical Region

Let us consider the following null and alternative hypotheses:

$$H0: \theta = \theta^* \quad \text{and} \quad H1: \ -\infty < \theta < +\infty, \tag{8.1}$$

where θ^* is some specified value of θ. In this case, one may use a test of the null hypothesis $H0$ called a two-tailed test. The test is constructed as follows:

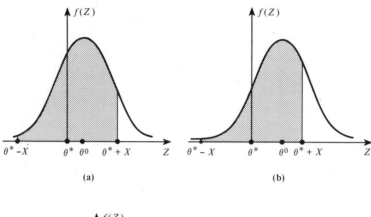

(a)　　　　　　　　　　　　　(b)

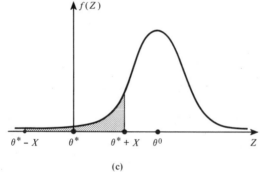

(c)

FIGURE 8.2   Probability of Type II Errors

1. Determine a probability α of a type I error that is acceptable. Usually one chooses α to be .05 or .01. The choice of a 5 percent ($\alpha = .05$) or a 1 percent ($\alpha = .01$) probability, however, is merely a convention.
2. Specify a number X and an interval $\{\theta^* - X \leq Z \leq \theta^* + X\}$ such that the probability of a type I error is in fact α. The interval $\{\theta^* - X \leq Z \leq \theta^* + X\}$ is called the region of acceptance. If Z lies outside this interval, it is in the region of rejection. The region of rejection is called the critical region.
3. Accept the null hypothesis if the calculated statistic Z lies in the region of acceptance.
4. Reject the null hypothesis if Z lies in the critical region.

Example 8.8.　Suppose the null hypothesis is true and the statistic Z is normally distributed with mean $\theta^* = 0$ and variance $\sigma^2 = 1$. The acceptable probability of a type I error is $\alpha = .01$. The critical region may be determined from Table A in the Appendix of Statistical Tables. Table A gives the area under a normal distribution curve (probability) for a random variable Z with a zero mean a unit variance.

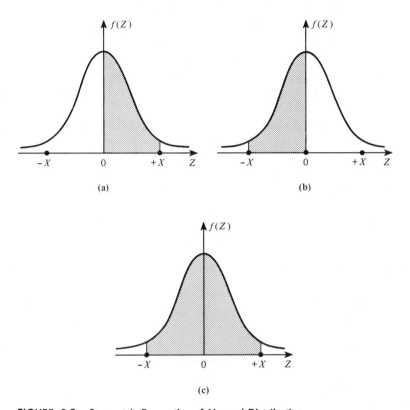

(a)

(b)

(c)

FIGURE 8.3   Symmetric Properties of Normal Distribution

Table A specifies the area between the mean 0 and X, where X takes on a number of different positive values. Such an area is shown in Fig. 8.3(a). From Table A we see that when $X = 2.57$, the area is approximately .495. Since the normal distribution is symmetric, the area between 0 and $-X = -2.57$ is also .495. Such an area is shown in Fig. 8.3(b). The total area between $-X$ and X is the sum of these two areas, or twice the area between 0 and X. This is $2 \cdot (.495) = .990$. Thus the probability that the random variable Z lies between $-X = -2.57$ and $X = 2.57$ is equal to .990.

The probability that Z lies outside the interval between $-X$ and X is equal to $1 - .990 = .01$. This is the probability that the variable Z lies in either the left or right "tail" of the normal distribution—that is, either of the unshaded areas in Fig. 8.3(c). Since this is the acceptable probability of a type I error, the region of acceptance is the interval

$$\{\theta^* - X \leq Z \leq \theta^* + X\} = \{-2.57 \leq Z \leq +2.57\}.$$

The probability that Z lies outside of this interval, if the null hypothesis $\theta = \theta^*$ is true, is .01—that is, .01 is the probability of a type I error. If the statistic Z lies in the critical region—that is, outside the interval $\{-2.57 \leq Z \leq +2.57\}$—the null hypothesis is rejected. Otherwise the null hypothesis is accepted.

The test we have just constructed for the hypothesis (8.1) is called a two-tailed test, since the critical region corresponds to the two "tails" of a probability distribution outside the interval $\{\theta^* - X \leq Z \leq \theta^* + X\}$. Such a test is appropriate whenever the alternative hypothesis $H1$ allows the parameter θ to take on any possible value—positive, negative, or zero. If we have some a priori reason to believe, however, that the parameter θ cannot be negative (or cannot be positive), then a *one-tailed test* may be constructed. Assume the following null and alternative hypotheses:

$$H0: \theta = \theta^* \quad \text{and} \quad H1: \theta \geqq \theta^*, \tag{8.2}$$

or

$$H0: \theta = \theta^* \quad \text{and} \quad H1: \theta \leqq \theta^*. \tag{8.3}$$

The set of hypotheses (8.2) is based on a priori knowledge that θ is not negative. We shall discuss a test of these hypotheses. The test of hypotheses (8.3) is analogous and follows in a straightforward manner.

The region of acceptance for a one-tailed test is an interval $\{-\infty < Z \leqq \theta^* + X\}$. Any point outside this region is in the critical region. If the statistic Z lies in the critical region, the null hypothesis is rejected. The probability that Z lies in the region of acceptance for a one-tailed test is the unshaded area in Fig. 8.4. The probability that Z lies in the critical region is the shaded area.

Example 8.9.   Suppose the null hypothesis in (8.2) is true and the statistic Z is distributed with mean $\mu = 0$ and variance $\sigma^2 = 1$. The acceptable type I error is $\alpha = .05$. From Table A in the Appendix of Statistical Tables we see that the probability that Z lies between 0 and $X = 1.64$ is .450. Since the normal distribution is symmetric, the probability that Z is negative is one-half (total probability is unity). Thus the probability that Z lies in the interval

$$\{-\infty < Z \leqq \theta^* + X\} = \{-\infty < Z \leqq 1.64\}$$

(that is, in the region of acceptance) is (i) the probability that Z is negative (lies in the interval $\{-\infty < Z \leqq 0\}$), (ii) *plus* the probability that Z lies between 0 and 1.64 (lies in the interval $\{0 \leq Z \leq 1.64\}$). The sum of these probabilities is $.500 + .450 = .950$. The probability that Z lies in the critical region is $1 - .950 = .05 = \alpha$, the probability of a type I error. Thus we reject the null hypothesis ($\theta = \theta^*$) if Z lies in the critical region (outside the interval $\{-\infty < Z \leqq 1.64\}$).

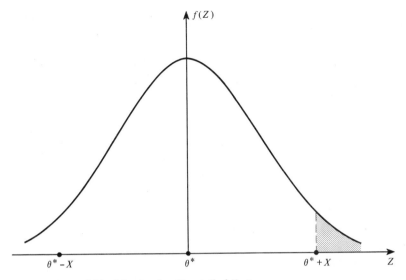

FIGURE 8.4   Critical Region for One-tailed Test

Tests of Means

Assume that the random variable Y is normally distributed with a theoretical mean μ and theoretical variance σ^2. From Theorem 7.1 of the last chapter we know that the sample mean $\bar{Y}$ is normally distributed with mean μ and variance σ^2/n. From Theorem 7.4 we know that ns^2/σ^2 has a chi-square distribution with $n - 1$ degrees of freedom. Finally, Theorem 7.6 implies that if we set $u = \bar{Y}$ and $v^2 = ns^2/\sigma^2$, the t ratio,

$$t = \frac{(\bar{Y} - \mu)\sqrt{n - 1}}{\dfrac{\sqrt{n}}{\sigma} \cdot s \cdot \dfrac{\sigma}{\sqrt{n}}} = \frac{(\bar{Y} - \mu)\sqrt{n - 1}}{s}, \tag{8.4}$$

has a Student's t distribution with $n - 1$ degrees of freedom. Thus the critical region α for a test of a hypothesis concerning the theoretical mean μ can be based on the Student's t distribution.

Example 8.10.   Suppose it is known that the average rate of return on capital invested in agriculture is 10 percent. We wish to determine whether investments in cattle ranches give a higher rate of return than the overall rate of return on capital. A sample of 10 cattle ranches is selected at random and the rate of return on capital in each ranch is determined. The data are given in the following table. The sample mean rate of return is $\bar{Y} = 8.38$

RANCH:	1	2	3	4	5	6	7	8	9	10
Rate of return:	9.9	15.3	12.6	9.4	11.7	1.8	0.9	9.6	6.6	6.0

and the sample variance is $s^2 = 18.96$. From (8.4), the t ratio is

$$t = \frac{(\bar{Y} - \mu) \sqrt{n-1}}{s} = \frac{(8.38 - \mu) \sqrt{9}}{4.35}. \tag{8.5}$$

The t ratio has a Student's t distribution with $n - 1 = 9$ degrees of freedom. The null hypothesis is

$$H0: \mu = 10.0. \tag{8.6}$$

If the critical region for a two-tailed test is $\alpha = .05$, then from Table C of the Appendix of Statistical Tables we see that the critical value of the t ratio for $\alpha = .05$ and 9 degrees of freedom is 2.262. If the null hypothesis is true ($\mu = 10.0$), the t ratio in (8.5) is only -1.117. This is smaller (in absolute value) than the critical value (2.262). Thus we must accept the null hypothesis that the rate of return on capital in cattle ranching equals the overall rate of return in agriculture.

Remember that if the degrees of freedom is large, the Student's t distribution is closely approximated by the normal distribution. Thus for large samples, tests of means may be made using the standard normal distribution tables. Furthermore, if the sample is large, we need not necessarily assume that the random variable Y is normally distributed. This derives from the central limit theorem (Theorem 7.2 of the last chapter), which asserts that the sample mean $\bar{Y}$ is approximately normally distributed regardless of the particular form of the distribution of Y.†

Thus for large samples, $\bar{Y}$ is approximately normally distributed with mean μ and variance σ^2/n. The standard normal variate (see Theorem 7.3) is

$$z = \frac{(\bar{Y} - \mu)}{\sigma/\sqrt{n}} = \frac{(\bar{Y} - \mu) \sqrt{n}}{\sigma}. \tag{8.7}$$

We do not in general know the *theoretical* standard deviation σ,.   It is appropriate to substitute the sample standard deviation s in (8.7). The result is

$$z = \frac{(\bar{Y} - \mu) \sqrt{n}}{s}. \tag{8.8}$$

Note then that the standard normal variate (8.8) is approximately equal to the t ratio in (8.4), the difference being that the numerator in (8.4) contains $\sqrt{n-1}$ while the numerator in (8.8) contains $\sqrt{n}$. Either formulation provides an approximation of the standard normal distribution and either may be used to perform tests of hypotheses.

Example 8.11.   Suppose a textile factory is thinking of installing a new type of loom but wishes to do so only if labor productivity can be assumed

---

† Provided that Y has a finite mean μ and a finite variance σ^2.

to increase. An experimental loom is installed for 200 working days. The mean daily gain in productivity of the sample loom is 2.3, while the sample standard deviation in daily productivity is 4.7. We wish to conduct a one-tailed test of the null hypothesis that the productivity gain is zero against the alternative hypothesis of a positive gain. The critical region is $\alpha = .01$. The critical value of the standard normal variate for a one-tailed test is 2.58 (see Table A in the Appendix of Statistical Tables). Using (8.8), we calculate the standard normal variate as

$$z = \frac{(\bar{Y} - \mu) \sqrt{n}}{s} = \frac{(2.3 - 0) \sqrt{200}}{4.7} = 6.925. \tag{8.9}$$

This easily exceeds the critical value, so that the null hypothesis is rejected in favor of the hypothesis that the new loom makes a significant difference in labor productivity.

Another kind of test of means arises when two samples are drawn from two different populations for which the means may be different but the theoretical variance is assumed to be the same. Let Y_1 and Y_2 represent two different random variables, each normally distributed with theoretical variance σ^2. The theoretical mean of Y_1, however, is μ_1 and the theoretical mean of Y_2 is μ_2. We wish to test the null hypothesis that the means of the two random variables are the same against the hypothesis that they are different.

$$H0: \mu_1 = \mu_2 \quad \text{and} \quad H1: \mu_1 \neq \mu_2. \tag{8.10}$$

It is possible to show then that the difference in sample means $u = \bar{Y}_1 - \bar{Y}_2$ is normally distributed with mean $\mu_1 - \mu_2$ and theoretical variance

$$\sigma_u^2 = \sigma_{\bar{Y}_1 - \bar{Y}_2}^2 = \sigma^2 \left(\frac{1}{n_1} + \frac{1}{n_2} \right), \tag{8.11}$$

where $\sigma_{\bar{Y}_1 - \bar{Y}_2}^2$ is the theoretical variance of the difference in means, σ^2 is the common theoretical variance of the random variables Y_1 and Y_2, and n_1 and n_2 are the respective sample sizes.

The sample variances of the two populations are

$$s_1^2 = \frac{\sum\limits_{i=1}^{n_1} (Y_{1i} - \bar{Y}_1)^2}{n_1}$$

and

$$s_2^2 = \frac{\sum\limits_{i=1}^{n_2} (Y_{2i} - \bar{Y}_2)^2}{n_2}. \tag{8.12}$$

It is possible to show that the variate $v^2 = (n_1 s_1 + n_2 s_2^2)/\sigma^2$ has a chi-square distribution with $n_1 + n_2 - 2$ degrees of freedom. Using Theorem 7.6 of the previous chapter, we see that the t ratio,

$$t = \frac{(u - \mu)\sqrt{f}}{v \cdot \sigma_u} = \frac{[(\bar{Y}_1 - \bar{Y}_2) - (\mu_1 - \mu_2)]\sqrt{n_1 + n_2 - 2}}{\frac{\sqrt{n_1 s_1^2 + n_2 s_2^2}}{\sigma} \cdot \sigma \sqrt{\frac{1}{n_1} + \frac{1}{n_2}}}$$

$$= \frac{[(\bar{Y}_1 - \bar{Y}_2) - (\mu_1 - \mu_2)]\sqrt{(n_1 + n_2 - 2) \cdot n_1 n_2}}{\sqrt{(n_1 s_1^2 + n_2 s_2^2)(n_1 + n_2)}}, \quad (8.13)$$

has a Student's t distribution with $n_1 + n_2 - 2$ degrees of freedom. The t ratio in (8.13), then, is a useful test statistic for hypotheses concerning differences in means.

Example 8.12.   Suppose a soap company is trying to decide whether to switch emphasis from television commercials to a major campaign of magazine advertising. First it wishes to test the magazine approach on a regional basis. If the increase in sales in the test region is significantly better than in all other regions, the decision will be made to go ahead with the magazine campaign. The critical region is $\alpha = .01$. Suppose the test results in a percentage increase in sales in the test region of 7.3 percent. The sample mean in the test region is $\bar{Y}_1 = 7.3$ (there is only one region in the sample). The sample variance $s_1^2 = 0$ in this case. In the twelve other regions of the country, the sales increases are

REGION:	1	2	3	4	5	6	7	8	9	10	11	12
Percent sales increase:	4.3	2.7	8.4	9.0	3.6	4.7	5.8	8.2	6.4	7.1	8.3	6.5

The sample mean increase in sales in all of these 12 regions is $\bar{Y}_2 = 6.25$. The sample variance $s_2^2 = 3.919$. The test statistic from (8.13) is

$$t = \frac{[(7.30 - 6.25) - 0]\sqrt{(1 + 12 - 2) \cdot 12}}{\sqrt{(1 \cdot 0 + 12 \cdot 3.919)(1 + 12)}} = .488, \quad (8.14)$$

where the hypothesis $\mu_1 = \mu_2$ is assumed to be true. For a two-tailed test with $n_1 + n_2 - 2 = 1 + 12 - 2 = 11$ degrees of freedom and $\alpha = .01$, the critical value using the Student's t distribution (see Table C in the Appendix of Tables) is 3.106. This is considerably greater than the t value of .488, and the hypothesis that magazine advertising is no more effective than television advertising is accepted. There is no significant difference between these two modes of advertising.

Tests of Variances

From Theorem 7.4 we know that $n \cdot s^2/\sigma^2$ is distributed as chi-square with $n - 1$ degrees of freedom, where n is the sample size, s^2 is the sample variance, and σ^2 is the theoretical variance. Hypotheses about the theoretical variance may be tested by using this result.

Example 8.13.   Suppose a manufacturer of steel rods wishes to achieve a product of fairly standard quality in terms of the tensile strength of the rods. In order to ensure that quality standards are being met, it conducts a test in which a sample of five steel rods is drawn from a day's output and the tensile strength is measured. The following table gives the tensile strength of the five rods from a day's sample:

ROD	1	2	3	4	5
Tensile strength:	1,045	985	1,067	975	928

A variance in strength less than $\sigma^2 = 1,000$ is tolerable. Any variance above 1,000 calls for the plant to be shut down temporarily and the machinery recalibrated and cleaned. The sample of five rods then is used to make a one-tailed test of the hypothesis,

$$H0: \sigma^2 = 1,000, \tag{8.15}$$

as opposed to the alternative hypothesis

$$H1: \sigma^2 \geqq 1,000. \tag{8.16}$$

The critical region adopted is $\alpha = .01$. The sample variance from the data above is $s^2 = 2509.6$. Assuming that the null hypothesis is true, the chi-square statistic,

$$\frac{n \cdot s^2}{\sigma^2} = \frac{5 \cdot 2509.6}{1,000} = 12.548, \tag{8.17}$$

has 4 degrees of freedom. From Table B of the Appendix of Statistical Tables, the critical value of the chi-square variate with $\alpha = .01$ and 4 degrees of freedom is 13.277. Thus the hypothesis that $\sigma^2 = 1,000$ is accepted, and no recalibration and cleaning of the machinery is considered necessary.

In some instances we may wish to test whether the variabilities of samples from two different populations are equal. In cases of this sort Theorem 7.7 of the last chapter is appropriate.

Denote the sample variance of the random variable Y_1 by s_1^2 and the sample variance of the random variable Y_2 by s_2^2. The theoretical variances are denoted by σ_1^2 and σ_2^2, respectively. The null and alternative hypotheses are

$$H0: \sigma_1^2 = \sigma_2^2 = \sigma^2 \quad \text{and} \quad H1: \sigma_1^2 \neq \sigma_2^2. \tag{8.18}$$

If the null hypothesis is true, the statistic $u = n_1 \cdot s_1^2 / \sigma^2$ has a chi-square distribution with $k_1 = n_1 - 1$ degrees of freedom where n_1 is the sample size for the random variable Y_1. Similarly, $v = n_2 \cdot s_2^2 / \sigma^2$ has a chi-square distribution with $k_2 = n_2 - 1$ degrees of freedom. From Theorem 7.7, we know that the F ratio,

$$F = \frac{u/k_1}{v/k_2} = \frac{n_1 \cdot s_1^2 / (n_1 - 1)}{n_2 \cdot s_2^2 / (n_2 - 1)}, \tag{8.19}$$

has an F distribution with $n_1 - 1$ degrees of freedom in the numerator and $n_2 - 1$ degrees of freedom in the denominator.

The critical region is shown in Fig. 8.5. Note, however, that the critical value F^* for the right-hand tail of the critical region is the reciprocal of the critical value $1/F^*$ for the left-hand tail of the critical region. Note that Table D in the Appendix of Statistical Tables gives only the right-hand critical value. The left-hand critical value is obtained by dividing the right-hand critical value into unity.

Example 8.14. Suppose that two different varieties of cotton are grown on adjacent plots under similar conditions. The two different varieties are grown over a ten-year period. Variety A has the higher average yield over the ten years, 750 pounds of cotton per acre compared to 680 pounds per acre for variety B. The sample variance of the yield of variety A, however, is $s_1^2 = 20,000$, while the sample variance of yield of variety B is $s_2^2 = 14,000$. Are these variances significantly different at the 5 percent level? That is, test the two hypotheses (8.18). From (8.19), the F ratio is

$$F = \frac{n_1 \cdot s_1^2 / (n_1 - 1)}{n_2 \cdot s_2^2 / (n_2 - 1)} = \frac{10 \cdot 20,000 / (10 - 1)}{10 \cdot 14,000 / (10 - 1)} = 1.4286 \tag{8.20}$$

with 10 degrees of freedom in both the numerator and denominator. With a 5 percent critical region using a two-tailed test, the right-hand critical value is $F^* = 3.7168$ while the left-hand critical value is $1/F^* = .2688$. Since 1.4286 lies between these critical values, we accept the null hypothesis that there is no significant difference in the variance of the yields of the two varieties of cotton.

Interval Estimation

Chapter 7 contained a discussion of estimates of parameters and the properties of various estimators. The estimators in Chapter 7 were all *point* estimates of population parameters. In this section we consider *interval* estimates of parameters. For example, if one says that the mean

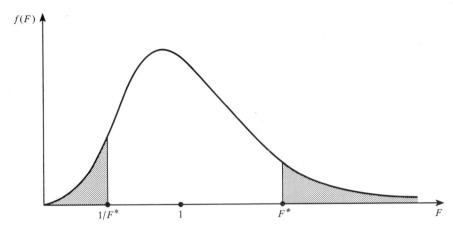

FIGURE 8.5   Critical Region—*F* Distribution

income of plumbers in Oregon is about $8,000, this is a point estimate of mean income. A statement in terms of an interval, such as "Mean income of plumbers is $8,000, plus or minus ($\pm$) $1,000," conveys in some sense the approximate nature of the estimation procedure.

One way to make interval estimates is in terms of *confidence intervals*. A β confidence interval is an interval about an estimator Z for which the probability is β that the interval contains the true population parameter β. For example, one might estimate that $8,000 \pm $1,000$ is a 95 percent confidence interval for the mean income of plumbers in Oregon. Thus the probability that the "true" mean income of plumbers lies in such a confidence interval is .95.

Confidence intervals and tests of hypotheses are related in the following way. If Z is an estimator of the population parameter and $\{\theta^* - X \leq Z \leq \theta^* + X\}$ is the region of acceptance for the null hypothesis that $\theta = \theta^*$, with a critical region α, then $Z \pm X$ is a $(1 - \alpha)$ confidence interval about the estimate Z.

For small samples, tests of hypotheses about theoretical means are based on the Student's t distribution. The same distribution may be used to construct a confidence interval about a sample mean. Suppose the critical values of the t ratio for a critical region α, using a two-tailed test of the mean μ, are $-X$ and $+X$. Then the region of acceptance is defined in terms of the t ratio (8.4) as

$$\left\{ -X \leqq \frac{(\bar{Y} - \mu) \sqrt{n - 1}}{s} \leqq X \right\}, \tag{8.21}$$

where n is the sample size. This interval may be analyzed in terms of two inequalities. The first is

$$-X \leqq \frac{(\bar{Y} - \mu) \sqrt{n - 1}}{s}. \tag{8.22}$$

Multiply both sides of the inequality by s and divide both sides by $\sqrt{n - 1}$ to obtain

$$\frac{-s \cdot X}{\sqrt{n - 1}} \leqq \bar{Y} - \mu. \tag{8.23}$$

Add $\mu + s \cdot X/\sqrt{n - 1}$ to both sides.

$$\mu \leqq \bar{Y} + \frac{s \cdot X}{\sqrt{n - 1}}. \tag{8.24}$$

The second inequality implied by the interval (8.21) is

$$\frac{(\bar{Y} - \mu) \sqrt{n - 1}}{s} \leqq X. \tag{8.25}$$

Multiply both sides by s, divide both sides by $\sqrt{n - 1}$, and add $\mu - s \cdot X/\sqrt{n - 1}$ to both sides to obtain

$$\bar{Y} - \frac{s \cdot X}{\sqrt{n - 1}} \leqq \mu. \tag{8.26}$$

Combining the inequalities (8.24) and (8.26), we obtain

$$\bar{Y} - \frac{s \cdot X}{\sqrt{n - 1}} \leqq \mu \leqq \bar{Y} + \frac{s \cdot X}{\sqrt{n - 1}}. \tag{8.27}$$

That is, the probability that the true mean μ lies in the interval $\bar{Y} \pm s \cdot X/\sqrt{n - 1}$ is $(1 - \alpha)$.

Example 8.15.   In Example 8.10 the sample mean was 8.38 and the sample standard deviation was 4.35, based on a sample of rates of return from 10 cattle ranches. Using a critical region of $\alpha = .05$, the null hypothesis was accepted if the t ratio in (8.5) was found to be between the critical values $+X = 2.262$ and $-X = -2.262$ of the t ratio. That is, the region of acceptance was

$$\left\{-2.262 \leqq \frac{(8.38 - \mu) \sqrt{9}}{4.35} \leqq +2.262\right\}. \tag{8.28}$$

The 95 percent confidence limits about the sample mean are

$$\bar{Y} \pm s \cdot X/\sqrt{n - 1} = 8.38 \pm (4.35)(2.262)/\sqrt{9}$$
$$= 8.38 \pm 3.2799. \tag{8.29}$$

The probability that the true mean lies between $8.38 + 3.2799 = 11.3337$ and $8.38 - 3.2799 = 5.1001$ is .95.

For large samples, one may use the standard normal variate of (8.8) in testing of hypotheses concerning the theoretical mean. If $+X$ and $-X$ are the critical values of the standard normal variate for a critical region α, the region of acceptance is

$$\left\{-X \leqq \frac{(\bar{Y} - \mu)\sqrt{n}}{s} \leqq +X\right\}.$$ (8.30)

Working with the two inequalities implied by (8.30) in the same manner as above, we obtain the $(1 - \alpha)$ confidence interval about the sample mean.

$$\bar{Y} \pm \frac{s \cdot X}{\sqrt{n}}.$$ (8.31)

Of particular interest are the 95 percent and 99 percent confidence intervals. Since we know that for a normal distribution, about 95 percent of the observations lie plus or minus two standard deviations and about 99 percent lie plus or minus three standard deviations, the 95 and 99 percent confidence intervals are

$$\bar{Y} \pm \frac{2 \cdot s}{\sqrt{n}},$$ (8.32)

and

$$\bar{Y} \pm \frac{3 \cdot s}{\sqrt{n}},$$ (8.33)

respectively.

Example 8.16.   In Example 8.11 the sample mean daily gain in productivity was 2.3 while the sample standard deviation for 200 daily trials was 4.7. The 95 percent confidence interval about the sample mean is

$$2.3 \pm \frac{(2)(4.7)}{\sqrt{200}} = 2.3 \pm .665.$$ (8.34)

The 99 percent confidence interval is

$$2.3 \pm \frac{(3)(4.7)}{\sqrt{200}} = 2.3 \pm .997.$$ (8.35)

Finally, let us consider the problem of constructing a confidence interval about a sample variance. The statistic ns^2/σ^2 has a chi-square distribution with $n - 1$ degrees of freedom. If χ_U^2 is the right hand critical value and χ_L^2 is the left hand critical value for a two-tailed critical region α, the region of acceptance is

$$\chi_L^2 \leqq ns^2/\sigma^2 \leqq \chi_U^2.$$ (8.36)

The first inequality implied by the region of acceptance is

$$\chi_L^2 \leqq ns^2/\sigma^2 \tag{8.37}$$

or

$$\sigma^2 \leqq ns^2/\chi_L^2. \tag{8.38}$$

The second inequality of (8.36) is

$$ns^2/\sigma^2 \leqq \chi_U^2 \tag{8.39}$$

or

$$ns^2/\chi_U^2 \leqq \sigma^2, \tag{8.40}$$

combining (8.38) and (8.40), we obtain

$$ns^2/\chi_U^2 \leqq \sigma^2 \leqq ns^2/\chi_L^2. \tag{8.41}$$

Thus the theoretical variance lies between ns^2/χ_U^2 and ns^2/χ_L^2 with a probability of $1 - \alpha$ where α is the critical region.

Example 8.17.   In Example 8.13 the sample variance in tensile strength of 5 rods was found to be $s^2 = 2509.6$. The right-hand critical value for the chi-square distribution (4 degrees of freedom) is 11.143 and the left-hand critical value is .4844 for a 5 percent, two-tailed critical region. (See Table B in the Appendix of Statistical Tables for $A = .975$ and $A = .025$, respectively.) Thus according to (8.41), the 95 percent confidence interval is

$$(5 \cdot 2509.6)/(11.143) \leqq \sigma^2 \leqq (5 \cdot 2509.6)/(.4844)$$

or

$$1126.08 \leqq \sigma^2 \leqq 2509.4.$$

Other Decision Rules

As we mentioned above, the choice of a 5 percent or 1 percent critical region is merely a convention. Similarly, 95 percent or 99 percent confidence intervals are used most often by convention. The choice of the critical region or a confidence interval need not be arbitrary, however. Modern statistical decision theory, based on two major decision rules for choosing a critical region, (1) the Bayesian decision rule, and (2) the minimax decision rule, theoretically provides a much more satisfactory approach. Both these decision rules assume that one may assign a numerical cost to type I and type II errors. Let C_1 be the cost of making a type I error and C_2 be the cost of a type II error.

The Bayesian approach to statistical decision theory is best illustrated if we assume that a theoretical parameter θ can assume one of only two different values, say θ_0 or θ_1. Let the null hypothesis be:

$H0: \theta = \theta_0.$ (8.42)

The alternative hypothesis is

$H1: \theta = \theta_1.$ (8.43)

In Bayesian theory one must assume the existence of a priori probabilities that one can assign to the two different hypotheses. Let q_0 stand for the a priori probability that the null hypothesis $H0$ is true and q_1 for the a priori probability that $H1$ is true. Then the *expected* cost C of making either a type I or type II error is

$C = q_0 \cdot C_1 \cdot p(\text{I}) + q_1 \cdot C_2 \cdot p(\text{II}),$ (8.44)

where $p(\text{I})$ is the probability of a type I error and $p(\text{II})$ the probability of a type II error. The probabilities $p(\text{I})$ and $p(\text{II})$ depend on the particular critical value of θ that is chosen. The Bayesian decision rule is to choose a critical value of θ so that the expected cost C is minimized.

The minimax decision rule does not require us to know the a priori probabilities q_0 and q_1 of the null and alternative hypotheses. Rather we choose a critical value of θ to minimize the maximum expected cost of error —that is, choose a critical value of θ so that the largest of the two expected costs

(1) $C_1 \cdot p(\text{I})$ and   (2) $C_2 \cdot p(\text{II})$

is minimized. For example, if one critical value θ^* produces expected costs of

(1) $C_1 \cdot p(\text{I}) = 5,$ (2) $C_2 \cdot p(\text{II}) = 10,$

and another critical value θ^{**} gives

(1) $C_1 \cdot p(\text{I}) = 8,$ (2) $C_2 \cdot p(\text{II}) = 4,$

the maximum expected cost with θ^* is 10 and with θ^{**} is 8. The minimum of these maxima is 8. Thus the minimax rule tells us to choose θ^{**} over θ^* as a critical value.

Minimax and Bayesian decision rules are rather sophisticated methods of choosing critical values. They are only rarely used in practice. Often it is not easy to determine the costs of either type I or type II errors or a priori probabilities, so that statisticians simply fall back on the usual 5 percent or 1 percent critical regions. One must realize, however, that these critical values may not be justifiable or reasonable in terms of decision-theory models.

PROBLEMS

1. A sample of 49 households in a rural county in Kentucky reveals a mean income of $3,630 and a sample standard deviation of $700. A family income of $3,500 is regarded as at the poverty level. (a) Is it reasonable to conclude that the

average income in the county is above the poverty level? (Use a 5 percent critical region.) (b) What is the 95 percent confidence interval about the mean income? (c) If the true average income in the county is $3,350, what is the probability of a type II error?

2. With five students from a small midwest college taking the Graduate Record Exam in economics, the following results are obtained: 540, 360, 720, 253, 437. The average score on a national basis is 500. (a) Do the five students on the average differ significantly from the national average? (Use a 5 percent critical region.) (b) Compute a 95 percent confidence interval about the mean score of the five students.

3. A school board wishes to test the effects of a new method of teaching reading to first grade students. For this purpose the board compares two classes, each containing 15 students matched for ability in terms of IQ and taught by the same teacher using the old method and the new method. The mean score on the reading test is 77 with a sample standard deviation of 6 for the class taught by the old technique. The mean score for students with the new method is 79 with a sample standard deviation of 8. (a) Can we conclude that the new method is significantly better than the old? (Use a 1 percent critical region.) (b) Does the new method produce a significantly greater variance in performance? (Use a 5 percent critical region.)

4. Suppose a sample of 400 households in New York reveals an average income of $7,562 with a sample standard deviation of $2,820. In Chicago, a sample of 225 households gives an average income of $7,150 with a sample standard deviation of $2,650. Is the average income in New York significantly greater than in Chicago? (Use a 5 percent critical region.)

5. A new automatic lathe is introduced into a factory for making machine parts. The tolerable standard deviation in circumference of the parts is 1/1000 of an inch. A sample of 10 parts is measured, giving the following results in terms of inches:

3.50031, 3.49979, 3.49996, 3.50001, 3.49989
3.49983, 3.50034, 3.50003, 3.49986, 3.50008

(a) Is the variance in these measurements significantly greater (using a 1 percent critical region) than the tolerable level? (b) Compute a 99 percent confidence interval about the estimated variance of the sample.

Outline of Further Reading for Part III

A. Elementary Introductions to Statistical Inference
   1. Bryant [1966], pp. 33–42 and 71–108
   2. Freund [1962], pp. 164–294
   3. Freund [1967], pp. 189–261
   4. Freund [1970], pp. 194–260
   5. Hoel [1962], pp. 131–159, 212–296
   6. Suits [1963], pp. 124–154
   7. Walker and Lev [1953], pp. 43–195
   8. Wallis and Roberts [1956], pp. 345–492

B. Intermediate Treatment of Statistical Inference
   1. Anderson and Bancroft, pp. 58–130
   2. Chou [1969], pp. 237–349, 372–398
   3. Christ [1966], pp. 243–297
   4. Cramér [1955], pp. 181–227
   5. Goldberger [1964], pp. 49–155
   6. Hoel [1966], pp. 117–191
   7. Kane [1968], pp. 172–216
   8. Mood and Graybill [1963], pp. 139–327
   9. Neter and Wasserman [1966], pp. 244–338, 361–471
   10. Neyman [1950], pp. 250–344
   11. Richmond [1964], pp. 158–304
   12. Wonnacott and Wonnacott [1969], pp. 128–194
   13. Yamane [1967], pp. 129–264, 502–641

C. Advanced Treatment of Statistical Inference
   1. Anderson [1958], pp. 44–59, 101–125, 154–177, 247–271
   2. Cramér [1946], pp. 323–535
   3. Hogg and Craig [1970], pp. 116–347
   4. Kendall and Stuart [1967], pp. 1–277
   5. Kendall and Stuart [1969], pp. 198–346, 369–378
   6. Wilks [1962], pp. 195–276

All of these references require a good deal of mathematical sophistication. Anderson concentrates on the estimation of parameters of the multivariate normal distribution. Hogg and Craig and Wilks contain full proofs of the central limit theorem and develop other aspects of asymptotic theory, that is, theory relating to the form which sampling distributions take as the number of observations increases without limit.

D. The Theory of Sampling
   1. Bryant [1966], pp. 244–262
   2. Chou [1969], pp. 350–371

3. Cochran [1963], *passim*
4. Neter and Wasserman [1966], pp. 339–361
5. Raj [1968], *passim*
6. Richmond [1964], pp. 324–339
7. Wallis and Roberts [1956], pp. 482–493

Cochran and Raj are books on sampling theory which discuss various nonrandom sampling methods, sampling from small finite populations, and the effects of nonrandomness and finite populations on estimation procedures and tests of hypotheses. The other references provide an elementary introduction to the type of material covered more extensively in Cochran and Raj.

E. Statistical Decision Theory
  1. Blackwell and Girschick [1954], *passim*
  2. Chernoff and Moses [1959], *passim*
  3. Hadley [1967], *passim*
  4. Neter and Wasserman [1966], pp. 472–511
  5. Raifa and Schlaifer [1961], *passim*
  6. Richmond [1964], pp. 227–289
  7. Savage [1962], *passim*
  8. Wald [1950], *passim*
  9. Weiss [1961], *passim*
  10. Wilks [1962], pp. 502–513
  11. Wonnacott and Wonnacott [1969], pp. 312–349

The classical approach to hypothesis testing involves an arbitrary limit on the probability of Type I errors and an arbitrarily specified critical region. There are two other approaches to hypothesis testing which take into account explicitly the costs of both Type I and Type II errors: the minimax approach, in which the researcher tries to minimize the maximum cost of either Type I or Type II errors, and the Bayesian approach, in which a priori probabilities are assigned to the two types of error. The development of decision rules for testing hypotheses is called statistical decision theory. The first important book in this area is by Wald. Blackwell and Girschick, Chernoff and Moses, Hadley, Neter and Wasserman, and Raifa and Schlaifer, and Wonnacott and Wonnacott provide elementary introductions to statistical decision theory. Savage is an excellent short monograph containing informal discussions by Savage and several colleagues about the Bayesian approach.

Part IV

Linear Models

Chapter **9**

Estimation and Hypothesis Testing with Linear Models

In Chapter 3 we indicated that two variables may be associated by way of a linear relationship,

$$Y = a + b \cdot X. \tag{9.1}$$

For example, the dependent variable Y might be consumption and the independent variable X income. If we have data on consumption and income for N different years, we may denote these observations by Y_1, $\ldots$, Y_N and X_1, $\ldots$, X_N. One way of estimating the linear relationship (9.1) is to find the values for a and b that minimize the sum of squares,

$$S = \sum_{i=1}^{N} (Y_i - a - b \cdot X_i)^2. \tag{9.2}$$

No rationale was given in Chapter 3 for this method of estimating a and b. If Y is a random variable, however, we can show that, given certain assumptions, the least-squares estimates have a number of desirable properties.

In particular, let us think of a particular observed value of the variable Y as a random variable composed of two parts:

$$Y_i = (a + b \cdot X_i) + e_i. \tag{9.3}$$

The first part $(a + b \cdot X_i)$ is a systematic linear component, and the second part e_i is a random component. The systematic part is determined by a linear relationship. The second or random component is often called the *error term*. The error term arises because we expect that the value of Y is

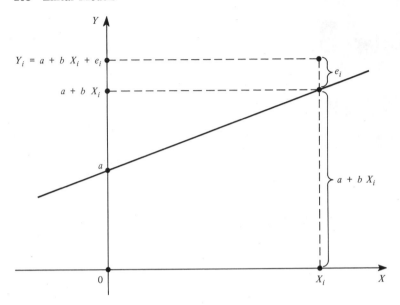

FIGURE 9.1   Systematic and Random Components of Y_i

not determined precisely by a linear relationship but that there is usually some error or deviation from the "true" exact linear relationship. The two components are diagramed in Fig. 9.1.

The probability distribution of the random variable Y_i depends on the parameters a, b, X_i and the error term e_i. In practice, we observe only the values of Y_i and X_i. The third of the sections that follow discusses ways in which the values of a, b, and e_i may be estimated given N sample values or observations of the Y and X variables. First, however, let us discuss some of the properties of the systematic and random components of the linear model.

The Systematic Component

We assume that a and b are fixed numbers that do not vary with the sample values of the X_i. One assumption often made is that sample values X_1, X_2, ... , X_N are fixed. That is, X_i is measured without error; it is nonrandom.

Perhaps the best way to illustrate the concept of the nonrandomness of the X_i values is to think of a physical experiment in which the N sample values of the X variable are determined beforehand by the experimenter. The experiment is performed N different times using each of the N values of X. The value of the Y variable is measured in each experiment. The measurement of the Y variable, however, is subject to errors of several sorts. There may be errors of measurement. There may be errors due to changes in experimental conditions that render the relationship between Y and X inexact. For example, temperature and humidity may vary from experiment to experiment, causing variations in the measured value of Y that are not accounted for by the variations in the experimental value of X.

Example 9.1. From classical physics we assume that the time t required for an object to fall a distance s in a complete vacuum is given by the formula:

$$t = t_0 + \sqrt{\frac{2 \cdot s}{g}}, \tag{9.4}$$

where t_0 is the initial time of the experiment and g is the so-called gravity constant. Suppose we are students of Newton and wish to estimate the value of g from a series of experiments.

Since we cannot conduct the experiment in a vacuum, and since we cannot be sure that the stopwatch used is completely accurate, we assume that under experimental conditions, the relationship will not be exact and that certain errors will give variable results for each experiment. Thus

$$t_i = t_0 + \sqrt{\frac{2}{g}} \cdot \sqrt{s_i} + e_i \tag{9.5}$$

or

$$Y_i = a + b \cdot X_i + e_i, \tag{9.6}$$

where $a = t_0$, $b = \sqrt{2/g}$, and $X_i = \sqrt{s_i}$. We then perform N experiments, measuring in each the time t (or Y) for an object to fall a different distance s (or $X = \sqrt{s}$). We then estimate the coefficient b (or $\sqrt{2/g}$) from the sample observations $Y_1, Y_2, \ldots, Y_N$ and the different values of X.

In the physical sciences, controlled experiments in which the values of the X variables can be specified beforehand are often easy to conduct. When dealing with problems involving economic and business statistics, it is often not possible to conduct controlled experiments. Thus it is sometimes assumed that the X variable is random but that the distribution of X has one very important property. This property is that the variable X_i and the error term e_i are independently distributed. They are uncorrelated. This assumption changes some of the properties of the estimated values of a and b. In particular, properties such as unbiasedness and consistency must be interpreted as being conditional upon the particular X values that occur in the sample. These distinctions between nonrandom X variables and random X variables are important in the theory of regression analysis, but for purposes of simplicity we can proceed "as if" the X variables were nonrandom.

Example 9.2.   Suppose an economist wishes to estimate the marginal propensity to consume. He may then specify a consumption function of the following form:

$$C = a + b \cdot Y, \tag{9.7}$$

where C is consumption, Y is income, and b the marginal propensity to consume. He assumes that consumption C is random. The relationship between consumption and income is not exact because of other factors that affect consumption expenditure (such as the interest rate, price inflation, changes in tastes, and so on) and because of errors in the measurement of consumption expenditure. Thus he assumes that

$$C_i = a + b \cdot Y_i + e_i \tag{9.8}$$

or

$$Y_i = a + b \cdot X_i + e_i, \tag{9.9}$$

where $Y_i = C_i$ and $X_i = Y_i$. Since controlled experiments are impossible, the economist uses instead a time series on personal income Y and consumption expenditure C for a certain period of time, say the last 20 years, as compiled by government statisticians. This gives him a set of values for income for 20 years which he assumes is uncorrelated with the error terms e_i.

In most applications of linear statistical techniques the dependent variable Y is related to a number of independent variables. As mentioned above, consumption may depend on income, the interest rate, and average family size among other possible independent variables. Imports may depend on consumption expenditure and investment expenditure. In such cases the linear model (9.3) may be generalized to the case of several independent variables.

$$Y_i = (a_0 + a_1 \cdot X_{1i} + a_2 \cdot X_{2i} + \cdots + a_K \cdot X_{Ki}) + e_i, \tag{9.10}$$

where there are K independent variables. X_{1i} is the ith observation or sample value of the first independent variable, X_{2i} the ith observation on the second variable, and so on. The systematic component of Y_i is now composed of $K + 1$ terms (including the constant a_0), and the error term e_i has the same interpretation as before.

The Error Term

Errors arise in experimental (or hypothetically experimental) situations for two reasons: (a) there may be errors of measurement in the dependent variable Y, or (b) the model may be misspecified. The first reason is self-explanatory. The second requires some elaboration.

Suppose consumption depends not only on income but on a host of other factors such as interest rates, the age structure of the population, the asset holdings of the population, past incomes, expected future incomes, the level of advertising expenditures in the economy, and so on. Furthermore, suppose the marginal propensity to consume is not constant but changes with the level of income. Then the specification (9.8), which implies that consumption is a linear function of income alone, is not a correct specification of the factors that affect consumption. These other influences on consumption may be regarded as being subsumed in the error term, e_i.

In order for the theory discussed in this chapter to be valid, the error term e_i must satisfy certain assumptions.

Assumption 1.   The error term e_i is a random variable distributed with a theoretical mean $\mu = 0$ and a finite variance $\sigma_{e_i}^2$.

The assumption of a mean $\mu = 0$ is usually not very stringent, since the constant term a_0 in (9.10) can always be assumed to take a value that ensures that the mean of the error term e_i is zero.

Assumption 2.   The sample values of the error terms e_i for $i = 1, \ldots, n$ are independently distributed.

Assumption 2 may be stated in two other equivalent ways. First, the expected value $E(e_i \cdot e_j) = 0$ for i not equal to j. $E(e_i \cdot e_j)$ is also called the covariance of e_i and e_j. Alternatively, we may say that the density function for the array of sample values, $(e_1, e_2, \ldots, e_N)$ is given by

$$f(e_1, e_2, \ldots, e_N) = f_1(e_1) \cdot f_2(e_2) \cdot \ldots \cdot f_N(e_N),$$

where $f_i(e_i)$ is the density function for the ith observed error term.

If Assumption 2 is violated, one says that the errors are autocorrelated or sometimes serially correlated. The error term in one period affects the probability distribution of the error term in other periods. Problems introduced by autocorrelation are discussed in a later chapter.

Example 9.3.   Autocorrelation may arise if the statistical model is misspecified. For example, a researcher assumes that investment in a particular industry is a function of the growth in sales.

$$I_t = a + b \cdot G_t + e_t, \tag{9.11}$$

where I_t is investment in period t, G_t is the growth in sales between period $t - 1$ and period t, and e_t is the error term in period t. Suppose, however, that interest rates also affect the rate of investment and that interest rates have been rising steadily through time. If higher interest rates depress investment, investment will be lower for a given growth in retail sales. That is, the error term in (9.11) will tend to be positive in earlier periods (investment at a higher level than that predicted by growth in sales) and decline to negative values in later periods. The error terms are not independently distributed; they are autocorrelated. A positive error term in an early period t will tend to be associated with a positive error term in period $t + 1$. Similarly, in later periods negative error terms tend to follow one another in succession.

Example 9.4.   Suppose that advertising expenditures E_t in period t are assumed to be a function of sales S_{t-1} in period $t - 1$ for a particular industry.

$$E_t = a + b \cdot S_{t-1} + e_t. \tag{9.12}$$

The "true" relationship between advertising expenditures and sales may not be linear as postulated in (9.12) but rather a quadratic relationship as in (9.13),

$$E_t = a + b \cdot S_{t-1} + c \cdot S_{t-1}^2 + e_t. \tag{9.13}$$

The linear relationship and the "true" quadratic relationship are shown in Fig. 9.2. The linear relationship is shown as a solid line and the quadratic relationship as a dotted line. The observed points tend to cluster about the quadratic relationship rather than the linear relationship. The observations in the early 1950s tend to be below the linear relationship [negative error terms in (9.12)], above the line in the middle 1950s (positive error terms), and below the line in the later 1950s. Again the error terms are autocorrelated, since they tend to follow a definite time sequence.

Assumption 3.   Each of the sample errors e_i is distributed with the same variance, σ_e^2.

This assumption is called the assumption of homoskedasticity. If it is violated, the errors are said to be heteroskedastic. The problems introduced by heteroskedasticity are also discussed in a later chapter.

Example 9.5.   In Example 9.2 we specified a linear relationship between consumption and income in equation (9.8). If consumption is measured with some error but the error is proportionate to the level of income, higher levels of income will be associated with higher absolute errors. If income grows through time, the variance of the error term will increase through time. The errors will be heteroskedastic, and Assumption 3 will be violated.

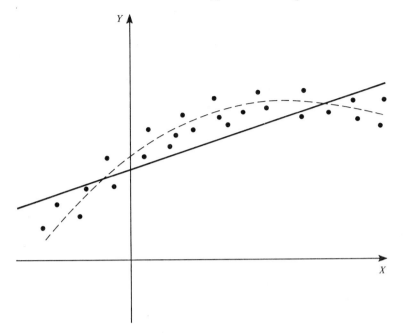

FIGURE 9.2   Autocorrelation of Residuals

Assumption 4.   Each of the error terms e_i is normally distributed.

This assumption is not always necessary for the theory of linear models that follows. The assumption of normally distributed error terms ensures the validity of tests of hypotheses for small samples. For large samples, appeals to various kinds of central limit theorems establish the approximate distributions of the various sample statistics.

Estimation Procedures

The parameters a and b of both the two-variable model (9.3) or a_0, a_1, ... , a_K of the multivariate model (9.10) are generally estimated by the method of least squares. (See Chapter 3.) If the independent X variables are uncorrelated with the error term and the error term satisfies Assumption 1 of the previous section, then the least-squares estimates are unbiased estimates of the parameters a_0, a_1, ... , a_K. Furthermore, if Assumptions 1, 2, and 3 are satisfied, the least-squares estimators are efficient.† If the

---

† More precisely, least-squares estimators are best linear estimators—that is, have the minimum variance among all linear estimators. Linear estimators are defined as linear functions of the observed values, Y_1, Y_2, ... , Y_N, of the dependent variables. In the appendix to this chapter we show that least-squares estimators are unbiased.

error terms also are normally distributed (Assumption 4), least-squares estimators are maximum-likelihood estimators. (See the appendix to this chapter for a demonstration.)

Let $\hat{a}_0, \hat{a}_1, \ldots, \hat{a}_K$ denote the least-squares estimates of the parameters $a_0, a_1, \ldots, a_K$. The estimated error term (residual) for the ith observation is determined by substituting the estimators $\hat{a}_0, \hat{a}_1, \ldots, \hat{a}_K$ into (9.10). That is,

$$\hat{e}_i = Y_i - (\hat{a}_0 + \hat{a}_1 X_{1i} + \cdots + \hat{a}_K X_{Ki}) = Y_i - \hat{Y}_i, \tag{9.14}$$

where $\hat{e}_i$ is the estimated value of the ith error term and $\hat{Y}_i = (\hat{a}_0 + \hat{a}_1 X_{1i} + \cdots + \hat{a}_K X_{Ki})$ is called the predicted value of Y_i from the linear regression. The variance of the error term σ_e^2 is estimated by

$$\hat{\sigma}_e^2 = \frac{\sum_{i=1}^{N} \hat{e}_i^2}{N - (K + 1)} = \frac{\sum_{i=1}^{N} (Y_i - \hat{Y}_i)^2}{N - (K + 1)}. \tag{9.15}$$

The estimator $\hat{\sigma}_e^2$ of the error variance is unbiased if the X's are nonrandom and Assumptions 1, 2, and 3 are satisfied. The numerator of (9.15),

$$\sum_{i=1}^{N} \hat{e}_i^2 = \sum_{i=1}^{N} (Y_i - \hat{Y}_i)^2,$$

is called the sum of the squared residuals. This sum plays an important role in the testing of hypotheses.

Linear Hypotheses

The random variable Y_i in (9.10) has a probability distribution that depends on (1) the probability distribution of the error term e_i, and (2) the parameters $a_0, a_1, \ldots, a_K$. That is, $a_0, a_1, \ldots, a_K$ can be regarded as parameters of the probability distribution of Y_i. We may wish to test a number of different null hypotheses concerning these parameters.

Example 9.6.   Suppose an economist is estimating a consumption function of the form

$$C = a + b \cdot Y. \tag{9.16}$$

He may wish to test the hypothesis that the average propensity to consume is constant. This is equivalent to testing whether the parameter $a = 0$. If $a = 0$, then the average propensity to consume is constant and equal to the marginal propensity to consume—that is,

$$\frac{C}{Y} = \frac{b \cdot Y}{Y} = b. \tag{9.17}$$

Example 9.7. Suppose imports M are specified as a linear function of consumption C and investment I:

$$M = a + b \cdot C + c \cdot I. \tag{9.18}$$

Suppose an economic policy maker wishes to know whether variations in investment expenditure have a more significant impact on the level of imports than consumption expenditures. He might then test the null hypothesis that $b = c$.

Example 9.8. A Cobb-Douglas production function has the general form

$$O = a \cdot K^b \cdot L^c, \tag{9.19}$$

where O stands for output, K is capital, and L is labor. The production function (9.21) is linear homogeneous† if

$$b + c = 1. \tag{9.20}$$

The production function (9.19) may be converted to a linear relationship by taking natural logarithms of both sides:

$$\log_e O = \log_e a + b \cdot \log_e K + c \cdot \log_e L \tag{9.21}$$

or, to simplify the notation,

$$Y = a_0 + a_1 X_1 + a_2 X_2, \tag{9.22}$$

where $Y = \log_e O$, $X_1 = \log_e K$, $X_2 = \log_e L$, $a_0 = \log_e a$, $a_1 = b$, and $a_2 = c$. We may test whether the production function is linear homogeneous by testing the null hypothesis that

$$b + c = a_1 + a_2 = 1. \tag{9.23}$$

Examples 9.6, 9.7, and 9.8 may all be treated as special cases of a *linear* null hypothesis of the form

$$H0: \gamma_0 a_0 + \gamma_1 a_1 + \cdots + \gamma_K a_K = r, \tag{9.24}$$

where $a_0, a_1, \ldots, a_K$ are the parameters of the linear regression model (9.10) and $\gamma_0, \gamma_1, \ldots, \gamma_K$ and r are constants.

Example 9.9. In Example 9.6 we proposed the null hypothesis that $a = 0$ in equation (9.16). The null hypothesis in terms of (9.24) implies that $\gamma_0 = 1, \gamma_1 = \gamma_2 = \cdots = \gamma_K = r = 0$. In Example 9.8, the null hypothesis specified in (9.23) implies that $\gamma_0 = \gamma_1 = r = 1$ and $\gamma_2 = \gamma_3 = \cdots = \gamma_K = 0$.

---

† A function $f(K, L)$ is linear homogeneous if $f(t \cdot K, t \cdot L) = t \cdot f(K, L)$. That is, an equal proportionate increase in the capital and labor inputs increases output in the same proportion.

In some cases we may want to test a null hypothesis involving more than one linear restriction on the parameters. If there are p such linear restrictions, the null hypothesis becomes

$$
\begin{aligned}
H0: \quad & \gamma_{01}a_0 + \gamma_{11}a_1 + \cdots + \gamma_{K1}a_K = r_1 \\
& \gamma_{02}a_0 + \gamma_{12}a_1 + \cdots + \gamma_{K2}a_K = r_2 \\
& \quad \vdots \\
& \gamma_{0p}a_0 + \gamma_{1p}a_1 + \cdots + \gamma_{Kp}a_K = r_p
\end{aligned}
\tag{9.25}
$$

Example 9.10.   In Example 9.7 we specified that imports depended on consumption and investment as in (9.18). One may wish to test not only the hypothesis that $b = c$ (or that consumption expenditure and investment expenditure affect imports equally) but the hypothesis that average and marginal effects on imports are the same—that is, $a = 0$. Both these hypotheses may be tested jointly by testing the null hypothesis

$$
H0: \quad a = 0, \quad b - c = 0.
\tag{9.26}
$$

The number p of linear restrictions in (9.25) is assumed to be less than the number $K + 1$ of parameters $a_0, a_1, \ldots, a_K$. If this were not true, the null hypothesis would generally be too restrictive; that is, no set of parameters could satisfy all the restrictions.

Estimation Under the Null Hypothesis

In the third section of this chapter, we suggested that the parameters $a_0, a_1, \ldots, a_K$ of the regression model (9.10) be estimated using a least-squares procedure. The resulting estimates are denoted by $\hat{a}_0, \hat{a}_1, \ldots, \hat{a}_K$, and the predicted or estimated value of Y_i is

$$
\hat{Y}_i = \hat{a}_0 + \hat{a}_1 X_{1i} + \cdots + \hat{a}_K X_{Ki}.
\tag{9.27}
$$

The error term e_i is estimated as the difference between the actual ith observed value of Y and the predicted value.

$$
\hat{e}_i = Y_i - \hat{Y}_i.
\tag{9.28}
$$

We may obtain an alternative set of estimates of the parameters $a_0, a_1, \ldots, a_K$ by forcing the parameters to satisfy the conditions of the null hypothesis. The resulting set of restricted estimators is denoted by $a_0^*, a_1^*, \ldots, a_K^*$. The estimated error term that results is

$$
e_i^* = Y_i - (a_0^* + a_1^* X_{1i} + \cdots + a_K^* X_{Ki}) = Y_i - Y_i^*,
\tag{9.29}
$$

where Y_i^* is the predicted value of Y_i using the restricted estimators, $a_0^*, \ldots, a_K^*$.

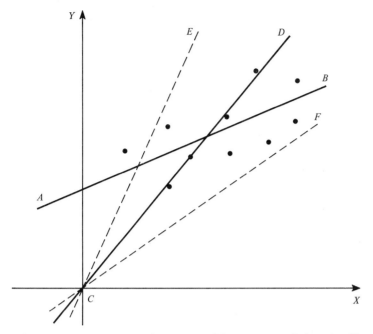

FIGURE 9.3 Restricted and Unrestricted Least-squares Regression Lines

For example, in Fig. 9.3, the line AB is a least-squares regression line, $Y = \hat{a} + \hat{b}X$, fitted to a scatter of points, where $\hat{a}$ and $\hat{b}$ denote *unrestricted* least-squares estimators. The line CD represents the least-squares regression line, $Y = a^* + b^*X$, where the intercept a is restricted to be zero. a^* and b^* denote *restricted* least-squares estimators. The line CD minimizes the sum of the squared vertical deviations from the regression line among all regression lines, such as CE and CF, that pass through the origin.

In Fig. 9.4, the line AB represents a more general linear restriction of the form $\gamma_0 a_0 + \gamma_1 a_1 = r$. The concentric ellipses about the point C represent combinations of the intercept a_0 and the slope a_1 of the regression line $Y = a_0 + a_1 X_1$ that result in equal sums of squared deviations about the regression line. The point C gives the combination of slope and intercept that minimizes the sum of squares, and the concentric ellipses represent increasing sums of squares as they become further removed from the point C. The point D indicates the slope and intercept that minimize the sum of squares for all points satisfying the linear restriction AB.

It is obvious that the sum of squared residuals using restricted estimators is greater than or equal to the sum of squared deviations using the unrestricted estimators. That is,

$$\sum_{i=1}^{N} \hat{e}_i^2 = \sum_{i=1}^{N} (Y_i - \hat{Y}_i)^2 \leq \sum_{i=1}^{N} e_i^{*2} = \sum_{i=1}^{N} (Y_i - Y_i^*)^2, \qquad (9.30)$$

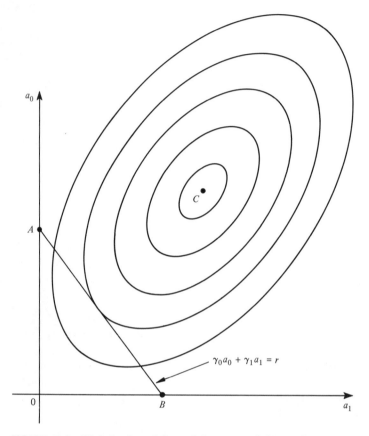

FIGURE 9.4   Minimization of Sum of Squares with Linear Restrictions on Coefficients

where Y_i^* is the predicted value of Y_i using the restricted estimators, and $\hat{Y}_i$ is the predicted value of Y_i with the unrestricted estimators.

Example 9.11.   In Example 9.6 we considered the consumption function of the form (9.16). If we test the hypothesis that the average propensity to consume is the same as the marginal propensity, we can estimate the parameters a and b subject to the restriction that $a = 0$. The estimate of a is, of course, $a^* = 0$. The estimate of b is obtained by minimizing the sum of squares

$$S_0 = \sum_{i=1}^{N} (C_i - bY_i)^2.$$

The value of b that minimizes this sum is

$$b^* = \sum_{i=1}^{N} C_i Y_i \Big/ \sum_{i=1}^{N} Y_i^2.$$

Example 9.12. In Example 9.7 we discussed the import function (9.18) and the null hypothesis that $b = c$. The unrestricted estimators are determined by minimizing the sum of squares

$$\sum_{i=1}^{N} (M_i - a - bC_i - cI_i)^2.$$

If $b = c$, the sum of squares becomes

$$S_0 = \sum_{i=1}^{N} [M_i - a - b(C_i + I_i)]^2.$$

The restricted estimators a^*, b^*, and $c^* = b^*$ are obtained by minimizing S_0. Note that this is equivalent to regressing M_i on the sum $(C_i + I_i)$, which is treated as one independent variable rather than two separate variables.

Example 9.13. In Example 9.8 the restricted estimators are obtained for equation (9.22) subject to the restriction (9.23) that $a_1 + a_2 = 1$. The unrestricted sum of squares is

$$\sum_{i=1}^{N} (Y_i - a_0 - a_1 X_{1i} - a_2 X_{2i})^2.$$

If the restriction is satisfied, we can write $a_1 = 1 - a_2$. Substituting this expression for a_1 into the sum of squares, we obtain

$$S_0 = \sum_{i=1}^{N} [(Y_i - X_{1i}) - a_0 - a_2(X_{2i} - X_{1i})]^2,$$

which is the sum of squares minimized to obtain the restricted estimators. Note that this is equivalent to treating $(Y_i - X_{1i})$ as the dependent variable and $(X_{2i} - X_{1i})$ as the single independent variable.

The Fundamental Identity

In this section we discuss a general test of the hypothesis (9.25). This test makes use of the *fundamental identity:*

$$\sum_{i=1}^{N} (Y_i - Y_i^*)^2 = \sum_{i=1}^{N} (Y_i - \hat{Y}_i)^2 + \sum_{i=1}^{N} (\hat{Y}_i - Y_i^*)^2$$

or   (9.31)

$$S_0 = S_1 + S_2.$$

That is, the sum of the squared residuals S_0 using the restricted estimators (null-hypothesis estimators) may be decomposed into two parts. The first S_1 is the sum of the squared residuals using the unrestricted estimators, and the second S_2 is the sum of the squared differences in the two sets of predicted values, one based on the unrestricted and the other based on the restricted estimators. The identity (9.31) is valid whenever the Y^* and $\hat{Y}_i$ are least-squares estimators.† The identity is derived in the appendix to this chapter.

Another way of writing (9.31) is

$$\sum_{i=1}^{N} e_i^{*2} - \sum_{i=1}^{N} \hat{e}_i^2 = \sum_{i=1}^{N} (\hat{Y}_i - Y_i^*)^2$$

or   (9.32)

$$S_0 - S_1 = S_2.$$

The difference in the sum of the squared residuals equals the sum of the squared differences in predicted values.

If Assumptions 1 through 4 concerning the error terms e_i are satisfied, it is possible to show that all three terms in the identity (9.31) have chi-square distributions.‡ The number of degrees of freedom of the term S_0 on the left-hand side of (9.31) is $N - (K + 1) + p$, where N is the number of observations; $K + 1$ is the number of parameters $a_0, a_1, \ldots, a_K$; and p is the number of linear restrictions in the null hypothesis. The first term S_1 on the right-hand side has $N - (K + 1)$ degrees of freedom, and the second term S_2 has p degrees of freedom. Note that the sum of the degrees of freedom on the right-hand side of (9.31) equals the degrees of freedom on the left-hand side.§

There is another way of looking at the number of degrees of freedom of each of the sums of squares in the fundamental identity (9.31). The degrees of freedom satisfy the identity:

degrees of freedom of S_0
$$= \text{degrees of freedom of } S_1 + \text{degrees of freedom of } S_2. \quad (9.33)$$

The sum of squares S_1 is the sum of squared residuals using the unrestricted parameter estimates, $a_0, a_1, \ldots, a_K$. There are $K + 1$ parameters estimated from N observations. The number of degrees of freedom of S_1 is the number of observations less the number of parameters being estimated or $N - (K + 1)$. The term S_0 is the sum of squared residuals using the $K + 1$

---

† In fact it is sufficient only that the $\hat{Y}_i$ be least-squares estimators.
‡ Strictly, this is true only if the null hypothesis is true. Otherwise the distributions are noncentral chi-square.
§ All this assumes that the rank of the matrix of γ's in (9.27) is equal to p, the number of restrictions.

restricted parameter estimates a_0^*, a_1^*, ... , a_K^*. Since these parameters, however, must satisfy the p linear restrictions of the null hypothesis (9.27), if any $(K + 1) - p$ of these parameters are known, the remaining p parameters can be determined by solving the linear equations (9.25). That is, only $(K + 1) - p$ independent parameters are being estimated. The number of degrees of freedom of S_0 then is the number of observations N less the number of independent parameters, $(K + 1) - p$, being estimated or $N - (K + 1) + p$. From the identity (9.33) concerning the number of degrees of freedom, we have

$$N - (K + 1) + p = N - (K + 1) + \text{degrees of freedom of } S_2$$

or

$$\text{degrees of freedom of } S_2 = [N - (K + 1) + p] - [N - (K + 1)] = p.$$
$$(9.34)$$

Testing the Null Hypothesis

If Assumptions 1 through 4 concerning the error term are satisfied, one can also show that the two terms on the right-hand side of (9.31) are distributed independently. Thus Theorem 7.7 of Chapter 7 applies and the ratio of the two terms has an F distribution.† That is, the ratio

$$F = \frac{\sum_{i=1}^{N} (\hat{Y}_i - Y_i^*)^2}{\sum_{i=1}^{N} (Y_i - \hat{Y}_i)^2} \cdot \frac{N - (K + 1)}{p} = \frac{S_2}{S_1} \cdot \frac{N - (K + 1)}{p} \qquad (9.35)$$

has an F distribution with p and $N - (K + 1)$ degrees of freedom.

From (9.32) we see that the S_2 term in the numerator of the F ratio is equal to the difference in the squared residuals under the null and alternative hypotheses. If the restrictions of the null hypothesis are really true, then relaxing them won't make much difference in the sum of squares, and the F ratio will be small. If, on the other hand, the null hypothesis is false, the difference in the sum of squares will be large and the F ratio will be large.

Example 9.14.   Suppose data have been gathered for 10 different textile firms on capital inputs, labor inputs, and output. These data are used by

---

† Strictly, the ratio has a noncentral F distribution, a generalized notion of the F distribution, which is not discussed in this volume. If the null hypothesis (9.24) or (9.25) is true, however, the ratio in (9.35) does have an F distribution, a special case of noncentral F. Thus tests of the null hypothesis can be based on the F distribution.

	FIRM									
	1	2	3	4	5	6	7	8	9	10
Capital	31.7	25.2	31.6	25.1	31.7	50.2	31.6	126.0	19.9	15.8
Labor	12.6	15.8	25.2	50.1	31.6	63.1	79.5	39.8	15.8	39.9
Output	15.9	20.0	25.1	39.8	25.2	63.1	50.1	79.4	15.8	31.6

an industry economist to estimate the parameters of a Cobb-Douglas production function.

$$O = a \cdot K^b \cdot L^c, \qquad (9.36)$$

where O is output, K is capital, and L is labor. If logarithms to the base e are taken of both sides, the production function becomes

$$\log_e O = \log_e a + b \cdot \log_e K + c \cdot \log_e L. \qquad (9.37)$$

The unrestricted least-squares estimators of the parameters using the above data are

$\log_e \hat{a}$	$\hat{b}$	$\hat{c}$
−.819	.4780	.6858

The economist wishes to test the null hypothesis that there are no economies of scale in textile production:

$$H0: b + c = 1. \qquad (9.38)$$

The sum of the squared residuals using the unrestricted estimators $\log_e \hat{a}$, $\hat{b}$, and $\hat{c}$ is calculated as follows:

FIRM (i)	LOG OF ACTUAL OUTPUT ($\log_e O_i$)	LOG OF PREDICTED OUTPUT ($\log_e \hat{O}_i$)	RESIDUAL ($\hat{c}_i$)	SQUARED RESIDUAL ($\hat{e}_i^2$)
1	1.2014	1.2496	.0482	.002323
2	1.3010	1.2465	−.0545	.002970
3	1.3997	1.4553	.0556	.003091
4	1.5999	1.5903	−.0096	.000092
5	1.4014	1.5239	.1225	.015006
6	1.8000	1.8013	.0013	.000002
7	1.6998	1.7265	.0267	.000713
8	1.8998	1.8554	−.0444	.001971
9	1.1987	1.1987	.0000	.000000
10	1.4997	1.4252	−.0745	.005550

Sum of squared residuals: $S_1 = \sum_{i=1}^{10} \hat{e}_i^2 = .031718.$

If one forces the parameters b and c to satisfy the linear restriction (9.38), the following estimates of the parameters are obtained:

$$\log_e a^* \quad b^* \quad c^*$$
$$-.004 \quad .405 \quad .595$$

The sum of squared residuals using the restricted estimators is calculated as follows:

FIRM (i)	LOG OF ACTUAL OUTPUT $(\log_e O_i)$	LOG OF PREDICTED OUTPUT $(\log_e O_i^*)$	RESIDUAL (e_i^*)	SQUARED RESIDUAL (e_i^{*2})
1	1.2014	1.2589	+.0575	.003306
2	1.3010	1.2769	−.0241	.000581
3	1.3997	1.4373	+.0376	.001414
4	1.5999	1.5742	−.0257	.000660
5	1.4014	1.4963	+.0949	.009006
6	1.8000	1.7558	−.0442	.001954
7	1.6998	1.7340	+.0342	.001170
8	1.8998	1.7989	−.1009	.010181
9	1.1987	1.2353	+.0366	.001340
10	1.4997	1.4339	−.0658	.004330

Sum of squared residuals: $S_0 = \sum_{i=1}^{10} e_i^{*2} = .033942.$

The sum of squared differences of predicted values may be calculated by taking the difference between S_0 and S_1.

$$S_2 = \sum_{i=1}^{10} (\log_e O_i^* - \log_e \hat{O}_i) = S_0 - S_1 = .033942 - .031718 = .002224.$$

The F ratio from (9.35) is

$$F = \frac{S_2}{S_1} \cdot \frac{N - (K + 1)}{p} = \frac{.0022}{.0317} \cdot \frac{10 - (2 + 1)}{1} = 0.49. \tag{9.39}$$

The least-squares estimation procedure produces unbiased estimates of the "true" parameters $a_0, a_1, \ldots, a_K$ in (9.10) (under Assumptions 1 through 3). If the true parameters $a_0, a_1, \ldots, a_K$ in fact satisfy the null hypothesis (9.24) or (9.25), then the unrestricted least-squares estimators $\hat{a}_0, \hat{a}_1, \ldots, \hat{a}_K$ tend to differ very little from the estimators $a_1^*, a_2^*, \ldots, a_K^*$

found by forcing the parameters to satisfy the linear restrictions of the null hypothesis. Thus Y_i and Y_i^*, the predicted values of Y_i under the two hypotheses, tend to differ very little. That is, the numerator in (9.35) tends to be very small if the null hypothesis (9.24) or (9.25) is true. If the null hypothesis is false, the predicted values Y_i and Y_i^* will tend to be quite different and the numerator of (9.35) will be large. Thus a small value for F indicates that the null hypothesis is true and a large value indicates that it is false. A one-tailed test is appropriate, using as the critical region the right-hand tail of the F distribution.

Table D in the Appendix of Statistical Tables gives critical values of F for one-tailed tests with critical regions of 5 percent, $2\frac{1}{2}$ percent, 1 percent, and $\frac{1}{2}$ percent for various degrees of freedom in the numerator and the denominator of the F ratio in (9.35). The degrees of freedom of the numerator are indicated in row heads and of the denominator in the column heads. Thus to use (9.35) to test the null hypothesis one looks across the row with p degrees of freedom and down the column with $N - (K + 1)$ degrees of freedom. An F value greater than the critical value results in rejection of the null hypothesis.

Example 9.15.   In Example 9.14 the F ratio was calculated in (9.39) to be 0.49. We wish to test the null hypothesis (9.38) that there are no returns to scale. The degrees of freedom in the numerator is $p = 1$. The degrees of freedom in the denominator is $N - (K + 1) = 10 - (2 + 1) = 7$. Using a 5 percent critical region, Table D of the Appendix of Statistical Tables gives a critical value of 5.59. Since the F ratio is smaller than this value, we accept the null hypothesis and conclude that there are no significant economies of scale.

Significance of a Single Independent Variable

In this section we consider in more detail a special case. This is the test of the null hypothesis that a single parameter is zero.

$$H0: a_j = 0. \tag{9.40}$$

A test of $H0$ is a test of whether the jth independent variable X_{ji} in (9.10) significantly explains variations in the dependent variable Y_i. Since the number of linear restrictions in $H0$ is $p = 1$, Theorem 7.8 of Chapter 7 applies, and the square root of F in (9.35) has a Student's t distribution with $N - (K + 1)$ degrees of freedom. Thus Table C in the Appendix of Statistical Tables may be used to test the null hypothesis. Note, however, that whereas a one-tailed test is appropriate with the F ratio, an equivalent critical t value is based on a two-tailed test.

Example 9.16.   In Example 9.6 we suggested that one might wish to test the null hypothesis that the average and marginal propensity to consume are the same. This is equivalent to the null hypothesis that $a = 0$ in the consumption function

$$C = a + b \cdot Y, \tag{9.41}$$

where C is consumption and Y is income. Suppose we have observations on consumption and income for 20 years. The unrestricted parameter estimates are $\hat{a} = .07$ and $\hat{b} = .92$. The restricted estimators are $a^* = 0$ and $b^* = .95$. The sum of squared residuals using the restricted estimators is $S_0 = \Sigma\, e_i^{*2} = 43$ and the sum of squared residuals using the unrestricted estimators is $S_1 = \Sigma\, \hat{e}_i^2 = 12$. The difference is $S_2 = 43 - 12 = 31$. The F ratio is

$$F = \frac{S_2}{S_1} \cdot \frac{[N - (K + 1)]}{p} = \frac{31}{12} \cdot (20 - 2) = 46.5. \tag{9.42}$$

The square root of F is the t ratio,

$$t = \sqrt{F} = \sqrt{46.5} = 6.8, \tag{9.43}$$

which has $N - (K + 1) = 18$ degrees of freedom. From Table C of the Appendix of Statistical Tables, the critical value of t for a two-tailed 1 percent critical region with 18 degrees of freedom is 2.878. We reject the null hypothesis that the marginal and average propensities to consume are the same.

Example 9.17. The following data are available on per capita sugar consumption (Y), sugar price (X_1), and the level of per capita disposable income (X_2) for Kenya over the ten-year period 1954 to 1963.

YEAR	SUGAR CONSUMPTION PER CAPITA (Y)	SUGAR PRICE (X_1)	DISPOSABLE INCOME PER CAPITA (X_2)
1954	16.1	70	22.97
1955	17.3	69	26.02
1956	19.9	62	26.43
1957	20.4	70	26.67
1958	20.5	62	26.11
1959	22.0	63	25.87
1960	24.0	62	26.34
1961	24.7	64	24.92
1962	25.7	65	25.49
1963	24.5	67	26.42

The estimated least-squares linear relationship is

$$Y = \hat{a}_0 + \hat{a}_1 X + \hat{a}_2 X$$
$$= 30.6 - (.386)X_1 + (.626)X_2. \qquad (9.44)$$

Suppose we wish to test the null hypothesis that price (X_1) does not significantly affect sugar consumption—that is, that $a_1 = 0$. If we estimate the least-squares regression line, restricting the parameter a_1 so that it equals zero, we obtain the relationship

$$Y = a_0^* + a_2^* X_2$$
$$= -5.2 + (1.039)X_2. \qquad (9.45)$$

The sum of the squared residuals using the unrestricted parameters is

$$S_1 = \sum_{i=1}^{10} \hat{e}_i^2 = 69.74. \qquad (9.46)$$

Using the restricted parameters, we obtain the sum of the squared residuals

$$S_0 = \sum_{i=1}^{10} \hat{e}_i^{*2} = 83.31. \qquad (9.47)$$

The sum of squared differences in predicted Y values is

$$S_2 = S_0 - S_1 = 13.57. \qquad (9.48)$$

The F ratio is

$$F = \frac{S_2}{S_1} \cdot \frac{N - (K+1)}{p} = (.195) \frac{10 - (2+1)}{1} = 1.365, \qquad (9.49)$$

which has $p = 1$ degrees of freedom in the numerator and $N - (K+1) = 7$ degrees of freedom in the denominator. The square root of F is the t ratio,

$$t^* = \sqrt{F} = 1.169, \qquad (9.50)$$

which has 7 degrees of freedom. Using a one-tailed 5 percent critical region, we obtain a critical value of 1.895 from the Appendix of Statistical Tables. Thus we accept the hypothesis that price has no significant effect on sugar consumption.

Another way of looking at the special case (9.40) is to investigate the sampling distribution of the unrestricted least-squares estimators $\hat{a}_0$, $\hat{a}_1$, ... , $\hat{a}_K$. For the general case of several independent variables (9.10) this is quite complicated. For the case (9.3) of a single independent variable, the unrestricted least-squares estimator $\hat{b}$ has a probability distribution with a variance

$$\sigma_{\hat{b}}^2 = \frac{\sigma_e^2}{\sum_{i=1}^{N} (X_i - \bar{X})^2},$$

where σ_e^2 is the variance of the error term, and

$$\bar{X} = \sum_{i=1}^{N} \frac{X_i}{N}$$

is the mean value of the independent variable. Note that the variance of the estimator $\hat{b}$ depends on the variance of the observed values X_i of the independent variable.

If Assumptions 1 through 4 are satisfied, then the error term e_i is normally distributed and the estimator $\hat{b}$ is normally distributed. If $H0$ is true, then, since $\hat{b}$ is unbiased, the theoretical mean of $\hat{b}$ is its hypothesized value of 0. Thus

$$t = \frac{\hat{b} - 0}{\sigma_{\hat{b}}} = \frac{\hat{b} - 0}{\sqrt{\sigma_e^2 / \sum_{i=1}^{N} (X_i - \bar{X})^2}} = \frac{\hat{b} \sqrt{\sum_{i=1}^{N} (X_i - \bar{X})^2}}{\sigma_e} \tag{9.51}$$

has a standard normal distribution. If we know σ_e, the theoretical standard deviation of the error term, then t in (9.51) could be used as a test statistic, where the critical value of t is obtained from the standard normal distribution in Table C of the Appendix of Statistical Tables.

σ_e is generally not known. If the sample is large, however, we may replace σ_e in (9.45) by its estimated value $\hat{\sigma}_e$. The sample statistic

$$t = \frac{\hat{b}}{\hat{\sigma}_{\hat{b}}} = \frac{\hat{b} \sqrt{\sum_{i=1}^{N} (X_i - \bar{X}_i)^2}}{\hat{\sigma}_e} \tag{9.52}$$

then is approximately normally distributed with mean 0, where

$$\hat{\sigma}_e^2 = \frac{\sum_{i=1}^{N} (Y_i - \hat{a} - \hat{b}X_i)^2}{N - 2} = \frac{\sum_{i=1}^{N} (Y_i - \hat{Y}_i)^2}{N - 2}. \tag{9.53}$$

$\hat{\sigma}_e^2$ is an unbiased estimator of the error variance $\hat{\sigma}_e^2$.†

---

† Note that sometimes $\hat{\sigma}_e^2$ is defined as

$$\hat{\sigma}_e^2 = \sum_{i=1}^{N} \frac{(Y_i - \bar{Y})^2}{N},$$

which is the maximum-likelihood estimate of σ_e^2. For large samples (large N), however, this difference is small.

If the sample is small, a different approach must be used. The variable t in (9.51) has a standard normal distribution. One can also show that the variable

$$v^2 = \frac{(N-2)\hat{\sigma}_e^2}{\sigma_e^2} \tag{9.54}$$

has a chi-square distribution with $N - 2$ degrees of freedom. Thus from Theorem 7.6 of Chapter 7

$$t^* = \frac{t \sqrt{N-2}}{v} = \frac{\hat{b} \cdot \sqrt{\sum_{i=1}^{N} (X_i - \bar{X})^2}}{\hat{\sigma}_e} \tag{9.55}$$

has a Student's t distribution with $N - 2$ degrees of freedom. Note that the expression (9.52) for the standard normal variate is the same as (9.55) for the t ratio. Thus the same computational procedure applies for large and small samples. The difference lies in the use of standard normal tables for large samples and the tables of the Student's t distribution in finding critical values for tests of hypotheses.

In either case, large samples or small, the square root of the F ratio in (9.35) produces the standard normal variate of (9.52) or the t ratio in (9.55). Thus the tests of hypotheses are equivalent whether one (i) uses the F-ratio approach, (ii) derives a t ratio (or standard normal variate) by taking the square root of the F ratio, or (iii) calculates the t ratio (the standard normal variate) directly from (9.52) or (9.55).

Example 9.18.   In Example 9.16 we estimated the consumption-function relationship (9.41). The unrestricted estimators were $\hat{a} = .07$ and $\hat{b} = .92$. The sum of the squared residuals using the unrestricted estimators was

$$S_1 = \sum_{i=1}^{20} (Y_i - \hat{a} - \hat{b}X_i)^2 = \sum_{i=1}^{20} \hat{e}_i^2 = 12. \tag{9.56}$$

The estimated error variance (9.53) is

$$\hat{\sigma}_e^2 = \frac{\sum_{i=1}^{20} (\hat{e}_i^2)}{20 - 2} = .667. \tag{9.57}$$

The estimated variance of the estimator $\hat{b} = .92$ is given by

$$\hat{\sigma}_{\hat{b}}^2 = \frac{\hat{\sigma}_e^2}{\sum_{i=1}^{20} (X_i - \bar{X})^2} = \frac{.667}{2350} = .000284. \tag{9.58}$$

The estimated standard deviation of $\hat{b}$ is the square root of the estimated variance or $\sqrt{.000284} = .0169$. If we divide this estimated standard deviation into the estimated value $\hat{b} = .92$, we obtain the t ratio in (9.52).

$$t^* = \frac{\hat{b}}{\hat{\sigma}_{\hat{b}}} = \frac{.92}{.0169} = 54.44. \tag{9.59}$$

Suppose we wish to test the null hypothesis that income has no significant effect on consumption. A one-tailed test seems appropriate, since we would hardly expect income to have a negative effect on consumption. For a 1 percent critical region and $20 - 2 = 18$ degrees of freedom the critical value of t is 2.552, which is smaller than the value of the t ratio in (9.59) of 54.44. Thus income has a significant effect on consumption—which is, of course, what one would expect.

This analysis can be generalized to the multivariate case in which

$$t = \frac{\hat{a}_j}{\hat{\sigma}_{\hat{a}_j}} \tag{9.60}$$

has the Student's t distribution (approximately a normal distribution for large samples), where $\hat{\sigma}_{\hat{a}_j}$ is the estimated variance of the estimator $\hat{a}_j$. The general expression for $\hat{\sigma}_{\hat{a}_j}$ is quite messy algebraically, although matrix notation may be used to simplify the expression considerably.

Example 9.19.   In Example 9.17 we calculated the regression line (9.44) relating per capita sugar consumption to price and income. Although the computational effort is a bit involved, one may calculate the estimated variance of the three estimators $\hat{a}_0 = 30.6$, $\hat{a}_1 = -.386$, and $\hat{a}_2 = .626$. These variances are 1088.36, .109, and 1.327, and the standard deviations are 32.99, .3298, and 1.1519. It is customary to present estimated standard deviations along with the estimators themselves in the following format:

$$Y = \underset{(32.99)}{30.6} \underset{(.3298)}{- (.386)X_1} + \underset{(1.1519)}{(.626)X_2} \tag{9.61}$$

The estimated standard deviations may be divided into the estimators themselves to produce a set of t ratios, one for each estimated parameter. They are sometimes presented in terms of the following format:

$$Y = \underset{(.928)}{30.6} \underset{(-1.169)}{- (.386)X_1} + \underset{(.543)}{(.626)X_2} \tag{9.62}$$

Each t ratio may be used to test the significance of each of the estimators $\hat{a}_0$, $\hat{a}_1$, and $\hat{a}_2$. The critical value for a 5 percent one-tailed test of significance using the Student's t distribution with 7 degrees of freedom is 1.895. Thus none of the coefficients, a_0, a_1, or a_2, is significant.

The variance approach to estimating the significance of a single regression parameter can be used also to construct a confidence interval about a regression parameter. To construct a 95 percent confidence interval, for example, we find the critical value of the t ratio (or standard normal variate in the large sample case) corresponding to a two-tailed 5 percent critical region. Let us denote this critical value by $t_{.05}$. Then a 95 percent confidence limit about the regression parameter a_j is given by

$$\hat{a}_j \pm t_{.05} \cdot \hat{\sigma}_{\hat{a}_j}. \tag{9.63}$$

Example 9.20.   In Example 9.18 we estimated the variance of the parameter $\hat{b}$ from the consumption function (9.41) to be .0169. For a 5 percent two-tailed critical region with 18 degrees of freedom, the critical value of the t ratio is 2.101. Since $\hat{b}$ is .92, the 95 percent confidence limit about $\hat{b}$ is

$$.92 \pm (2.101)(.0169) \quad \text{or} \quad .92 \pm .036.$$

Significance of the Regression

Of considerable interest is the null hypothesis

$$H0: \quad a_1 = 0$$
$$a_2 = 0$$
$$\cdot$$
$$\cdot \tag{9.64}$$
$$\cdot$$
$$a_K = 0$$

That is, one tests the null hypothesis that the parameters for all K independent variables are 0. This test is sometimes called a test of the significance of the regression.

In this special case, the number p of linear restrictions in (9.64) is equal to K the number of independent variables. The F ratio of (9.35) becomes

$$F = \frac{S_2}{S_1} \cdot \frac{N - (K + 1)}{K}. \tag{9.65}$$

Of particular interest is the set of restricted parameter estimates. It is possible to show that the estimator of a_0 is

$$a_0^* = \bar{Y}, \tag{9.66}$$

where $\bar{Y}$ is the sample mean of the dependent variable. The other restricted estimators are, of course, $a_1^* = \cdots = a_K^* = 0$. The predicted value of Y_i using the restricted estimators is $Y_i^* = \bar{Y}$; it is the same for all observations and equal to the sample mean.

The fundamental identity (9.31) then has a special interpretation, which is illustrated in Figure 9.5. The identity may be written

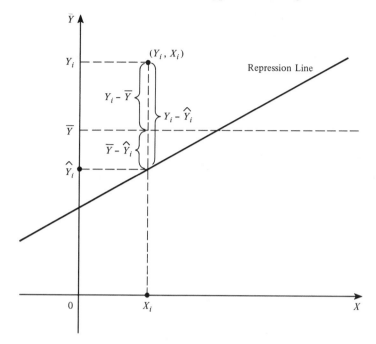

FIGURE 9.5   Interpretation of fundamental identity

$$\sum_{i=1}^{N} (Y_i - \bar{Y})^2 = \sum_{i=1}^{N} (Y_i - \hat{Y}_i)^2 + \sum_{i=1}^{N} (\hat{Y}_i - \bar{Y})^2$$

or

$$S_0 = S_1 + S_2.$$

(9.67)

The term S_0 on the left-hand side is the sum of squared deviations about the mean. The first term S_1 on the right-hand side is the sum of squared residuals about the regression line, and the second term S_2 is the sum of squared residuals of the regression line about the mean. In terms of Figure 9.5, the square root of the ith term in S_0 is the distance between the ith observation and the mean $\bar{Y}$. The square root of the ith term in S_1 is the vertical distance between the ith observation and the regression line. The ith term in S_2 is the squared distance between the regression line and the mean $\bar{Y}$.

Another way to write the sum of squares S_0, S_1, and S_2 is as follows:

$$S_0 = \sum_{i=1}^{N} y_i^2,$$

(9.68)

$$S_1 = \sum_{i=1}^{N} y_i^2 - \left(\hat{a}_1 \sum_{i=1}^{N} y_i x_{1i} + \cdots + \hat{a}_K \sum_{i=1}^{N} y_i x_{Ki} \right)$$

(9.69)

$$S_2 = \hat{a}_1 \sum_{i=1}^{N} y_i x_{1i} + \cdots + \hat{a}_K \sum_{i=1}^{N} y_i x_{Ki}, \tag{9.70}$$

where

$$y_i = Y_i - \bar{Y}, \quad x_{1i} = X_{1i} - \bar{X}_1, \ldots, x_{Ki} = X_{Ki} - \bar{X}_K. \tag{9.71}$$

This formulation is particularly useful for computational purposes.

The sum of squares S_0 is sometimes called the total variation of the dependent variable Y (about the mean). S_2, then, is considered to be the variation in Y explained by the regression on X_1 and X_2, and S_1 is the unexplained variation in Y. In these terms, the fundamental identity may be written as

total variation (S_0)
$$= \text{unexplained variation } (S_1) + \text{explained variation } (S_2). \tag{9.72}$$

The F ratio then may be written

$$F = \frac{\text{explained variation } (S_2)}{\text{unexplained variation } (S_1)} \cdot \frac{N - (K + 1)}{K}. \tag{9.73}$$

If the F-ratio is significant (greater than the critical value), one says that the independent variables, X_1, X_2, $\ldots$, X_K, significantly explain the variation in the dependent variable Y.

Example 9.21.   Let us consider again the estimated relationship between per capita sugar consumption (Y) and price (X_1) and per capita income (X_2) discussed in Examples 9.17 and 9.19. The total variation in per capita sugar consumption is

$$S_0 = \sum_{i=1}^{9} (Y_i - \bar{Y})^2 = \sum_{i=1}^{9} y_i^2 = 94.95.$$

The unexplained variation is

$$S_1 = \sum_{i=1}^{9} (Y_i - \hat{Y}_i)^2 = \sum_{i=1}^{9} (Y_i - \hat{a}_0 - \hat{a}_1 X_1 - \hat{a}_2 X_2)^2$$

$$= \sum_{i=1}^{9} y_i^2 - \hat{a}_1 \sum y_i x_{1i} - \hat{a}_2 \sum y_i \cdot x_{2i} = 69.74.$$

The variation explained by the regression is

$$S_2 = S_0 - S_1 = 94.95 - 69.74 = 25.21.$$

The F ratio then is explained variation S_2 divided by the unexplained variation S_1 multiplied by a factor $[N - (K + 1)]/K$.

$$F = \frac{S_2}{S_1} \cdot \frac{N - (K + 1)}{K} = (.361) \frac{10 - (2 + 1)}{2} = 1.263.$$

There are 2 degrees of freedom in the numerator and 7 degrees of freedom in the denominator. For a 1 percent critical region in a one-tailed test, the critical value of F is 9.95, which is larger than the F we calculated. Thus the regression is not significant, and price (X_1) and per capita income (X_2) do not significantly explain the variation in Y.

Note also that the coefficient R^2 as defined in equation (4.7) of Chapter 4 is related to the F statistic and to the fundamental identity. The definition is

$$R^2 = \frac{\text{explained variation}}{\text{total variation}} = \frac{S_2}{S_0}. \tag{9.74}$$

Since $S_0 = S_1 + S_2$ we may write

$$R^2 = \frac{S_2}{S_1 + S_2} = \frac{1}{1 + \dfrac{S_2}{S_1}} = \frac{1}{1 + F\left[\dfrac{K}{N - (K + 1)}\right]}. \tag{9.75}$$

Alternatively (9.75) may be written to express the F ratio in terms of the coefficient of determination

$$\begin{aligned}
F &= \frac{S_2}{S_1} \cdot \frac{N - (K + 1)}{K} = \frac{S_2}{S_0 - S_2} \cdot \frac{N - (K + 1)}{K} \\
&= \frac{S_2/S_0}{1 - (S_2/S_0)} \cdot \frac{N - (K + 1)}{K} \\
&= \frac{R^2}{1 - R^2} \cdot \frac{N - (K + 1)}{K}.
\end{aligned} \tag{9.76}$$

From (9.75) and (9.76) one may show that the F ratio and the coefficient of determination R^2 vary directly. When $R^2 = 0$, then $F = 0$ also. As R^2 becomes larger, F becomes larger also. As R^2 approaches unity, the F ratio approaches infinity. Thus a test of the significance of the regression is also a test of the significance of the coefficient of determination.

The Case of a Single Restriction

In any case where there is only a single restriction in the null hypothesis, the degrees of freedom in the numerator of (9.35) is $p = 1$. Theorem 7.8 of Chapter 7 applies, and the square root of F in (9.35) has a Student's t distribution with $N - (K + 1)$ degrees of freedom. In such cases one may conduct both one-tailed and two-tailed tests, since the Student's t statistic has a symmetric distribution.

In general, the null hypothesis is

$$H0:\ \gamma_0 a_0 + \gamma_1 a_1 + \cdots + \gamma_K a_K = r, \tag{9.77}$$

where $a_0, a_1, \ldots, a_K$ are the parameters of the linear regression model (9.10) and the γ's and r are constants. Suppose we wish to test the null hypothesis against the alternative hypothesis:

$$H1\colon \gamma_0 a_0 + \gamma_1 a_1 + \cdots + \gamma_K a_K \geqq r. \tag{9.78}$$

This requires a one-tailed test as opposed to a two-tailed test. The square root of F in (9.35) will be positive if the unrestricted estimators $\hat{a}_0, \hat{a}_1, \ldots, \hat{a}_K$ satisfy the inequality

$$\gamma_0 \hat{a}_0 + \gamma_1 \hat{a}_1 + \cdots + \gamma_K \hat{a}_K > r. \tag{9.79}$$

We reject the null hypothesis if the positive square root of F is greater than the critical value of the t ratio (or the standard normal variate in the case of large samples). Of course, if the left-hand side of (9.79) is less than r, the square root of F in (9.35) will be negative. In this case we cannot reject the null hypothesis.

Another (equivalent) way of viewing the t-ratio test of a single linear restriction of the form (9.77) is to form the t ratio as follows:

$$t = \frac{(\gamma_0 \hat{a}_0 + \gamma_1 \hat{a}_1 + \cdots + \gamma_K \hat{a}_K) - r}{\text{estimated standard deviation } (\gamma_0 \hat{a}_0 + \gamma_1 \hat{a}_1 + \cdots + \gamma_K \hat{a}_K)}. \tag{9.80}$$

The term in the denominator of this ratio is the estimated standard deviation of $(\gamma \hat{a}_0 + \gamma_1 \hat{a}_1 + \cdots + \gamma_K \hat{a}_K)$. The estimated standard deviation is the square root of the estimated variance, which may be determined from the estimated variances of the individual $\hat{a}_0, \hat{a}_1, \ldots, \hat{a}_K$ *and* the estimated covariances of these estimated parameters.

Let us consider the null hypothesis when only the first three parameters are included in the linear restriction (9.77). Then the variance of $\gamma_0 \hat{a}_0 + \gamma_1 \hat{a}_1 + \gamma_2 \hat{a}_2$ is given by

$$\begin{aligned}
\text{Var}\,(\gamma_0 \hat{a}_0 + \gamma_1 \hat{a}_1 + \gamma_2 \hat{a}_2) &= E(\gamma_0 \hat{a}_0 + \gamma_1 \hat{a}_1 + \gamma_2 \hat{a}_2 - \gamma_0 a_0 - \gamma_1 a_1 - \gamma_2 a_2)^2 \\
&= E[\gamma_0(\hat{a}_0 - a_0) + \gamma_1(\hat{a}_1 - a_1) + \gamma_2(\hat{a}_2 - a_2)]^2 \\
&= \gamma_0^2 E(\hat{a}_0 - a_2)^2 + \gamma_1^2 E(\hat{a}_1 - a_1)^2 + \gamma_2^2 E(\hat{a}_2 - a_2)^2 \\
&\quad + 2\gamma_0\gamma_1 E(\hat{a}_0 - a_0)(\hat{a}_1 - a_1) \\
&\quad + 2\gamma_1\gamma_2 E(\hat{a}_1 - a_1)(\hat{a}_2 - a_2) \\
&\quad + 2\gamma_0\gamma_2 E(\hat{a}_0 - a_0)(\hat{a}_2 - a_2). \tag{9.81}
\end{aligned}$$

In terms of variances and covariances of $\hat{a}_0$, $\hat{a}_1$, and $\hat{a}_2$, we have

$$\begin{aligned}
\text{Var}\,(\gamma_0 \hat{a}_0 + \gamma_1 \hat{a}_1 + \gamma_2 \hat{a}_2) &= \gamma_0 \, \text{Var}\,(a_0) + \gamma_1 \, \text{Var}\,(a_1) + \gamma_2 \, \text{Var}\,(a_2) \\
&\quad + 2\gamma_0\gamma_1 \, \text{Cov}\,(\hat{a}_0 \hat{a}_1) + 2\gamma_1\gamma_2 \, \text{Cov}\,(\hat{a}_1 \hat{a}_2) \\
&\quad + 2\gamma_0\gamma_2 \, \text{Cov}\,(\hat{a}_0 \hat{a}_2). \tag{9.82}
\end{aligned}$$

The formulae for the variances and covariances are quite messy unless matrix algebra is used. In any case, most packaged computer regression programs print out a matrix of variances and covariances of the parameters. Usually then the variance of the linear restriction may be computed from (9.82) or its generalization to an indefinite number K of parameters.

Example 9.22. Let us consider a rather simple hypothesis relating to the consumption function (9.41). In particular let us test the null hypothesis that the marginal propensity to consume is unity against the alternative hypothesis that it is less than unity. The t ratio in (9.80) becomes

$$t = \frac{\hat{b} - 1}{\hat{\sigma}_{\hat{b}}} = \frac{.92 - 1}{.0169} = -4.734,$$

where $\hat{b}$ was estimated in Example 9.16 and the estimated standard deviation of $\hat{b}$ was estimated in Example 9.18 to be $\hat{\sigma}_{\hat{b}} = .0169$. Using a one-tailed test and a 5 percent critical region with 18 degrees of freedom, we find the critical value of t to be -1.734. Since -4.734 is much larger in absolute value than the critical value, we reject the null hypothesis that the marginal propensity to consume is unity.

Example 9.23. In Example 9.14, we indicated that a Cobb-Douglas production function could be estimated as a linear regression equation of the form (9.37). Suppose we wish to test the null hypothesis (9.38) that there are no economies of scale in production as opposed to the alternative hypothesis that there are economies of scale. The alternative hypothesis can be written

$H1: b + c > 1.$

A one-tailed test is appropriate. Suppose the estimated values of $\hat{b}$ and $\hat{c}$, the variance $\hat{\sigma}_{\hat{b}}^2$ of $\hat{b}$ and $\hat{\sigma}_{\hat{c}}^2$ of $\hat{c}$, and the covariance $\hat{\sigma}_{\hat{b}\hat{c}}$ are determined from 50 observations and given as follows:

$\hat{b} = .4, \quad \hat{c} = .9, \quad \hat{\sigma}_{\hat{b}}^2 = .12, \quad \hat{\sigma}_{\hat{c}}^2 = .05, \quad \hat{\sigma}_{\hat{b}\hat{c}} = -.07.$

From (9.82), we can deduce

$\text{Var}(\hat{b} + \hat{c}) = \text{Var}(\hat{b}) + \text{Var}(\hat{c}) + 2\,\text{Cov}(\hat{b}\hat{c}),$

or the estimated variance of $\hat{b} + \hat{c}$ is

$$\begin{aligned}
\hat{\sigma}_{\hat{b}+\hat{c}}^2 &= \hat{\sigma}_{\hat{b}}^2 + \hat{\sigma}_{\hat{b}}^2 + 2\hat{\sigma}_{\hat{b}\hat{c}} \\
&= .12 + .05 - .14 \\
&= .03.
\end{aligned}$$

The estimated standard deviation of $\hat{b} + \hat{c}$ is the square root of .03 or

$\hat{\sigma}_{\hat{b}+\hat{c}} = .175.$

The t ratio in (9.80) is

$$t = \frac{(\hat{b} + \hat{c}) - 1}{\hat{\sigma}_{\hat{b}+\hat{c}}} = \frac{(.4 + .9) - 1}{.175} = \frac{.3}{.175} = 1.714.$$

Since there are 50 observations, one may use the standard normal tables rather than the Student's t distribution. For a one-tailed test with a 5 percent critical region, the critical value of the standard normal variate is 1.64. Since this is smaller than 1.714, we reject the null hypothesis and conclude that there are significant economies of scale.

APPENDIX   Further Discussion

In this appendix we will further expand on a number of points raised in the main text of the chapter. First, we show that if the error terms e_i are normally and independently distributed with zero mean and constant variance, least-squares estimates are maximum-likelihood estimates. Second, we show that least-squares estimators are unbiased. Finally, we provide a proof of the fundamental identity (9.31) for a simple case, which may be extended easily to more complex situations.

For convenience, let us rewrite here the linear regression equation

$$Y_i = a_0 + a_1 X_{1i} + \cdots + a_K X_{Ki} + e_i \tag{9.83}$$

for $i = 1, \ldots, N$. The normal distribution function for e_i is

$$f(e_i) = \frac{1}{\sqrt{2\pi}\,\sigma} \exp\left(-\frac{e_i^2}{\sigma^2}\right). \tag{9.84}$$

The likelihood function for a random sample of size N is

$$
\begin{aligned}
L &= f(e_1, e_2, \ldots, e_N) \\
&= f(e_1) \cdot f(e_2) \cdot \ldots \cdot f(e_N) \\
&= \frac{1}{\sqrt{2\pi}\,\sigma} \exp\left(-\frac{e_1^2}{\sigma^2}\right) \cdot \frac{1}{\sqrt{2\pi}\,\sigma} \exp\left(-\frac{e_2^2}{\sigma^2}\right) \cdot \ldots \cdot \frac{1}{\sqrt{2\pi}\,\sigma} \exp\left(-\frac{e_N^2}{\sigma^2}\right).
\end{aligned}
\tag{9.85}
$$

If we take logarithms of both sides of (9.85), we obtain

$$
\begin{aligned}
\log L &= -N \log\left(\sqrt{2\pi}\,\sigma\right) - \frac{e_1^2}{\sigma^2} - \frac{e_2^2}{\sigma^2} - \cdots - \frac{e_N^2}{\sigma^2} \\
&= -N \log\left(\sqrt{2\pi}\,\sigma\right) - \frac{1}{\sigma^2} \sum_{i=1}^{N} e_i^2.
\end{aligned}
\tag{9.86}
$$

Maximizing $\log L$ is equivalent to maximizing the likelihood function L. For a given sample size N and a given variance σ^2, maximization of $\log L$ is equivalent to minimizing the sum of squares:

$$\sum_{i=1}^{N} e_i^2 = \sum_{i=1}^{N} (Y_i - a_0 - a_1 X_{1i} - \cdots - a_K X_{Ki})^2. \tag{9.87}$$

Next let us prove that least-squares estimators are unbiased for the simple two-variable case, which may be extended to the multivariate case. The regression equation is

$$Y_i = a + b X_i + e_i. \tag{9.88}$$

The least-squares estimate of b is

$$\hat{b} = \frac{\Sigma \, (Y_i - \bar{Y})(X_i - \bar{X})}{\Sigma \, (X_i - \bar{X})^2}. \tag{9.89}$$

For simplicity, let us assume that the X_i are nonrandom. Then the expected value of $\hat{b}$ is

$$E(\hat{b}) = \frac{E[\Sigma \, (Y_i - \bar{Y})(X_i - \bar{X})]}{\Sigma \, (X_i - \bar{X})^2}. \tag{9.90}$$

The denominator of $\hat{b}$ has been factored out, since it is nonrandom and can be treated as a constant. Thus we need only evaluate the numerator of (9.90). Substitute the expression (9.82) for Y_i into the numerator of (9.90).

$$E[\Sigma \, (Y_i - \bar{Y})(X_i - \bar{X})] = E \, [\Sigma \, (a + bX_i + e_i - \bar{Y})(X_i - \bar{X})]. \tag{9.91}$$

By summing over (9.88) and dividing by N, we can derive an expression for $\bar{Y}$.

$$\Sigma \, Y_i = N \cdot a + b \, \Sigma \, X_i + \Sigma \, e_i \tag{9.92}$$

or

$$\bar{Y} = a + b\bar{X} + \bar{e}. \tag{9.93}$$

If we substitute this expression for $\bar{Y}$ into (9.91), we obtain

$$\begin{aligned} E[\Sigma \, (Y_i - \bar{Y})(X_i - \bar{X})] &= E[\Sigma \, (bX_i - b\bar{X} + e_i - \bar{e})(X_i - \bar{X})] \\ &= E[b \, \Sigma \, (X_i - \bar{X})^2 + \Sigma \, (e_i - \bar{e})(X_i - \bar{X})]. \end{aligned} \tag{9.94}$$

Since b and the X variables are nonrandom, (9.94) can be written as

$$E[\Sigma \, (Y_i - \bar{Y})(X_i - \bar{X})] = b \, \Sigma \, (X_i - \bar{X})^2 + \Sigma \, (X_i - \bar{X})E(e_i - \bar{e}). \tag{9.95}$$

If the error term has a zero mean, however, we have

$$E(e_i - \bar{e}) = E(e_i) - E(\bar{e}) = 0. \tag{9.96}$$

Thus (9.95) becomes

$$E[\Sigma \, (Y_i - \bar{Y})(X_i - \bar{X})] = b \, \Sigma \, (X_i - \bar{X})^2. \tag{9.97}$$

Substituting this expression into (9.90), we obtain

$$E(\hat{b}) = b. \tag{9.98}$$

Thus the estimator of b is unbiased.

Finally, let us prove the validity of the fundamental identity in the two-variable case (9.88). We rewrite the fundamental identity:

$$\Sigma (Y_i - Y_i^*)^2 = \Sigma (Y_i - \hat{Y}_i)^2 + \Sigma (\hat{Y}_i - Y_i^*)^2. \tag{9.99}$$

First let us work with the expression on the left-hand side of (9.99). It can be written

$$\begin{aligned}
\Sigma (Y_i - Y_i^*)^2 &= \Sigma (Y_i - \hat{Y}_i + \hat{Y}_i - Y_i^*)^2 \\
&= \Sigma (Y_i - \hat{Y}_i)^2 + 2 \Sigma (Y_i - \hat{Y}_i)(\hat{Y}_i - Y_i^*) \\
&\qquad\qquad + \Sigma (\hat{Y}_i - Y_i^*)^2. \tag{9.100}
\end{aligned}$$

Comparing (9.100) with (9.99) we see that the fundamental identity is proved if we can show that the middle term on the right-hand side of (9.100) is identically equal to zero. That is,

$$\Sigma (Y_i - \hat{Y}_i)(\hat{Y}_i - Y_i^*) = 0. \tag{9.101}$$

Let us rewrite the left-hand side of (9.101) in terms of the restricted estimators a^* and b^* and the nonrestricted estimators $\hat{a}$ and $\hat{b}$.

$$\begin{aligned}
\Sigma (Y_i - \hat{Y}_i)(\hat{Y}_i - Y_i^*) &= \Sigma (Y_i - \hat{a} - \hat{b}X_i)(\hat{a} + \hat{b}X_i - a^* - b^*X_i) \\
&= \Sigma (Y_i - \hat{a} - \hat{b}X_i)(\hat{a} - a^*) \\
&\qquad + \Sigma (Y_i - \hat{a} - \hat{b}X_i)(\hat{b} - b^*)X_i. \tag{9.102}
\end{aligned}$$

We can take $\hat{a} - a^*$ and $\hat{b} - b^*$ out of the summation signs to obtain

$$\begin{aligned}
\Sigma (Y_i - \hat{Y}_i)(\hat{Y}_i - Y_i^*) &= (\hat{a} - a^*)[\Sigma (Y_i - \hat{a} - \hat{b}X_i)] \\
&\quad + (\hat{b} - b^*)[\Sigma (Y_iX_i - \hat{a}X_i - \hat{b}X_i^2)] \\
&= (\hat{a} - a^*)[\Sigma Y_i - N \cdot \hat{a} - \hat{b} \Sigma X_i] \\
&\quad + (\hat{b} - b^*)[\Sigma Y_iX_i - \hat{a} \Sigma X_i - \hat{b} \Sigma X_i^2]. \tag{9.103}
\end{aligned}$$

Now if $\hat{a}$ and $\hat{b}$ are least-squares estimators, they must satisfy the following normal equations.

$$\begin{aligned}
\hat{a} \cdot N + \hat{b} \Sigma X_i &= \Sigma Y_i, \\
\hat{a} \Sigma X_i + \hat{b} \Sigma X_i^2 &= \Sigma Y_iX_i
\end{aligned} \tag{9.104}$$

or

$$\begin{aligned}
\Sigma Y_i - N \cdot \hat{a} - \hat{b} \Sigma X_i &= 0, \\
\Sigma Y_iX_i - \hat{a} \Sigma X_i - \hat{b} \Sigma X_i^2 &= 0. \tag{9.105}
\end{aligned}$$

But this means that both terms on the right-hand side of (9.103) are identically equal to zero. Thus the middle term on the right-hand side of (9.100) is zero, and the fundamental identity (9.99) must hold.

The proof of the fundamental identity is easily generalized to the multi-variate case.

PROBLEMS

1. Consider the following data on Y and X:

Y	X
4	5
7	4
8	2
3	6
3	8

(a) Compute the regression line using the model

$$Y_i = a + b \cdot X_i + e_i.$$

(b) Compute the sum of squared residuals.
(c) Consider the null hypothesis

$$H0: a = 0.$$

Compute the regression line subject to the restriction $a = 0$. [*Hint:* The sum of squares $\Sigma (Y_i - b \cdot X_i)^2$ is minimized. The solution is $b^* = \Sigma Y_i X_i / \Sigma X_i^2$.]
(d) Compute the sum of squared residuals of the restricted regression line.
(e) Using a 5 percent critical region, test the null hypothesis using the relevant F statistic.
(f) Perform the same test using the relevant t ratio.

2. Consider the general regression model

$$Y_i = a_0 + a_1 X_{1i} + \cdots + a_K X_{Ki} + e_i$$

and the null hypothesis

$$H0: a_1 = a_2 = \cdots = a_K.$$

(a) What are the degrees of freedom of the sums of squares S_0, S_1, and S_2?
(b) How could you obtain restricted estimates of the parameters $a_1, \ldots, a_K$?

3. Using the data from problem 1, (a) compute the estimated standard deviation of the error term, (b) determine the estimated standard deviation of the estimated parameter $\hat{b}$, (c) calculate the t ratio and test whether $b = 0$ using a two-tailed 5 percent critical region.

4. Consider the following data on per capita sugar consumption, sugar price, and income per capita for Tanzania.

YEAR	SUGAR CONSUMPTION Y	SUGAR PRICE X_1	INCOME PER CAPITA X_2
1954	9.1	57	16.20
1955	9.8	63	16.68
1956	10.3	70	17.69
1957	10.5	70	17.23
1958	10.2	57	16.59
1959	11.6	62	18.64
1960	12.0	56	17.45
1961	12.7	56	17.51
1962	13.5	60	18.30

(a) Compute the least-squares regression line, $Y = a_0 + a_1X_1 + a_2X_2$.

(b) Test the significance of the regression using a 5 percent critical region.

5. Consider the model

$$Y_i = a_0 + a_1X_{1i} + a_2X_{2i} + a_3X_{3i} + a_4X_{4i}$$

and the null hypothesis

$H0$: $a_0 = 5,$ $2a_0 + 3a_1 = 4,$ $a_0 + 2a_1 + a_3 = 7.$

(a) If there are 20 observations, what are the degrees of freedom of S_0, S_1, and S_2?

(b) How would one estimate the restricted parameters?

Chapter 10

Analysis of Variance and Covariance

In the preceding chapter we discussed tests of hypotheses concerning a general linear model. The theory was then applied to some special cases. This chapter may be viewed as a further discussion of special cases. The first of these refers to models in which the independent variables, X_1, ... , X_K, assume only the value of 0 or 1. This special case is often called the *analysis of variance*.

A more complicated situation arises when some of the values of the independent variables are restricted to be 0 or 1. This is called the *analysis of covariance*.

Analysis of Variance—One-Way Classification

In this section we develop two different approaches to the analysis of variance. The first is called the *ratio-of-variance* approach, the second the *regression* approach. Both arrive at the same result. The first approach relies on Corollary 7.7, discussed in Chapter 7; the second relies on the general theory of linear models developed in Chapter 9.

THE RATIO-OF-VARIANCE APPROACH

Suppose there are K different random variables Y_1, ... , Y_K, which are normally distributed with theoretical means μ_1, ... , μ_K and a common theoretical variance σ^2. Further assume that a sample of N_i observations is taken on the ith random variable, Y_i. The jth observation of the ith random variable is denoted by Y_{ij}. We wish to test the null hypothesis that the means are all equal.

$$H0: \mu_1 = \mu_2 = \cdots = \mu_K = \mu. \tag{10.1}$$

Example 10.1. Suppose that random samples of dairy farmers' incomes are taken in three different states, Wisconsin, Connecticut, and California. A sample of size 21 is taken from Wisconsin, 18 from Connecticut, and 23 from California. The sample incomes of dairy farms in Wisconsin is denoted by $Y_{1,1}$, $Y_{1,2}$, ... , $Y_{1,21}$, in Connecticut by $Y_{2,1}$, $Y_{2,2}$, ... , $Y_{2,18}$, and in California by $Y_{3,1}$, $Y_{3,2}$, ... , $Y_{3,23}$. We wish to use these data to test the hypothesis that average farmers' incomes are the same in each of the three states.

A reasonable first step in the testing of this hypothesis is to calculate the sample means within each of the K groups of observations. The sample mean of the ith group, denoted $\bar{Y}_i$, is given by

$$\bar{Y}_i = \frac{\sum_{j=1}^{N_i} Y_{ij}}{N_i}. \tag{10.2}$$

The grand sample mean $\bar{Y}$ is the sum of all observations in all groups divided by the total number of observations.

$$\bar{Y} = \frac{\sum_{j=1}^{N_1} Y_{1j} + \sum_{j=1}^{N_2} Y_{2j} + \cdots + \sum_{j=1}^{N_K} Y_{Kj}}{(N_1 + N_2 + \cdots + N_K)}. \tag{10.3}$$

Another way of calculating this grand mean is to take a weighted average of the group means, where the weights are the number of observations in each group.

$$\bar{Y} = \frac{N_1\bar{Y}_1 + N_2\bar{Y}_2 + \cdots + N_K\bar{Y}_K}{(N_1 + N_2 + \cdots + N_K)}. \tag{10.4}$$

By substitution of (10.2) into (10.4), it is simple to show that (10.3) and (10.4) are equivalent.

Example 10.2. Suppose a survey of current expenditures in a number of school districts in three states reveals the following data on expenditures per pupil. The sum of observations in each group is divided by the number of observations in each group to obtain the three group mean expenditures per pupil.

NEW JERSEY	OHIO	MISSISSIPPI
$Y_{11} = 940$	$Y_{21} = 830$	$Y_{31} = 543$
$Y_{12} = 735$	$Y_{22} = 656$	$Y_{32} = 198$
$Y_{13} = 610$	$Y_{23} = 324$	$Y_{33} = 687$
$Y_{14} = 1035$	$Y_{24} = 783$	$Y_{34} = 498$
$Y_{15} = 916$	$Y_{25} = 597$	$Y_{35} = 433$
$Y_{16} = 821$	$Y_{26} = 348$	$Y_{36} = 397$
$Y_{17} = 715$	$Y_{27} = 698$	$Y_{37} = 416$
$Y_{18} = 843$	$Y_{28} = 549$	$Y_{38} = 587$
	$Y_{29} = 683$	$Y_{39} = 469$
	$Y_{2,10} = 715$	$Y_{3,10} = 476$
	$Y_{2,11} = 732$	

New Jersey:
$$\bar{Y}_1 = \frac{\sum_{j=1}^{8} Y_{1j}}{8} = \frac{6,615}{8} = 826.9$$

Ohio:
$$\bar{Y}_2 = \frac{\sum_{j=1}^{11} Y_{2j}}{11} = \frac{6,915}{11} = 628.6 \qquad (10.5)$$

Mississippi:
$$\bar{Y}_3 = \frac{\sum_{j=1}^{10} Y_{3j}}{10} = \frac{4,704}{10} = 470.4.$$

The grand mean expenditure per pupil is the weighted sum,

$$\bar{Y} = \frac{8 \cdot \bar{Y}_1 + 11 \cdot \bar{Y}_2 + 10 \cdot \bar{Y}_3}{8 + 11 + 10} = \frac{18,234.0}{29} = 628.8. \qquad (10.6)$$

The sample variances of observations within each group are also useful statistics. The sample variance *within* the ith group is denoted by s_i^2 and

$$s_i^2 = \frac{\sum_{j=1}^{N_i} (Y_{ij} - \bar{Y}_i)^2}{N_i}. \qquad (10.7)$$

A test of the null hypothesis also requires us to calculate the statistic s_Y, the sample variance of group means about the grand mean, defined by

$$s_Y^2 = \frac{\sum_{i=1}^{K} N_i \cdot (\bar{Y}_i - \bar{Y})^2}{N}. \qquad (10.8)$$

The statistic s_Y^2 is often called the sample variance *among* groups.

Finally, we compute an F statistic that is a function of the ratio of the sample variance among groups to a weighted sum of the within group sample variances.

$$F = \frac{N \cdot s_{\bar{Y}}^2/(K - 1)}{(N_1 s_1^2 + N_2 s_2^2 + \cdots + N_K s_K^2)/(N - K)}. \tag{10.9}$$

The numerator of this F ratio is an unbiased estimate of the theoretical variance σ^2, which is assumed to be equal in all groups. The denominator of F, which is a kind of weighted average of the variances within groups, is also an unbiased estimate of σ^2. Both these estimators are independently distributed. As the ratio of independent unbiased estimates of variance of normally distributed random variables, the F ratio has an F distribution with $K - 1$ degrees of freedom in the numerator and $N - K$ degrees of freedom in the denominator (see Corollary 7.7 of Chapter 7).

Now if the null hypothesis is true, the theoretical means for all groups are the same, and we would expect that the sample means within groups would be quite similar. That is, we would expect the variance of the group means about the grand mean, $s_{\bar{Y}}^2$, to be small relative to the variance within groups. Thus we would expect the numerator in (10.9) to be small relative to the denominator and the F statistic to be small if the null hypothesis is true. A one-tailed test then seems appropriate, and a large value of F would result in rejection of the hypothesis. The critical value of F, of course, depends on the size of the critical region chosen.

Example 10.3.   From Example 10.2, we calculated the sample means within groups to be

$$\bar{Y}_1 = 826.9, \qquad \bar{Y}_2 = 628.6, \qquad \bar{Y}_3 = 470.4.$$

The grand mean is $\bar{Y} = 628.8$. The variance among groups then is

$$s_{\bar{Y}}^2 = \frac{8 \cdot (\bar{Y}_1 - \bar{Y})^2 + 11 \cdot (\bar{Y}_2 - \bar{Y})^2 + 10 \cdot (\bar{Y}_3 - \bar{Y})^2}{29} = \frac{564,854}{29}$$

$$= 19,477.7.$$

The sample variances within groups are

$$s_1^2 = \frac{\sum_{j=1}^{8} (Y_{1j} - \bar{Y}_1)^2}{8} = \frac{132,343}{8} = 16,542.9,$$

$$s_2^2 = \frac{\sum_{j=1}^{11} (Y_{2j} - Y_2)^2}{11} = \frac{269,937}{11} = 24,539.7, \tag{10.10}$$

$$s_3^2 = \frac{\sum_{j=1}^{10} (Y_{3j} - Y_3)^2}{10} = \frac{150,524}{10} = 15,052.4.$$

Thus the F statistic is

$$F = \frac{(29 \cdot 19{,}477.7)/(3-1)}{(8 \cdot 16{,}542.9 + 11 \cdot 24{,}539.7 + 10 \cdot 15{,}052.4)/(29-3)} = 13.283$$

(10.11)

with 2 degrees of freedom in the numerator and 26 degrees of freedom in the denominator. If we choose a 5 percent critical region, the critical value of F is approximately 3.331. Since F as calculated in (10.11) is greater than this critical value, we reject the null hypothesis and conclude that there is a significant difference in per pupil expenditures among the three states.

The test we have just performed could be more powerful if we took into account the number of pupils in each school district sampled. Each observation would have to be weighted by the number of pupils per district, and the test statistics would have to be appropriately modified.

THE REGRESSION APPROACH

Another way of looking at the test above is to postulate that the jth observation on the ith random variable Y_i is composed of a systematic part, the theoretical mean μ_i, and an error component e_{ij}.

$$Y_{ij} = \mu_i + e_{ij} \qquad \text{for } i = 1, \ldots, K, j = 1, \ldots, N_i.$$

(10.12)

The error term e_{ij} is assumed to have a zero mean and variance σ^2. Let us now define a new random variable Y_t, a set of K independent variables $X_{1t}, \ldots, X_{Kt}$, and a new error term e_t in the following way:

$$
\begin{aligned}
Y_t &= Y_{1,t} & &\text{for } t = 1, \ldots, N_1, \\
Y_t &= Y_{2,t-N_1} & &\text{for } t = N_1 + 1, \ldots, N_1 + N_2,
\end{aligned}
$$

.

.

.

$$
\begin{aligned}
Y_t &= Y_{K,t-N+N_K} & &\text{for } t = N - N_K + 1, \ldots, N, \\[4pt]
X_{1t} &= 1 & &\text{for } t = 1, \ldots, N_1, \\
X_{1t} &= 0 & &\text{for } t = N_1 + 1, \ldots, N, \\[4pt]
X_{2t} &= 0 & &\text{for } t = 1, \ldots, N_1, \\
X_{2t} &= 1 & &\text{for } t = N_1 + 1, \ldots, N_1 + N_2, \\
X_{2t} &= 0 & &\text{for } t = N_1 + N_2 + 1, \ldots, N,
\end{aligned}
$$

(10.13)

.

.

.

$$
\begin{aligned}
X_{Kt} &= 0 & &\text{for } t = 1, \ldots, N - N_K, \\
X_{Kt} &= 1 & &\text{for } t = N - N_K + 1, \ldots, N,
\end{aligned}
$$

where $N = N_1 + N_2 + \cdots + N_K$. The independent variables assume only the values of 0 or 1. Using these undefined variables, we can rewrite (10.12) as

$$Y_t = \mu_1 X_{1t} + \mu_2 X_{2t} + \cdots + \mu_K X_{Kt} + e_t \qquad \text{for } t = 1, \ldots, N. \qquad (10.14)$$

That is, for $t = 1, \ldots, N_1$, only $X_{1t} = 1$ and all the other independent variables are equal to zero. Thus if t is somewhere between 1 and N_1, we have

$$Y_t = \mu_1 + e_t. \qquad (10.15)$$

Thus (10.14) and (10.12) are equivalent for t between 1 and N_1. Similarly, the X's are set so that they extract the relevant equations from (10.12) for all other values of t.

Example 10.4.   Suppose there are three groups with five observations on Y_1, the random variable from group 1, four observations on Y_2, and three observations on Y_3. The observations on the Y's are transformed, and we define the observations on the independent variables X_1, X_2, and X_3 as follows:

$Y_1 = Y_{1,1}$	$X_{1,1} = 1$	$X_{2,1} = 0$	$X_{3,1} = 0$
$Y_2 = Y_{1,2}$	$X_{1,2} = 1$	$X_{2,2} = 0$	$X_{3,2} = 0$
$Y_3 = Y_{1,3}$	$X_{1,3} = 1$	$X_{2,3} = 0$	$X_{3,3} = 0$
$Y_4 = Y_{1,4}$	$X_{1,4} = 1$	$X_{2,4} = 0$	$X_{3,4} = 0$
$Y_5 = Y_{1,5}$	$X_{1,5} = 1$	$X_{2,5} = 0$	$X_{3,5} = 0$
$Y_6 = Y_{2,1}$	$X_{1,6} = 0$	$X_{2,6} = 1$	$X_{3,6} = 0$
$Y_7 = Y_{2,2}$	$X_{1,7} = 0$	$X_{2,7} = 1$	$X_{3,7} = 0$
$Y_8 = Y_{2,3}$	$X_{1,8} = 0$	$X_{2,8} = 1$	$X_{3,8} = 0$
$Y_9 = Y_{2,4}$	$X_{1,9} = 0$	$X_{2,9} = 1$	$X_{3,9} = 0$
$Y_{10} = Y_{3,1}$	$X_{1,10} = 0$	$X_{2,10} = 0$	$X_{3,10} = 1$
$Y_{11} = Y_{3,2}$	$X_{1,11} = 0$	$X_{2,11} = 0$	$X_{3,11} = 1$
$Y_{12} = Y_{3,3}$	$X_{1,12} = 0$	$X_{2,12} = 0$	$X_{3,12} = 1$

$$(10.16)$$

Then the linear model is

$$Y_t = \mu_1 X_{1t} + \mu_2 X_{2t} + \mu_3 X_{3t} + e_t. \qquad (10.17)$$

Because of the way in which the X's are defined, this becomes

$$
\begin{aligned}
Y_t &= \mu_1 + e_t && \text{for } t = 1, \ldots, 5, \\
Y_t &= \mu_2 + e_t && \text{for } t = 6, \ldots, 9, \\
Y_t &= \mu_3 + e_t && \text{for } t = 10, \ldots, 12.
\end{aligned}
\qquad (10.18)
$$

The first five observations ($t = 1, \ldots, 5$) are assumed to come from group 1 with a theoretical mean μ_1; the next four observations ($t = 6, \ldots, 9$) are assumed to come from group 2 with a mean μ_2; and the last three observations from group 3 with a mean μ_3.

The linear model in (10.14) is equivalent to the general linear model (9.10) discussed in the last chapter. The parameters a_1, a_2, ... , a_K correspond to the theoretical means μ_1, μ_2, ... , μ_K. The only difference is that in this chapter we assume that $a_0 = 0$. It is easy to show, however, that this makes only a slight difference in the analysis and is quite simple to take into account.

The parameters μ_1, ... , μ_K in (10.14) may be estimated by the least-squares technique. Not surprisingly, it is possible (though messy without matrix algebra) to show that the least-squares estimate of the parameter μ_i, the theoretical mean of Y in the ith group, is the sample mean $\bar{Y}_i$ in the ith group—that is,

$$\hat{\mu}_i = \bar{Y}_i = \frac{\sum_{j=1}^{N_i} Y_{ij}}{N_i} \qquad \text{for } i = 1, \dots, K. \tag{10.19}$$

If we ignore the error term and substitute the estimated values of the μ_i parameters into (10.14), we obtain a set of predicted values for Y_t.

$$\hat{Y}_t = \hat{\mu}_1 X_{1t} + \hat{\mu}_2 X_{2t} + \cdots + \hat{\mu}_K X_{Kt} \qquad \text{for } t = 1, \dots, N. \tag{10.20}$$

Because of the way in which the X's are defined, the predicted values within each of the K groups are

$$\hat{Y}_t = \hat{\mu}_1 = \bar{Y}_1 \qquad \text{for } t = 1, \dots, N_1,$$
$$\hat{Y}_t = \hat{\mu}_2 = \bar{Y}_2 \qquad \text{for } t = N_1 + 1, \dots, N_1 + N_2,$$

$$\tag{10.21}$$

$$\hat{Y}_t = \hat{\mu}_K = \bar{Y}_K \qquad \text{for } t = N_1 + \cdots + N_{K-1} + 1, \dots, N.$$

If the null hypothesis (10.1) is true, all the theoretical means are equal. Under the null hypothesis (10.14) becomes

$$Y_t = \mu X_{1t} + \mu X_{2t} + \cdots + \mu X_{Kt} + e_t \tag{10.22}$$
$$= \mu(X_{1t} + X_{2t} + \cdots + X_{Kt}) + e_t,$$

where μ is the common theoretical mean. Since only one of the X's is equal to unity for any specific value of t and the rest of the X's are equal to zero, the composite variable $(X_{1t} + X_{2t} + \cdots + X_{Kt}) = 1$ for all values of t. Thus we may write (10.22) as

$$Y_t = \mu X_t + e_t, \tag{10.23}$$

where X_t is a new variable which assumes the value of unity for all t. The parameter μ may be estimated by least-squares techniques. The estimated value of μ, as one might expect, is the grand sample mean.

$$\mu^* = \frac{\sum\limits_{t=1}^{N} Y_t}{N} = \bar{Y}. \tag{10.24}$$

The predicted value of Y_t in all groups then is

$$Y_t^* = \mu^* = \bar{Y} \qquad \text{for } t = 1, \ldots, N. \tag{10.25}$$

From Chapter 9 we have the fundamental identity:

$$\sum_{t=1}^{N} (Y_t - Y_t^*)^2 = \sum_{t=1}^{N} (Y_t - \hat{Y}_t)^2 + \sum_{t=1}^{N} (\hat{Y}_t - Y_t^*)^2. \tag{10.26}$$

From (10.21) and (10.25), this identity can be written

$$\sum_{t=1}^{N} (Y_t - \bar{Y})^2 = \sum_{t=1}^{N} (Y_t - \bar{Y}_1)^2 + \cdots + \sum_{t=N-N_K+1}^{N} (Y_t - \bar{Y}_K)^2$$
$$+ N_1(\bar{Y}_1 - \bar{Y})^2 + \cdots + N_K(\bar{Y}_K - \bar{Y})^2. \tag{10.27}$$

In terms of our original group notation (10.27) may be written as

$$\left[\sum_{j=1}^{N_1} (Y_{1j} - \bar{Y})^2 + \cdots + \sum_{j=1}^{N_K} (Y_{Kj} - \bar{Y})^2 \right]$$
$$= \left[\sum_{j=1}^{N_1} (Y_{1j} - \bar{Y}_1)^2 + \cdots + \sum_{j=1}^{N_K} (Y_{Kj} - \bar{Y}_K)^2 \right] \tag{10.28}$$
$$+ [N_1(Y_1 - Y)^2 + \cdots + N_K(Y_K - \bar{Y})^2].$$

In analysis of variance the first term in brackets on the right-hand side of (10.28) is called the variation *within* groups; the second term in brackets is called the variation *among* groups. The first term has a chi-square distribution with $N - K$ degrees of freedom, since it is the sum of squared residuals resulting when K parameters $\mu_1, \ldots, \mu_K$ are estimated from N observations.

The degrees of freedom of the second term in brackets on the right-hand side of (10.28) can be determined as follows. The term on the *left*-hand side is the sum of squared residuals resulting when one parameter, the common mean μ, is estimated from a sample of size N. Therefore its degrees of freedom is $N - 1$. Since the sum of degrees of freedom on the left-hand side must equal those on the right-hand side, the degrees of freedom of the second term on the right-hand side denoted by d can be determined from the following equation:

$$(N - 1) = (N - K) + d \quad \text{or} \quad d = K - 1. \tag{10.29}$$

Thus the statistic

$$F = \frac{[N_1(\bar{Y}_1 - \bar{Y})^2 + \cdots + N_K(\bar{Y}_K - \bar{Y})^2]/(K - 1)}{\left[\sum_{j=1}^{N_1} (Y_{1j} - \bar{Y}_1)^2 + \cdots + \sum_{j=1}^{N_K} (Y_{Kj} - \bar{Y}_K)^2\right]/(N - K)} \qquad (10.30)$$

has an F distribution with $K - 1$ degrees of freedom in the numerator and $N - K$ degrees of freedom in the denominator.

Example 10.5. Suppose the following observations are obtained on three different groups.

$Y_{11} = 1.5$	$Y_{21} = 5.0$	$Y_{31} = 3.0$
$Y_{12} = 3.0$	$Y_{22} = 8.5$	$Y_{32} = 2.5$
$Y_{13} = 1.5$	$Y_{23} = 3.0$	$Y_{33} = 3.5$
	$Y_{24} = 7.0$	$Y_{34} = 3.0$
$\sum_{j=1}^{3} Y_{1j} = 6.0$	$Y_{25} = 6.5$	
	$\sum_{j=1}^{5} Y_{2j} = 30.0$	$\sum_{j=1}^{4} Y_{3j} = 12.0$

The three sample means are

$$\bar{Y}_1 = \frac{6.0}{3} = 2.0, \qquad \bar{Y}_2 = 6.0, \qquad \bar{Y}_3 = 3.0.$$

The grand mean is

$$\bar{Y} = \frac{6.0 + 30.0 + 12.0}{12} = 4.0.$$

The numerator of F in (10.30) is

$$[3(2 - 4)^2 + 5(6 - 4)^2 + 4(3 - 4)^2]/2 = [12 + 20 + 4]/2 = 36/2 = 18.0.$$

The denominator is

$$[1.5 + 17.5 + .5]/(12 - 3) = 19.5/9 = 2.17.$$

The F ratio is

$$F = \frac{18.00}{2.17} = 8.29$$

with 2 degrees of freedom in the numerator and 9 degrees of freedom in the denominator. From Table D of the Appendix of Statistical Tables the critical value of F with a 5 percent critical region is 4.257. Since 7.78 is greater than 4.257, we reject the null hypothesis that the means of the three groups are equal.

Note that the F ratio in (10.30) and the F ratio proposed in (10.9) are equivalent. That is, if we substitute (10.7) and (10.8) into (10.9), we obtain the ratio (10.30). Thus the two tests are equivalent.

Finally, note that if there are only two groups ($K = 2$), then the numerator of F in (10.30) has $K - 1 = 2 - 1 = 1$ degree of freedom. In this case we are testing the difference of two means μ_1 and μ_2, and the F test reduces to a Student's t test. This special case was discussed in Chapter 7.

Analysis of Variance—Two-Way Classification

Suppose samples of average salaries of computer programmers are drawn in a number of different cities. Each of the city samples, however, contains a number of male programmers as well as female programmers. The sample may be divided into cells on the basis of a two-way classification, by city and by sex. If there are three cities, the two-way classification of cells may be illustrated as in Fig. 10.1. The first row refers to male programmers. The

1,1	1,2	1,3
2,1	2,2	2,3

FIGURE 10.1

second row refers to female programmers. The three columns refer to each of the three cities. For example, salaries of female programmers in the third city are entered in the cell labeled 2, 3 in Fig. 10.1.

Figure 10.2 refers to the number of observations in each cell. For example, N_{12} is the number of observations of salaries of male computer programmers in the second city.

N_{11}	N_{12}	N_{13}
N_{21}	N_{22}	N_{23}

FIGURE 10.2

A two-way classification enables us to test whether there are significant differences in salaries among cities as well as between male and female computer programmers. Tests of this sort may be performed with the following model:

$$Y_{ijk} = \mu + a_i + b_j + e_{ijk}, \tag{10.31}$$

where Y_{ijk} is the kth observation in cell i, j. a_i is called the cell row effect and b_j the cell column effect. μ is the theoretical grand mean. μ, a_i, and b_i are fixed parameters, while e_{ijk} is a random-error term that has a zero mean and constant theoretical variance σ^2. In terms of our two-way classification by cities and sex, b_j represents the city differential and a_i the sex differential.

The model (10.31) may be converted to a linear model of the form (9.10) discussed in the previous chapter by an appropriate definition of independent variables, which assume the value of 0 or 1. The details of this conversion are not important here; suffice it to say that in principle it can be done. If the conversion is made, the sum of squared deviations can be minimized as before. For simplicity let us assume that the number of observations in each cell is the same; that is, $N_{ij} = N$ for all i and all j. Then a set of least-squares estimates of the parameters μ, a_i, and b_j are

$$\hat{\mu} = \bar{Y},$$

$$\hat{a}_1 = \frac{\bar{Y}_{11} + \bar{Y}_{12} + \bar{Y}_{13}}{3} - \bar{Y} = \bar{Y}_{1\cdot} - \bar{Y},$$

$$\hat{a}_2 = \frac{\bar{Y}_{21} + \bar{Y}_{22} + \bar{Y}_{23}}{3} - \bar{Y} = \bar{Y}_{2\cdot} - \bar{Y},$$

$$\hat{b}_1 = \frac{\bar{Y}_{11} + \bar{Y}_{21}}{2} - \bar{Y} = \bar{Y}_{\cdot 1} - \bar{Y}, \qquad (10.32)$$

$$\hat{b}_2 = \frac{\bar{Y}_{12} + \bar{Y}_{22}}{2} - \bar{Y} = \bar{Y}_{\cdot 2} - \bar{Y},$$

$$\hat{b}_3 = \frac{\bar{Y}_{13} + \bar{Y}_{23}}{2} - \bar{Y} = \bar{Y}_{\cdot 3} - \bar{Y},$$

where $\bar{Y}_{ij}$ is the mean of observations in cell i, j; and $\bar{Y}$ is the grand mean of all observations. The mean of all observations in the first row is $\bar{Y}_{1\cdot\cdot}$. $\bar{Y}_{2\cdot}$ is the mean of all observations in the second row. Similarly, $\bar{Y}_{\cdot 1}$, $\bar{Y}_{\cdot 2}$, and $\bar{Y}_{\cdot 3}$ are the means of all observations in columns 1, 2, and 3, respectively.

The predicted values of Y_{ijk} differ by cell. From (10.31) and (10.32) they are

$$\begin{aligned}
\hat{Y}_{11k} &= \hat{\mu} + \hat{a}_1 + \hat{b}_1 = \bar{Y}_{1\cdot} + \bar{Y}_{\cdot 1} - \bar{Y} && \text{for } k = 1, \ldots, N, \\
\hat{Y}_{12k} &= \hat{\mu} + \hat{a}_1 + \hat{b}_2 = \bar{Y}_{1\cdot} + \bar{Y}_{\cdot 2} - \bar{Y} && \text{for } k = 1, \ldots, N, \\
\hat{Y}_{13k} &= \hat{\mu} + \hat{a}_1 + \hat{b}_3 = \bar{Y}_{1\cdot} + \bar{Y}_{\cdot 3} - \bar{Y} && \text{for } k = 1, \ldots, N, \\
\hat{Y}_{21k} &= \hat{\mu} + \hat{a}_2 + \hat{b}_1 = \bar{Y}_{2\cdot} + \bar{Y}_{\cdot 1} - \bar{Y} && \text{for } k = 1, \ldots, N, \\
\hat{Y}_{22k} &= \hat{\mu} + \hat{a}_2 + \hat{b}_2 = \bar{Y}_{2\cdot} + \bar{Y}_{\cdot 2} - \bar{Y} && \text{for } k = 1, \ldots, N, \\
\hat{Y}_{23k} &= \hat{\mu} + \hat{a}_2 + \hat{b}_3 = \bar{Y}_{2\cdot} + \bar{Y}_{\cdot 3} - \bar{Y} && \text{for } k = 1, \ldots, N.
\end{aligned} \qquad (10.33)$$

The sum of squared residuals is

$$S_1 = \sum_{k=1}^{N} (Y_{11k} - \hat{\mu} - \hat{a}_1 - \hat{b}_1)^2 + \sum_{k=1}^{N} (Y_{12k} - \hat{\mu} - \hat{a}_1 - \hat{b}_2)^2$$

$$+ \sum_{k=1}^{N} (Y_{13k} - \hat{\mu} - \hat{a}_1 - \hat{b}_3)^2 + \sum_{k=1}^{N} (Y_{21k} - \hat{\mu} - \hat{a}_2 - \hat{b}_1)^2$$

$$+ \sum_{k=1}^{N} (Y_{22k} - \hat{\mu} - \hat{a}_2 - \hat{b}_2)^2 + \sum_{k=1}^{N} (Y_{23k} - \hat{\mu} - \hat{a}_2 - \hat{b}_3)^2. \qquad (10.34)$$

This sum has a chi-square distribution. Since six parameters, μ, a_1, a_2,

b_1, b_2, b_3, are being estimated, one might think that S_1 has $6 \cdot N - 6$ degrees of freedom, where $6 \cdot N$ is the total number of observations in all cells. In fact, however, not all the parameter estimates are independent. If we sum $\hat{a}_1$ and $\hat{a}_2$, and $\hat{b}_1$, $\hat{b}_2$, and $\hat{b}_3$, we obtain

$$\hat{a}_1 + \hat{a}_2 = 0, \tag{10.35}$$

$$\hat{b}_1 + \hat{b}_2 + \hat{b}_3 = 0. \tag{10.36}$$

That is, if we know $\hat{a}_1$, then from (10.35) we can determine $\hat{a}_2$. Similarly if we know $\hat{b}_1$ and $\hat{b}_2$, then $\hat{b}_3$ can be determined from (10.36). In effect, two of the six parameters are known once the four others are specified. Thus S_1 in (10.34) has $6 \cdot N - 4$ degrees of freedom.

In general, for any two-way classification for which there are K_1 rows of cells and K_2 columns, there are $K_1 \cdot K_2 \cdot N - (K_1 + K_2) + 1$ degrees of freedom. This derives from the fact that the weighted sum of the row effects and the column effects are both zero. Two parameters are not independent of the rest.

If we wish to test whether the mean salary of computer programmers in all three cities is the same, except for possible male-female differentials, this is equivalent to the null hypothesis that all the city effects, b_1, b_2, and b_3, are the same—that is,

$$H0: b_1 = b_2 = b_3 = b. \tag{10.37}$$

If the null hypothesis is true, we can restrict the parameters b_1, b_2, and b_3 to satisfy (10.33) and estimate the row effects a_1 and a_2, the common column effect b, and the theoretical grand mean μ. A set of least-squares estimates of these parameters is

$$\mu^* + b^* = \bar{Y},$$

$$a_1^* = \hat{a}_1 = \frac{\bar{Y}_{11} + \bar{Y}_{12} + \bar{Y}_{13}}{3} - \bar{Y} = \bar{Y}_{1.} - \bar{Y},$$

$$a_2^* = \hat{a}_2 = \frac{\bar{Y}_{21} + \bar{Y}_{22} + \bar{Y}_{23}}{3} - \bar{Y} = \bar{Y}_{2.} - \bar{Y}. \tag{10.38}$$

The predicted values of Y_{ijk} differ only by the cell row. That is, the predicted value of Y_{ijk} is the same for all cells in a given row. The predicted values are

$$\hat{Y}_{11k} = \hat{Y}_{12k} = \hat{Y}_{13k} = \mu^* + b^* + a_1^* = \bar{Y}_{1.},$$

$$\hat{Y}_{21k} = \hat{Y}_{22k} = \hat{Y}_{23k} = \mu^* + b^* + a_2^* = \bar{Y}_{2..} \tag{10.39}$$

The sum of the squared residuals is

$$S_0 = \sum_{k=1}^{N} (Y_{11k} - \mu^* - b^* - a_1^*)^2 + \sum_{k=1}^{N} (Y_{12k} - \mu^* - b^* - a_1^*)^2$$

$$+ \sum_{k=1}^{N} (Y_{13k} - \mu^* - b^* - a_1^*)^2 + \sum_{k=1}^{N} (Y_{21k} - \mu^* - b^* - a_2^*)^2$$

$$+ \sum_{k=1}^{N} (Y_{22k} - \mu^* - b^* - a_2^*)^2 + \sum_{k=1}^{N} (Y_{23k} - \mu^* - b^* - a_2^*)^2. \qquad (10.40)$$

This sum S_0 also has a chi-square distribution. There are four estimated parameters, μ, b, a_1, and a_2. μ and b, however, are not really distinguishable as separate parameters; rather the sum $\mu + b$ is estimated as in (10.34). Furthermore, the two estimators a_1^* and a_2^* satisfy the following linear relationship:

$$a_1^* + a_2^* = 0. \qquad (10.41)$$

Thus there are, in effect, only two independent parameters being estimated. The sum of squares S_0 in (10.40), therefore, has $6 \cdot N - 2$ degrees of freedom.

In general, for any two-way classification in which one set of effects, say the column effects, is considered to be identical, the sum of squares S_0 has $K_1 \cdot K_2 \cdot N - K_1$ degrees of freedom, where K_1 is the number of rows of cells.

The difference in the sum of squares—that is,

$$S_2 = S_0 - S_1 \qquad (10.42)$$

—also has a chi-square distribution. The number of degrees of freedom is the difference in the degrees of freedom in S_0 and S_1. In the general case the number of degrees of freedom is

$$[K_1 \cdot K_2 \cdot N - K_1] - [K_1 \cdot K_2 \cdot N - (K_1 + K_2) + 1] = K_2 - 1. \qquad (10.43)$$

In the special case where the number of rows $K_1 = 2$ and the number of columns is $K_2 = 3$ the number of degrees of freedom is $3 - 1 = 2$.

The F ratio then is

$$F = \frac{S_2}{S_1} \cdot \frac{[K_1 \cdot K_2 \cdot N - (K_1 + K_2) + 1]}{[K_2 - 1]}, \qquad (10.44)$$

which has an F distribution with $K_2 - 1$ degrees of freedom in the numerator and $K_1 \cdot K_2 \cdot N - (K_1 + K_2) + 1$ degrees of freedom in the denominator. The general formulae for the sums of squares are:

$$S_0 = \sum_{i=1}^{K_1} \sum_{j=1}^{K_2} \sum_{k=1}^{N} (Y_{ijk} - \mu^* - b^* - a_i^*)^2 = \sum_{i,j,k} (Y_{ijk} - \bar{Y}_{i\cdot})^2,$$

$$S_1 = \sum_{i=1}^{K_1} \sum_{j=1}^{K_2} \sum_{k=1}^{N} (Y_{ijk} - \hat{\mu} - \hat{a}_i \hat{b}_j)^2$$

$$= \sum_{i,j,k} (Y_{ijk} - \bar{Y}_{i\cdot} - \bar{Y}_{\cdot j} + \bar{Y})^2, \qquad (10.45)$$

$$S_2 = S_0 - S_1 = \sum_{i,j,k} (\bar{Y} - \bar{Y}_{\cdot j})^2,$$

where the restricted estimators are given by

$$\hat{\mu}^* + \hat{b}^* = \bar{Y},$$

$$a_i^* = \frac{\sum_{j=1}^{K_2} \bar{Y}_{ij}}{K_1} - \bar{Y} = \bar{Y}_{i\cdot} - \bar{Y} \qquad (10.46)$$

and the unrestricted estimators are given by

$$\hat{\mu} = \bar{Y}$$

$$\hat{a}_i = \frac{\sum_{j=1}^{K_2} \bar{Y}_{ij}}{K_2} - \bar{Y} = \bar{Y}_{i\cdot} - \bar{Y} \qquad \text{for } i = 1, \ldots, K_1$$

$$\qquad (10.47)$$

$$\hat{b}_j = \frac{\sum_{i=1}^{K_1} \bar{Y}_{ij}}{K_1} - \bar{Y} = \bar{Y}_{\cdot j} - \bar{Y} \qquad \text{for } j = 1, \ldots, K_2.$$

The grand mean of all cells is denoted by $\bar{Y}$; $\bar{Y}_{ij}$ represents the mean of observations in cell i, j; $\bar{Y}_{i\cdot}$ is the ith row mean; and $\bar{Y}_{\cdot j}$ is the jth column mean.

Example 10.6. Suppose there are two rows of cells and three columns of cells as shown in the following table:

4, 3, 2	2, 3, 1	5, 7, 9
8, 4, 3	4, 3, 2	9, 7, 5

Let us test the hypothesis that there are no significant differences in the columns. The cell means are entered in the corresponding cells.

3.0	2.0	7.0	4.0
5.0	3.0	7.0	5.0
4.0	2.5	7.0	

The two row means are $Y_1. = 4.0$ and $Y_2. = 5.0$, while the three column means are $Y._1 = 4.0$, $Y._2 = 2.5$, and $Y._3 = 7.0$. The grand mean of all observations is $\bar{Y} = 4.5$.

From (10.47), we have the unrestricted least-squares estimators

$\hat{\mu} = \bar{Y} = 4.5$,
$\hat{a}_1 = \bar{Y}_1. - \bar{Y} = 4.0 - 4.5 = -.5$,
$\hat{a}_2 = \bar{Y}_2. - \bar{Y} = 5.0 - 4.5 = .5$,
$\hat{b}_1 = \bar{Y}._1 - \bar{Y} = 4.0 - 4.5 = -.5$,
$\hat{b}_2 = \bar{Y}._2 - \bar{Y} = 2.5 - 4.5 = -2.0$,
$\hat{b}_3 = \bar{Y}._3 - \bar{Y} = 7.0 - 4.5 = 2.5$.

Let us calculate the sum of squares S_1 in (10.45) by writing in each cell the sum of squared deviations for each cell.

$(4 - 4.5 + .5 + .5)^2$ $+ (3 - 4.5 + .5 + .5)^2$ $+ (2 - 4.5 + .5 + .5)^2$ $= 2.75$	$(2 - 4.5 + .5 + 2.0)^2$ $+ (3 - 4.5 + .5 + 2.0)^2$ $+ (3 - 4.5 + .5 + 2.0)^2$ $= 2.0$	$(5 - 4.5 + .5 - 2.5)^2$ $+ (7 - 4.5 + .5 - 2.5)^2$ $+ (9 - 4.5 + .5 - 2.5)^2$ $= 8.75$
$(8 - 4.5 - .5 + .5)^2$ $+ (4 - 4.5 - .5 + .5)^2$ $+ (3 - 4.5 - .5 + .5)^2$ $= 14.75$	$(4 - 4.5 - .5 + 2.0)^2$ $+ (3 - 4.5 - .5 + 2.0)^2$ $+ (2 - 4.5 - .5 + 2.0)^2$ $= 2.00$	$(9 - 4.5 - .5 - 2.5)^2$ $+ (7 - 4.5 - .5 - 2.5)^2$ $+ (5 - 4.5 - .5 - 2.5)^2$ $= 8.75$

The sum of all these squared deviations is $S_1 = 39.0$.

From (10.46) the restricted estimators are

$\mu^* + b^* = \bar{Y} = 4.5$,
$a_1^* = \bar{Y}_1. - \bar{Y} = 4.0 - 4.5 = -.5$,
$a_2^* = \bar{Y}_2. - \bar{Y} = 5.0 - 4.5 = .5$.

We can calculate the sum of squares S_0 in (10.45) by writing the sum of squared deviations in each cell.

$(4 - 4.5 + .5)^2$ $+ (3 - 4.5 + .5)^2$ $+ (2 - 4.5 + .5)^2$ $= 5$	$(2 - 4.5 + .5)^2$ $+ (3 - 4.5 + .5)^2$ $+ (1 - 4.5 + .5)^2$ $= 14$	$(5 - 4.5 + .5)^2$ $+ (7 - 4.5 + .5)^2$ $+ (9 - 4.5 + .5)^2$ $= 35$
$(8 - 4.5 - .5)^2$ $+ (4 - 4.5 - .5)^2$ $+ (3 - 4.5 - .5)^2$ $= 14$	$(4 - 4.5 - .5)^2$ $+ (3 - 4.5 - .5)^2$ $+ (2 - 4.5 - .5)^2$ $= 14$	$(9 - 4.5 - .5)^2$ $+ (7 - 4.5 - .5)^2$ $+ (5 - 4.5 - .5)^2$ $= 20$

The sum of all these squared deviations is $S_0 = 102$, and $S_2 = S_0 - S_1 = 102 - 39 = 63$. To test the null hypothesis that the column effects are the same, we compute the F ratio

$$F = \frac{S_2}{S_1} \cdot \frac{N \cdot K_1 \cdot K_2 - (K_1 + K_2) + 1}{K_2 - 1} = \frac{63}{39} \cdot \frac{18 - 5 + 1}{3 - 1} = 11.308$$

with 2 degrees of freedom in the numerator and 14 degrees of freedom in the denominator. With a 1 percent critical region, the critical value of F is approximately 6.548. (See Table D in the Appendix of Statistical Tables.) Thus the column effects are significantly different.

Analysis of Covariance

Another kind of linear model is that of analysis of covariance. This kind of model is often useful, for example, when attempts are made to pool cross-section and time-series data. Thus one may have data on household expenditure on food (Y), total income (X_1), and an index of food prices (X_2) paid by different households over a period of years. Let Y_{ij}, X_{1ij}, and X_{2ij} represent, respectively, observations on food expenditure, total income, and food prices for the ith household in year j. Suppose the data cover N_1 households for N_2 years.

Suppose a regression relationship is determined for the N_1 households in year j. The regression relationship is denoted by

$$Y_{ij} = a_j + b_{1j}X_{1ij} + b_{2j}X_{2ij} + e_{ij} \qquad \text{for } i = 1, \ldots, N_1. \tag{10.48}$$

The least-squares estimators for year j are denoted by $\hat{a}_j$, $\hat{b}_{1j}$, and $\hat{b}_{2j}$. The predicted values of Y_{ij} in year j are

$$\hat{Y}_{ij} = \hat{a}_j + \hat{b}_{1j}X_{1ij} + \hat{b}_{2j}X_{2ij} \qquad \text{for } i = 1, \ldots, N_1. \tag{10.49}$$

The sum of squared errors in year j is

$$S_{1j} = \sum_{i=1}^{N_1} (Y_{1j} - \hat{a}_j - \hat{b}_{1j}X_{1ij} - \hat{b}_{2j}X_{2ij})^2 = \sum_{i=1}^{N_1} (\hat{e}_{ij})^2. \tag{10.50}$$

The total sum of squared error terms for all years is

$$S_1 = S_{11} + S_{12} + \cdots + S_{1N_2}. \tag{10.51}$$

This sum of squared error terms was determined by using least-squares techniques to estimate separate linear regressions for each year. We may ask the question, however, whether it is legitimate to pool all of the data, time-series, and cross-section, to obtain one grand regression relationship:

$$Y_{ij} = a + b_1X_{1ij} + b_2X_{2ij} + e_{ij} \qquad \text{for } i = 1, \ldots, N_1; j = 1, \ldots, N_2. \tag{10.52}$$

In order to test the legitimacy of this procedure, we are essentially testing the null hypothesis:

$$
\begin{aligned}
H0: a_1 &= a_2 = \cdots = a_{N_2} = a \\
b_{11} &= b_{12} = \cdots = b_{1N_2} = b_1 \\
b_{21} &= b_{22} = \cdots = b_{2N_2} = b_2
\end{aligned}
\tag{10.53}
$$

If this hypothesis is valid, the regression coefficients for each year are the same and it is justified to pool the data and obtain one grand regression to estimate the common parameters a, b_1, and b_2. These estimators are denoted by a^*, b_1^*, and b_2^*. They are restricted estimators in the sense that they are restricted to satisfy the linear relationships (10.53).

The predicted values of Y_{ij} are

$$
Y_{ij}^* = a^* + b_1^* X_{1ij} + b_2^* X_{2ij} \qquad \text{for } i = 1, \ldots, N_1; j = 1, \ldots, N_2.
\tag{10.54}
$$

Likewise, the sum of squared error terms is

$$
S_0 = \sum_{i=1}^{N_1} \sum_{j=1}^{N_2} (Y_{ij} - a^* - b_1^* X_{1ij} - b_2^* X_{2ij})^2 = \sum_{i=1}^{N_1} \sum_{j=1}^{N_2} (e_{ij}^*)^2.
\tag{10.55}
$$

Both sums of squared error terms S_1 and S_0 have chi-square distributions. The number of degrees of freedom of S_1 is

$$
N_1 \cdot N_2 - 3 \cdot N_2 = (N_1 - 3) \cdot N_2.
\tag{10.56}
$$

That is, there are a total of $N_1 \cdot N_2$ observations (N_1 households over N_2 years) and three estimated parameters a_j, b_{1j}, and b_{2j} for each of N_2 years for a total of $3 \cdot N_2$ estimated parameters. Thus the number of degrees of freedom of S_1 is the total number of observations less the total number of estimated parameters that enter into the total sum of squared errors.

Similarly, the number of degrees of freedom for S_0 is

$$
N_1 \cdot N_2 - 3,
\tag{10.57}
$$

since only three estimated parameters, a^*, b_1^*, and b_2^* enter the sum of squared errors.

The sum of squares

$$
S_2 = S_0 - S_1 = \sum_{i=1}^{N_1} \sum_{j=1}^{N_2} (Y_{ij} - Y_{ij}^*)^2
\tag{10.58}
$$

also has a chi-square distribution with degrees of freedom

$$
(N_1 \cdot N_2 - 3) - (N_1 - 3)N_2 = 3N_2 - 3 = 3(N_2 - 1).
\tag{10.59}
$$

Thus the F ratio

$$
F = \frac{S_2}{S_1} \cdot \frac{(N_1 - 3) \cdot N_2}{3 \cdot (N_2 - 1)}
\tag{10.60}
$$

has an F distribution with $3 \cdot (N_2 - 1)$ degrees of freedom in the numerator and $(N_1 - 3) \cdot N_2$ degrees of freedom in the denominator. This F ratio may be used to test the hypothesis (10.53) that permits a pooling of the data in order to obtain a regression relationship.

In the more general case in which there are K independent variables, $X_1, X_2, \ldots , X_K$, the F ratio is of the form

$$F = \frac{S_2}{S_1} \frac{(N_1 - K) \cdot N_2}{K \cdot (N_2 - K)},$$ (10.61)

where

$$S_0 = \sum_{i=1}^{N_1} \sum_{j=1}^{N_2} (Y_{ij} - a^* - b_1^* X_{1ij} - \cdots - b_K^* X_{Kij})^2$$

$$S_1 = \sum_{i=1}^{N_1} \sum_{j=1}^{N_2} (Y_{ij} - \hat{a}_j - \hat{b}_{1j} X_{1ij} - \cdots - \hat{b}_{Kj} X_{Kij})^2$$ (10.62)

$$S_2 = S_0 - S_1$$

and the F ratio has $K \cdot (N_2 - K)$ degrees of freedom in the numerator and $(N_1 - K) \cdot N_2$ degrees of freedom in the denominator.

Alternatively, one might wish to pool the data to estimate just some of the parameters. For example, one might wish to pool the data to estimate the effect of price on food expenditure but use only the cross-sectional data in the estimate of the effect of income. In the general case, suppose we wish to pool the data to estimate b_K, the parameter of the Kth independent variable. All other parameters $a_j, b_{1j}, \ldots , b_{K-1,j}$ are estimated on an annual basis from cross-section data. The null hypothesis is

$$H0: b_{K1} = b_{K2} = \cdots = b_{KN} = b_K.$$ (10.63)

First, one may obtain the unrestricted estimators, $\hat{a}_j, \hat{b}_{1j}, \ldots , \hat{b}_{K-1,j}, \hat{b}_{Kj}$ for each year j by running regressions on each cross-section of data. The restricted or pooled estimators, $a_j^*, b_{1j}^*, \ldots , b_{K-1,j}^*$ and b_K^*, are obtained by minimizing the following sum of squares:

$$\sum_{i=1}^{N_1} \sum_{j=1}^{N_2} (Y_{ij} - a_j - b_{1j} X_{1ij} - \cdots - b_{K-1,j} X_{K-1,ij} - b_K X_{Kij})^2.$$ (10.64)

The sum of squared errors using the nonrestricted (cross-section) estimators and the restricted (partially pooled) estimators are

$$S_1 = \sum_{i=1}^{N_1} \sum_{j=1}^{N_2} (Y_{ij} - \hat{a}_j - \hat{b}_{1j} X_{1ij} - \cdots - \hat{b}_{K-i,j} X_{K-1,ij} - \hat{b}_{K,j} X_{K,ij})^2$$

$$S_0 = \sum_{i=1}^{N_1} \sum_{j=1}^{N_2} (Y_{ij} - a_j^* - b_{1j}^* X_{1ij} - \cdots - b_{K-1,j}^* X_{K-1,ij} - b_K^* X_{K,ij})^2.$$

(10.65)

The first of these sums S_1 has

$$(N_1 \cdot N_2 - K \cdot N_2) = (N_1 - K) \cdot N_2 \tag{10.66}$$

degrees of freedom, while S_0 has

$$N_1 \cdot N_2 - (K - 1) \cdot N_2 - 1, \tag{10.67}$$

since $(K - 1) \cdot N_2 + 1$ parameters are estimated in the sum of squares S_0. The degrees of freedom of $S_2 = S_0 - S_1$ is

$$[N_1 \cdot N_2 - (K - 1) \cdot N_2 - 1] - (N_1 \cdot N_2 - K \cdot N_2) = N_2 - 1. \tag{10.68}$$

The F ratio is

$$F = \frac{S_2}{S_1} \cdot \frac{(N_1 - K) \cdot N_2}{(N_2 - 1)} \tag{10.69}$$

with $N_2 - 1$ degrees of freedom in the numerator and $(N_1 - K) \cdot N_2$ degrees of freedom in the denominator.

Dummy Variables in Regression

In this section let us consider the use of dummy variables in regression analysis. The insertion of dummy variables in a regression relationship may be treated as a special case of the covariance analysis discussed in the last section, but this special case is of such practical significance that some additional comments are useful.

Suppose, for example, that one wishes to estimate a relationship among tea consumption, income, the price of tea, the price of sugar, and the price of coffee by using time-series data from 1932 to 1968. Since this period covers World War II, in which the economy was severely disrupted and food rationing was imposed, the regression relationship might be assumed to be altered significantly during the war years 1942–1945. One way to take account of the war-year differences is to insert a dummy variable into the relationship. This dummy variable assumes the value of unity during the war years and the value of zero for other years. Thus we have the relationship:

$$Y_t = a + b_1 X_{1t} + b_2 X_{2t} + b_3 X_{3t} + b_4 X_{4t} + c Z_t + e_t, \tag{10.70}$$

where Y is tea consumption, X_1 is income, X_2 is the price of tea, X_3 is the price of sugar, X_4 is the price of coffee, and Z is the dummy variable. The value of the dummy coefficient c tells us how much consumption is altered upward (if c is positive) or downward (if c is negative) as the result of wartime restrictions. Of course, in this particular case, we would expect c to be negative to reflect the fact that wartime restrictions would tend to reduce tea consumption. Note that the use of a dummy variable assumes that tea consumption is altered by some constant amount and that the coefficients of the other independent variables remain the same in the war years as the nonwar years. (Differences in the other parameters between war years and nonwar years can be handled by the covariance techniques discussed above.)

Example 10.7. Suppose a regression relationship is estimated to be

$$Y_t = 2.000 + .060X_{1t} - .031X_{2t} - .004X_{3t} + .002X_{4t} - .987Z_t + e_t, \tag{10.71}$$

where Y_t is annual per capita tea consumption in pounds, X_{1t} is income per capita, X_{2t} is price of tea, X_{3t} is price of sugar, and X_{4t} is price of coffee in year t. The dummy variable $Z_t = 0$ for $t = 1932$ to $t = 1941$, $Z_t = 1$ for $t = 1942$ to $t = 1945$, and $Z_t = 0$ for $t = 1946$ to 1968. The coefficient $-.987$ of the dummy variable Z_t indicates that annual per capita tea consumption was reduced by nearly one pound as a result of wartime restrictions.

Suppose we wish to test whether wartime restrictions had a significant effect on tea consumption. Formally, we wish to test the null hypothesis:

$$H0: c = 0. \tag{10.72}$$

The null hypothesis, then, is that the war had no significant effect on consumption. Let $\hat{a}$, $\hat{b}_1$, $\hat{b}_2$, $\hat{b}_3$, $\hat{b}_4$, and $\hat{c}$ represent the least-squares estimators of the regression relationship (10.70). If the null hypothesis is true, then $c = 0$ and the dummy variable drops out of the regression relationship (10.70). By leaving out the dummy variable and running the regression, we obtain a set of restricted estimators, a^*, b_1^*, b_2^*, b_3^*, and b_4^*. The respective sums of squared deviations are

$$S_0 = \sum_{t=1}^{N} (Y_t - a^* - b_1^* X_{1t} - b_2^* X_{2t} - b_3^* X_{3t} - b_4^* X_{4t})^2, \tag{10.73}$$

$$S_1 = \sum_{t=1}^{N} (Y_t - \hat{a} - \hat{b}_1 X_{1t} - \hat{b}_2 X_{2t} - \hat{b}_3 X_{3t} - \hat{b}_4 X_{4t} - \hat{c} Z_t)^2.$$

The test of the null hypothesis (10.72) is performed by using the F ratio

$$F = \frac{S_2}{S_1} \cdot \frac{N-6}{1}, \tag{10.74}$$

where $S_2 = S_0 - S_1$. The degrees of freedom in the denominator is $N - 6$, where N is the total number of observations and there are six parameters, $\hat{a}$, $\hat{b}_1$, $\hat{b}_2$, $\hat{b}_3$, $\hat{b}_4$, and $\hat{c}$, being estimated. The sum of squares S_0 has $N - 5$ degrees of freedom so that $S_2 = S_0 - S_1$ has $(N - 6) - (N - 5) = 1$ degree of freedom. Thus the numerator of F has one degree of freedom.

Since the numerator of F has one degree of freedom and the denominator has $N - 6$ degrees of freedom, $\sqrt{F} = t$ has a Student's t distribution with

$N - 6$ degrees of freedom. It is often desirable to use the Student's t distribution to test the null hypothesis (10.72). The use of the Student's t distribution enables a one-tailed test of the null hypothesis. This is particularly useful in the tea-consumption example, since one would expect the war years to show reduced tea consumption; that is, we would expect c to be negative and would want to use a critical region in the left-hand tail of the Student's t distribution.

To indicate how one might perform a one-tailed test, let us assert that the square root of the F-ratio in (10.74) can be written as

$$\sqrt{F} = t = \frac{\hat{c}}{\hat{\sigma}_{\hat{c}}}, \tag{10.75}$$

where $\hat{c}$ is the least-squares estimator of the parameter c of the dummy variable and $\hat{\sigma}_{\hat{c}}^2$ is an unbiased estimator of the variance of the estimator $\hat{c}$. The exact expression for $\hat{\sigma}_{\hat{c}}$ in terms of the observations on the dependent and independent variables is complicated, and the derivation of the formula (10.75) is messy without the use of matrix algebra. This formulation, however, indicates that the t ratio is positive if $\hat{c}$ is positive and negative if $\hat{c}$ is negative. For a one-tailed test using the left-hand tail of the Student's t distribution, the critical value of t is negative. One may calculate the F ratio in (10.74) and take as the t ratio the negative square root if $\hat{c}$ is negative and the positive square root if $\hat{c}$ is positive. We reject the null hypothesis if t is negative and larger in absolute value than the critical value of t.

A somewhat more general approach must be used whenever there is more than one dummy variable. For example, one may wish to include two different dummy variables in a regression relationship, one for World War II and one for the Korean War. In general terms, let us consider the regression relationship with L different dummy variables:

$$Y_i = a + b_1 X_{1i} + \cdots + b_K X_{Ki} + c_1 Z_{1i} + \cdots + c_L Z_{Li} + e_i, \tag{10.76}$$

where $X_{1i}, \ldots, X_{Ki}$ are K independent variables and $Z_{1i}, \ldots, Z_{Li}$ are L dummy variables. Let us test the hypothesis that none of the dummy variables has a significant effect on the dependent variable Y_i; that is, the null hypothesis is

$$H0: c_1 = c_2 = \cdots = c_L = 0. \tag{10.77}$$

The test may be performed by using the F ratio

$$F = \frac{S_2}{S_1} \cdot \frac{N - (K + 1) - L}{L}, \tag{10.78}$$

where S_2 and S_1 are determined from the following sums of squares:

$$S_0 = \sum_{i=1}^{N} (Y_i - a^* - b_1^* X_{1i} - \cdots - b_K^* X_{Ki})^2,$$

$$S_1 = \sum_{i=1}^{N} (Y_i - \hat{a} - \hat{b}_1 X_{1i} - \cdots - \hat{b}_K X_{Ki} - \hat{c}_1 Z_{1i} - \cdots - \hat{c}_L Z_{Li})^2,$$

$$S_2 = S_0 - S_1. \tag{10.79}$$

The numerator of F has L degrees of freedom and the denominator has $N - (K + 1) - L$ degrees of freedom.

Alternatively, we might wish to test the hypothesis that only one of the dummy variables has a significant effect. The null hypothesis is

$$H0: c_j = 0. \tag{10.80}$$

For example, we might wish to test the significance of the Korean War alone whereas the hypothesis (10.77) is a test of the significance of both the Korean War and World War II.

This test may be performed by using the F-ratio

$$F = \frac{S_2}{S_1} \frac{[N - (K + 1) - L]}{1}, \tag{10.81}$$

where S_2 and S_1 are determined from the following sums of squares:

$$S_0 = \sum_{i=1}^{N} (Y_i - a^* - b_1^* X_{1i} - \cdots + b_K^* X_{Ki} - c_1^* Z_{1i} - \cdots$$
$$- c_{j-1}^* Z_{ji} - c_{j+1}^* Z_{j+1,i} - \cdots - c_L^* Z_{Li})^2,$$

$$S_1 = \sum_{i=1}^{N} (Y_i - \hat{a} - \hat{b}_1 X_{1i} - \cdots - \hat{b}_K X_{Ki} - \hat{c}_1 Z_{1i} - \cdots - \hat{c}_L Z_{Li})^2,$$

$$S_2 = S_0 - S_1 \tag{10.82}$$

The numerator of F in (10.81) has one degree of freedom while the denominator has $N - (K + 1) - L$ degrees of freedom. Since the numerator has one degree of freedom, $\sqrt{F}$ has a Student's t distribution with $N - (K + 1) - L$. Thus the hypothesis (10.80) may be tested using the Student's t distribution.

Furthermore in this case, we may write

$$\sqrt{F} = \frac{\hat{c}_j}{\hat{\sigma}_{\hat{c}_j}}. \tag{10.83}$$

That is, the t ratio may be viewed as the least-squares estimate (unrestricted) divided by an estimator $\hat{\sigma}_{\hat{c}_j}$ of the standard deviation of the coefficient of the jth dummy variable.

Conclusions

This chapter has considered only some of the various tests in the theory of analysis of variance and covariance. For example, the various models considered above may be combined to produce additional types of testing procedures. The same basic principles still apply, however. That is, the F ratio is the ratio of the sums of squares S_2/S_1 multiplied by the inverse of the ratio of the respective degrees of freedom. Various computational schemes may be used to compute the sums of squares. Most of them rely on the basic identity $S_0 = S_1 + S_2$, where S_0 is the sum of squared residuals using the restricted parameter estimates, S_1 is the sum of squared residuals using the unrestricted parameter estimates, and S_2 is the sum of squared differences in the predicted values of Y_i using the restricted and unrestricted parameter estimates. This same identity may be used to determine the degrees of freedom. That is,

degrees of freedom of S_0
$$= \text{degrees of freedom of } S_1 + \text{degrees of freedom of } S_2. \qquad (10.84)$$

The number of degrees of freedom of S_0 is the number of observations N less the number of restricted parameters. S_1 has degrees of freedom equal to N less the number of unrestricted parameters. The number of degrees of freedom of S_2 is the difference between the degrees of freedom of S_0 and S_1 or alternatively the number of linear restrictions in the null hypothesis that apply to the total set of parameters.

PROBLEMS

1. Suppose a sample of building lots in four suburban communities gives the following prices per acre:

COMMUNITY 1	COMMUNITY 2	COMMUNITY 3	COMMUNITY 4
$7,642	$1,438	$4,320	$16,004
6,876	2,343	5,157	13,876
5,936	4,387	3,120	
9,041	540		
8,375			

(a) Using a 5 percent critical region, test whether the mean price per acre is the same in all four communities. (b) Perform the same test using a 1 percent critical region.

2. Let Y_{ijk} stand for the kth observation in the cell of the ith row and jth column.

Suppose there are K_1 rows ($i = 1, \ldots, K_1$), K_2 columns ($j = 1, \ldots, K_2$), and N observations in each cell ($k = 1, \ldots, N$). Let the model be

$$Y_{ijk} = \mu + a_i + b_j + c_{ij}, \tag{10.85}$$

where a_i is the row effect, b_j is the column effect, and c_{ij} is called the interaction (cell) effect. The least-squares estimates of the parameters are

$$\hat{\mu} = \bar{Y}, \tag{10.86}$$

$$\hat{a}_i = \sum_{j=1}^{K_2} \sum_{k=1}^{N} Y_{ijk} - \bar{Y} = \bar{Y}_{i\cdot} - \bar{Y} \quad \text{for } i = 1, \ldots, K_1, \tag{10.87}$$

$$\hat{b}_j = \sum_{i=1}^{K_1} \sum_{k=1}^{N} Y_{ijk} - \bar{Y} = \bar{Y}_{\cdot j} - \bar{Y} \quad \text{for } j = 1, \ldots, K_2, \tag{10.88}$$

$$\hat{c}_{ij} = \bar{Y}_{ij} + \bar{Y} - \bar{Y}_{i\cdot} - \bar{Y}_{\cdot j}, \tag{10.89}$$

where $\bar{Y}$ is the grand mean, $Y_{i\cdot}$ is the mean of observations in the ith row, $\bar{Y}_{\cdot j}$ is the mean of observations in the jth column, and $\bar{Y}_{ij}$ is the mean of observations in the cell of the ith row and jth column. The estimated parameters satisfy the following restrictions:

$$\sum_{i=1}^{K_1} \hat{a}_i = 0 \tag{10.90}$$

$$\sum_{j=1}^{K_2} \hat{b}_j = 0 \tag{10.91}$$

$$\sum_{i=1}^{K_1} \hat{c}_{ij} = 0 \quad \text{for } j = 1, \ldots, K_2, \tag{10.92}$$

$$\sum_{j=1}^{K_2} \hat{c}_{ij} = 0 \quad \text{for } i = 1, \ldots, K_1. \tag{10.93}$$

Suppose the null hypothesis is that the interaction effects are zero. That is,

$$c_{ij} = 0 \quad \text{for } i = 1, \ldots, K_1, \quad j = 1, \ldots, K_2. \tag{10.94}$$

Under the null hypothesis the restricted least-squares estimates are given by (10.86), (10.87), and (10.88), and the restricted estimates satisfy (10.90) and (10.91). (a) Denoting the restricted estimators by a_i^* (for $i = 1, \ldots, K_1$) and b_j^* (for $j = 1, \ldots, K_2$), write the expression for the restricted sum of squares S_0 and the nonrestricted sum of squares S_1. (b) What is the F ratio and how many degrees of freedom are in the numerator and in the denominator? (c) Given the following observations in cells:

5, 7	6, 8	2, 2
9, 5	10, 14	4, 6

what are the estimated values of the parameters μ, a_1, a_2, b_1, b_2, b_3, c_{11}, c_{12}, c_{13}, c_{21}, c_{22}, and c_{23}? (d) Test the hypothesis that the interaction effects are zero with a 5 percent critical region. (e) Test the hypothesis that the interaction effects and the column effects are zero using a 5 percent critical region.

3. Suppose a sample survey of rice farmers in Thailand gives data on yields per acre (Y), labor input per acre (X_1), capital input per acre (X_2), educational level, and ethnic background. Suppose the farmers are grouped into two educational levels, primary school graduate and non-primary school graduate, and three ethnic categories. The model used is

$$Y_{ijk} = \mu + a_i + b_j + c_1 X_{1ijk} + c_2 X_{2ijk} + e_{ijk} \tag{10.95}$$

for $i = 1, 2$; $j = 1, 2, 3$; and $k = 1, 2, \ldots, N_{ij}$; where i refers to educational group, j refers to ethnic group, and k indicates the kth observation. For example, Y_{ijk} is the kth observation on farmers with educational level i and in ethnic category j. The a and b parameters must satisfy

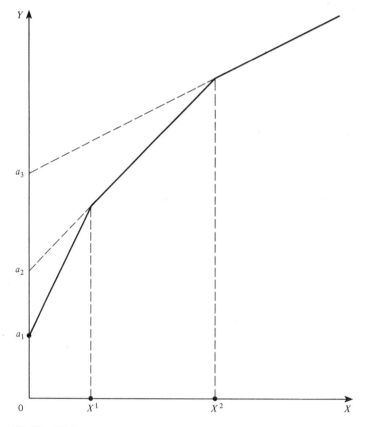

FIGURE 10.3

$$a_1 + a_2 = 0, \tag{10.96}$$

$$b_1 + b_2 + b_3 = 0. \tag{10.97}$$

(a) Show how you would test the hypothesis that ethnic grouping made no difference in the level of output per acre. Write expressions for the appropriate sums of squares and the F ratio. Give degrees of freedom of the F ratio. (b) In the same way, show how you would test the hypothesis that neither educational background nor ethnic category affects the level of output per acre.

4. Suppose a survey of manufacturing firms gives data on rates of return on capital over a period of ten years. We wish to determine whether there is a relationship between average rate of return on capital (Y) and the variance in the rate of return on capital (X). The relationship is assumed to be nonlinear, which we approximated by a piecewise linear function as shown in Fig. 10.3. The firms are grouped into three categories: those with low variability in rate of return ($X \leq X^1$), those with a medium variability in rate of return ($X^1 < X \leq X^2$), and those with high variability in the rate of return ($X > X^2$). A linear relationship between the two variables, Y and X, is fitted for each of the three groups

$$
\begin{aligned}
Y_{1i} &= a_1 + b_1 X_{1i} && \text{for } i = 1, \ldots, N_1, \\
Y_{2i} &= a_2 + b_2 X_{2i} && \text{for } i = 1, \ldots, N_2, \\
Y_{3i} &= a_3 + b_3 X_{3i} && \text{for } i = 1, \ldots, N_3,
\end{aligned}
\tag{10.98}
$$

where N_j (for $j = 1, 2, 3$) is the number of firms in the jth group. In order that the three linear pieces meet at the transition points, $X = X^1$ and $X = X^2$ (see Fig. 10.3), the coefficients in (10.97) must satisfy the following constraints:

$$
\begin{aligned}
a_1 + b_1 X^1 &= a_2 + b_2 X^1, \\
a_2 + b_2 X^2 &= a_3 + b_3 X^3.
\end{aligned}
\tag{10.99}
$$

Show how you would test the hypothesis that the curve is linear rather than piecewise linear. That is, test the null hypothesis

$$a_1 = a_2 = a_3 = a, \qquad b_1 = b_2 = b_3 = b. \tag{10.100}$$

Outline of Further Reading for Part IV

A. The Linear Regression Model
   1. Johnston [1960], pp. 106–144
   2. Kane [1968], pp. 255–282
   3. Kendall and Stuart [1967], pp. 26–30, 278–464
   4. Malinvaud [1966], pp. 139–240
   5. Mood and Graybill [1963], pp. 328–359
   6. Walker and Lev [1953], pp. 230–260, 315–347

These references treat the linear regression model and use the analysis of variance approach to testing hypotheses concerning the parameters.

B. Simple Introduction to Analysis of Variance
   1. Bryant [1966], pp. 147–181
   2. Chou [1969], pp. 399–438
   3. Freund [1962], pp. 328–350
   4. Freund [1967], pp. 302–315
   5. Hope [1968], pp. 23–32
   6. Richmond [1964], pp. 304–323
   7. Walker and Lev [1953], pp. 196–229, 348–386
   8. Wonnacott and Wonnacott [1969], pp. 195–219
   9. Yamane [1967], pp. 642–671

C. Introduction to Analysis of Covariance
   1. Chow [1960]
   2. Hope [1968], pp. 138–156
   3. Walker and Lev [1953], pp. 487–512

The Chow reference applies analysis of covariance to combining time-series and cross-section data.

D. Advanced Treatment of Analysis Variance and Covariance
   1. Anderson [1958], pp. 178–229
   2. Anderson and Bancroft [1952], pp. 153–378
   3. Kendall and Stuart [1966], pp. 1–118
   4. Malinvaud [1966], pp. 215–240, 242–250
   5. Scheffé [1957]

The Scheffé and Anderson and Bancroft references emphasize least-squares estimation techniques in analysis of variance.

E. Dummy Variables in Regression Analysis
   1. Malinvaud [1966], pp. 241–242
   2. Suits [1957]

F. Integrated Treatment of Analysis of Variance and Regression Analysis
   1. Graybill [1961], pp. 93–222
   2. Yamane [1967], pp. 752–844

The Yamane reference shows in an elementary fashion the relationship between regression analysis and analysis of variance. The Graybill book uses a very general approach to regression analysis and the analysis of variance. Much of the discussion in Chapter 9 of the present volume is based on Graybill's approach. The Graybill work, however, is advanced.

G. Restricted Regression
   1. Chipman and Rao [1964]
   2. Goldberger [1964], pp. 255–259
   3. Zellner [1961b]

Goldberger discusses estimation of regression parameters when the regression parameters are expected to satisfy linear constraints. Zellner deals with the case where the constraints are inequalities. Chipman and Rao discuss both estimation procedures and tests of hypotheses with linear restrictions.

H. Quadratic Forms and the F Distribution
   1. Box [1954]
   2. Graybill [1961], pp. 74–92
   3. Hogg and Craig [1970], pp. 384–394
   4. Kendall and Stuart [1969], pp. 347–368

I. Experimental Design
   1. Cochran and Cox [1957], pp. 1–390
   2. Graybill, pp. 223–336
   3. Kendall and Stuart [1966], pp. 119–238
   4. Mood and Graybill [1963], pp. 360–382
   5. Wallis and Roberts [1956], pp. 475–482

In setting up a scientific research experiment the data can be gathered and tabulated in ways which permit efficient use of analysis of variance techniques. The science of setting up experiments is called experimental design. Wallis is a very elementary introduction and Mood and Graybill a more advanced introduction to experimental design. Cochran and Cox, Graybill, and Kendall and Stuart are advanced and comprehensive treatments of experimental design models.

Part V

Problems in Econometrics

Chapter **11**

Estimation Problems

In Chapter 9 we discussed four different assumptions concerning the error term e_i in a linear regression relationship of the form

$$Y_i = a_0 + a_1 \cdot X_{1i} + a_2 \cdot X_{2i} + \cdots + a_K \cdot X_{Ki} + e_i. \tag{11.1}$$

The second and third of these assumptions concerning the error term were as follows:

Assumption 2   The sample values of the error terms e_i for $i = 1, \ldots, n$ are independently distributed.

Assumption 3   Each of the sample error terms e_i is distributed with the same variance σ_e^2.

If Assumption 2 is violated, the error terms are said to be autocorrelated. If Assumption 3 is violated, the error terms are said to be heteroskedastic. If either of these assumptions is violated, the estimation procedures and the tests of hypotheses used and discussed in Chapters 8, 9, and 10 are no longer universally valid. The estimation procedures discussed in Chapters 7 through 10 do not always provide efficient and unbiased estimates. Modified estimation and hypothesis-testing procedures must be used when these two assumptions are violated.

Autocorrelation

Autocorrelation of error terms arises under a number of different circumstances—for example, when variables are missing from a relationship or when a nonlinear relationship is specified as a linear relationship. The manner in which these circumstances give rise to autocorrelation was discussed in Examples 9.3, 9.4, and 9.5 of Chapter 9.

Autocorrelation alone does not generally produce biased estimates of the coefficients a_0, a_1, ... , a_K.† The estimates of the variance of the estimators, however, can be seriously biased downward. In such cases, it is illegitimate to use the usual Student's t and F tests of significance. A parameter a_i may be significant according to a t test, but truly nonsignificant because of an underestimate of the variance of a_i. Furthermore, ordinary least-squares estimation procedures are inefficient. Other methods of estimation exist that produce less sampling variance in the estimators.

TESTING FOR AUTOCORRELATION

How does one know when autocorrelation is present in the error terms? A quick way of determining the presence of autocorrelation is to note the pattern of the residuals from a regression line. This can be done most expeditiously by looking at a scatter diagram of the residuals about the regression line. If the residuals seem to follow a pattern, the error term is likely to be autocorrelated. For example, in Fig. 11.1 the residuals seem

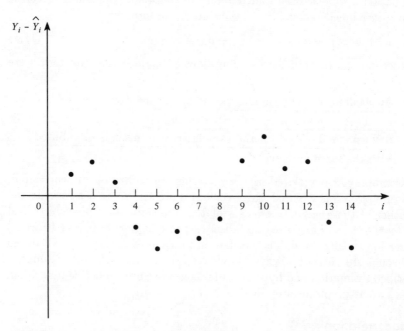

FIGURE 11.1   Cyclical Patterns of Residuals

---

† When lagged values of the dependent variable Y appear as independent variables, however, ordinary least-squares estimates are both biased and inconsistent.

to follow a cyclical pattern about the regression line. In Fig. 11.2 the
residuals follow an oscillating pattern; a negative residual is followed by a
positive residual and a positive residual by a negative residual. The error
terms in both these cases are likely to be autocorrelated.

There are more exact tests for autocorrelation, but these tests usually
require one to make a specific assumption about the *form* of the autocorrelation. A simple form of autocorrelation is *linear first-order autocorrelation*
(most often called simply first-order autocorrelation). First-order autocorrelation is present whenever the error term e_i is linearly dependent on the
error term e_{i-1} in the following way:

$$e_i = \rho e_{i-1} + u_i, \quad \text{for } i = 2, \ldots, N, \tag{11.2}$$

where ρ is a constant called the coefficient of autocorrelation and u_i is an
error term. The error term u_i is assumed to satisfy Assumptions 1, 2, and
3; that is, it is not autocorrelated nor heteroskedastic and has a zero mean
and finite variance. One also assumes that the coefficient ρ has an absolute
value less than unity.

$$|\rho| < 1. \tag{11.3}$$

The coefficient ρ may be positive or negative, however. If ρ is negative,
the errors e_i tend to oscillate between positive and negative values, and the
errors are said to possess negative first-order autocorrelation. A positive ρ
corresponds to positive first-order autocorrelation. If the absolute value

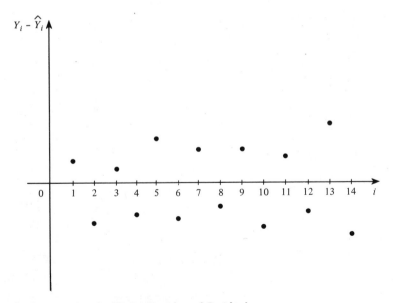

FIGURE 11.2 Oscillating Patterns of Residuals

of ρ were greater than unity, the errors would tend to become larger and larger through time. An economic process with this kind of autocorrelation would be inherently unstable.

One test for first-order autocorrelation is based on the Durbin-Watson statistic d. (This statistic is sometimes called the von Neumann ratio.) The formula for this statistic is

$$d = \frac{\sum_{i=2}^{N} (\hat{e}_i - \hat{e}_{i-1})^2}{\sum_{i=1}^{N} \hat{e}_i^2}, \tag{11.4}$$

where $\hat{e}_i$ is the estimated error term. The estimated error term is

$$\hat{e}_i = Y_i - (\hat{a}_0 + \hat{a}_1 X_{1i} + \cdots + \hat{a}_K X_{Ki}), \tag{11.5}$$

where $\hat{a}_0, \hat{a}_1, \ldots, \hat{a}_K$ are least-squares estimates of the parameters a_0, $a_1, \ldots, a_K$.

Example 11.1.   Suppose the following equation has been estimated:

$$Y = \hat{a} + \hat{b} \cdot X = -4.0 + 4.0X. \tag{11.6}$$

The data and the calculation of the Durbin-Watson statistic are as follows:

Y_i	X_i	$\hat{Y}_i = \hat{a} + \hat{b}X_i$	$\hat{e}_i = Y_i - \hat{Y}_i$	$\hat{e}_i - \hat{e}_{i-1}$	$(\hat{e}_i - \hat{e}_{i-1})^2$	$(\hat{e}_i)^2$
18	4	12	6	—	—	36
14	3	8	6	0	0	36
26	8	28	−2	−8	64	4
24	6	20	4	6	36	16
18	5	16	2	−2	4	4
5	3	8	−3	−5	25	9
27	9	32	−5	−2	4	25
29	10	36	−7	−2	4	49
33	12	44	−11	−4	16	121
43	14	52	−9	2	4	81
47	14	52	−5	4	16	25
66	16	60	6	11	121	36
63	15	56	7	1	1	49
49	13	48	1	−6	36	1
78	18	68	10	9	81	100
					412	592

$$d = \frac{\sum_{i=2}^{15} (\hat{e}_i - \hat{e}_{i-1})^2}{\sum_{i=1}^{15} (\hat{e}_i)^2} = \frac{412}{592} = .696.$$

If there is no first-order autocorrelation, the expected value of the Durbin-Watson statistic is 2. A high degree of positive autocorrelation (ρ close to $+1$) means that the expected value of the Durbin-Watson statistic is very nearly zero. A high degree of negative autocorrelation (ρ close to -1) implies that the expected value of d is close to 4. Let us summarize these properties:

$\rho = 0$: $E(d) = 2$.
$\rho \cong +1$: $E(d) \cong 0$.
$\rho \cong -1$: $E(d) \cong 4$.

The Durbin-Watson statistic is never less than 0 nor greater than 4. The exact sampling distribution of the statistic d depends on the particular values of the independent variables $X_{1i}, \ldots, X_{Ki}$ for $i = 1, \ldots, N$. For a given set of values, one could determine the sampling distribution of d and design a two-tailed test for autocorrelation with critical regions as shown in Fig. 11.3. If the value of d falls below d^* or above d^{**}, one accepts the hypothesis that the error terms are autocorrelated.

In practical terms, however, it would be extremely difficult to perform

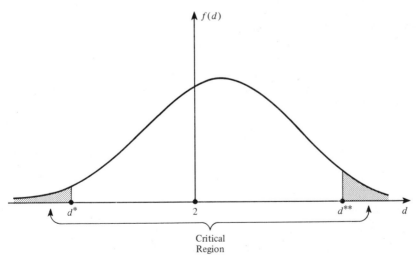

FIGURE 11.3   Critical Region for Durbin-Watson Distribution

this test. Since the sampling distribution of d depends on the values assumed by the independent variables, a large number of tables would have to be constructed for a large number of different combinations of values of the independent variables. It is possible, however, to construct a nonexact test for first-order autocorrelation based on the Durbin-Watson statistic. While the Durbin-Watson statistic has a variable distribution depending on the values of the X variables, the distribution lies within certain limits. For example, if there are 30 observations on the data and one independent variable and if the critical region is 5 percent, the lower critical value d^* is never less than 1.25 nor more than 1.38 regardless of the values of the X variables. The upper critical value d^{**} is never greater than 2.75 or less than 2.62. In more general terms, given the number of observations and the number of independent variables, for a two-tailed critical region α, the lower critical value is never less than d_1 nor greater than d_2, and the upper critical value is never less than $4 - d_2$ nor greater than $4 - d_1$. This is illustrated in Fig. 11.4. Distribution 1 illustrates the lower limit d_1 for the lower critical value and the lower limit $4 - d_2$ for the higher critical value. Distribution 2 illustrates the upper limit d_2 of the lower critical value and the upper limit $4 - d_1$ of the higher critical value. All other distributions of the Durbin-Watson statistic produce critical values that lie between the two limits at either end of the distribution.

These limiting properties of the Durbin-Watson distribution enable one to make a test of a hypothesis concerning first-order autocorrelation that is sometimes conclusive and sometimes not. If the Durbin-Watson statistic d is greater than d_2 and less than $4 - d_2$, one accepts the hypothesis of no significant autocorrelation. If the Durbin-Watson statistic is less than d_1

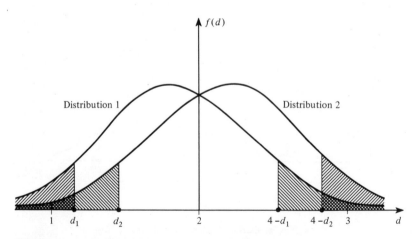

FIGURE 11.4   Limiting Durbin-Watson Distributions

or greater than $4 - d_1$, one concludes that there is significant first-order autocorrelation. If, however, the statistic lies between d_1 and d_2 or between $4 - d_2$ and $4 - d_1$, then the hypothesis of zero first-order autocorrelation can neither be accepted nor rejected. This analysis is summarized as follows:

DURBIN-WATSON STATISTIC	SIGNIFICANT AUTOCORRELATION
$d < d_1 < d_2$	Yes (positive)
$d > 4 - d_1 > 4 - d_2$	Yes (negative)
$d > d_2 > d_1$	No
$d < 4 - d_2 < 4 - d_1$	No
$d_1 < d < d_2$	Indeterminate
$4 - d_2 < d < 4 - d_1$	Indeterminate

Example 11.2. Let us use the Durbin-Watson statistic calculated in Example 11.1 to test the hypothesis of no first-order autocorrelation. Let us choose a two-tailed critical region of 5 percent. Table E of the Appendix of Statistical Tables gives values of d_1 and d_2 for two-tailed critical regions. Note that d_1 and d_2 depend both on the number of observations and on the number K of independent variables. Since there is only one independent variable and 15 observations, the value of d_1 and d_2 for a 5 percent critical region are

$$d_1 = .95, \qquad d_2 = 1.23.$$

Since the estimated Durbin-Watson statistic $d = .696$ is lower than both d_1 and d_2, we conclude that there is significant positive autocorrelation.

If, however, d was between .95 and 1.23, we could neither accept nor reject the hypothesis concerning first-order autocorrelation.

ESTIMATION WITH AUTOCORRELATION

In order to determine efficient estimates with autocorrelated error terms, it is necessary to know the precise nature of the autocorrelation. This is not usually possible. Thus the usual procedure is to assume that the autocorrelation follows a first-order scheme of the form (11.2). Let us assume for the moment that in fact we know the value of ρ, the coefficient of autocorrelation. Later on we will discuss ways to estimate ρ from a set of data, but the explanation of an efficient estimation scheme for a regression equation with autocorrelated error terms is facilitated if we assume that ρ is already known.

Although the error terms e_i are autocorrelated, we assume that the error terms u_i are not autocorrelated and satisfy our basic Assumptions 1, 2, and 3. Then an efficient estimation procedure is one that minimizes the sum of the squared error terms u_i, where u_i is given by (11.2).

$$\sum_{i=2}^{N} u_i^2 = \sum_{i=2}^{N} (e_i - \rho e_{i-1})^2, \tag{11.7}$$

where

$$e_i = Y_i - (a_0 + a_1 X_{1i} + \cdots + a_K X_{Ki}) \tag{11.8}$$

and

$$e_i - \rho e_{i-1} = (Y_i - \rho Y_{i-1})$$
$$- [a_0(1 - \rho) + a_1(X_{1i} - \rho X_{1,i-1}) + \cdots + a_K(X_{Ki} - \rho X_{K,i-1})]. \tag{11.9}$$

It is easy to see from (11.9) that minimization of the sum of squared error terms in (11.7) is equivalent to minimizing a sum of squared residuals by transforming the original variables Y_i and $X_{1i}, \ldots, X_{Ki}$ into new variables Y_i' and $X_{1i}', \ldots, X_{Ki}'$, where

$$Y_i' = Y_i - \rho Y_{i-1} \qquad \text{for } i = 2, \ldots, N, \tag{11.10}$$

and

$$X_{1i}' = X_{1i} - \rho X_{1,i-1}$$
$$. $$
$$. \tag{11.11}$$
$$. $$
$$X_{Ki}' = X_{Ki} - \rho X_{K,i-1} \qquad \text{for } i = 2, \ldots, N.$$

The transformed variables are then used to estimate the regression equation

$$Y_i' = a_0(1 - \rho) + a_1 X_{1i}' + \cdots + a_K X_{Ki}' + u_i \tag{11.12}$$

by the ordinary least-squares procedures. Note that we assume that the error term u_i is neither autocorrelated nor heteroskedastic. Thus the transformed equation (11.12) satisfies the usual assumptions, while the original equation (11.1) does not.

How does one determine ρ. One way is to estimate the original equation (11.1) in the nontransformed variables and obtain a set of estimated residuals $\hat{e}_1, \hat{e}_2, \ldots, \hat{e}_N$. The Durbin-Watson statistic may then be used to test for the presence of autocorrelation. If autocorrelation is significant, then one may use the estimated error terms $\hat{e}_i$ to estimate the linear relationship (11.2) between successive error terms. The estimated value of ρ in (11.2) is given by

$$\hat{\rho} = \frac{\displaystyle\sum_{i=2}^{N} \hat{e}_i \hat{e}_{i-1}}{\displaystyle\sum_{i=2}^{N} \hat{e}_i^2}. \tag{11.13}$$

This estimated value of ρ may then be used to transform the variables as in (11.10) and (11.11). A new regression equation (11.12) can then be estimated with the transformed variables.†

Example 11.3.   Let us use the data from Example 11.1. The test using the Durbin-Watson statistic indicates the presence of autocorrelation, as we showed in Example 11.2. The estimated value of ρ is

$$\hat{\rho} = \frac{374}{556} = .673.$$

The transformed variables are computed as follows:

Y_i	ρY_{i-1}	$Y'_i = Y_i - \rho Y_{i-1}$	X_i	ρX_{i-1}	$X'_i = X_i - \rho X_{i-1}$
18	—	—	4	—	—
14	12.114	1.886	3	2.692	.308
26	9.422	16.578	8	2.019	5.981
24	17.498	6.502	6	5.384	.616
18	16.152	1.848	5	4.038	.962
5	12.114	−7.114	3	3.365	−.365
27	3.365	23.635	9	2.019	6.981
29	18.171	10.829	10	6.057	3.943
33	19.517	13.483	12	6.730	5.270
43	22.209	20.791	14	8.076	5.924
47	28.939	18.061	14	9.422	4.578
66	31.631	34.369	16	9.422	6.578
63	44.418	18.582	15	10.768	4.232
49	42.399	6.601	13	10.095	2.905
78	32.977	45.023	18	8.749	9.251

The transformed variables may then be used to estimate the following equation

$$Y' = \hat{a}(1 - \hat{\rho}) + \hat{b} \cdot X',$$
$$Y' = -2.8166 + 4.3931 \cdot X'. \tag{11.14}$$

The coefficient $\hat{a}$ is given by

$$\hat{a} = \frac{-2.8166}{(1 - \hat{\rho})} = \frac{-2.8166}{.327} = -8.6135.$$

Thus in terms of the original Y and X variables, the estimated regression equation is

$$Y = -8.6135 + 4.3931X.$$

---

† Note from (11.4) that an approximate estimate of ρ is $\tilde{\rho} = \frac{1}{2}(2 - d)$, where d is the Durbin-Watson statistic. Thus if d has already been estimated, it is simple to obtain a quick estimate of ρ.

This method of estimation with autocorrelated error terms is generally more efficient than ordinary least-squares. It assumes, however, that the estimate of ρ given by (11.13) is a reasonably accurate estimate of the true ρ. A more efficient procedure, which attempts to obtain successively better estimates of the true ρ, is to use the estimated $a_0, a_1, \ldots, a_K$ obtained by minimizing the sum of squares (11.7) to obtain new estimates of the error terms e_i for $i = 1, \ldots, N$. That is, the new estimated error term is

$$\tilde{e}_i = Y_i - \tilde{a}_0 - \tilde{a}_1 X_{1i} - \cdots - \tilde{a}_K X_{Ki}, \qquad (11.15)$$

where $\tilde{a}_0, \tilde{a}_1, \ldots, \tilde{a}_K$ are the estimated values of $a_0, a_1, \ldots, a_K$ obtained by estimating the sum of squares (11.7) (not the usual sum of squares). A new estimate of ρ, call it $\tilde{\rho}$, may be obtained from formula (11.13) by substituting $\tilde{e}_i$ for $\hat{e}_i$. This procedure is repeated until successive estimates of ρ are arbitrarily close. This procedure is called the Cochran-Orcutt iterative technique.

So far we have assumed that the autocorrelation between error terms e_i is first-order and that the error terms u_i are not autocorrelated. If the error terms u_i are in fact autocorrelated, then we may have a case of linear autocorrelation of a higher order. A method of determining whether higher-order autocorrelation exists is to estimate the Durbin-Watson statistic with respect to the estimated error terms $\hat{u}_i = \hat{e}_i - \rho \hat{e}_{i-1}$. If there is significant autocorrelation of the u_i, we may assume at least second-order autocorrelation. In general, for pth-order autocorrelation, the error term e_i is assumed to be correlated not only with e_{i-1} but also with $e_{i-2}, \ldots, e_{i-p}$. The relationship among error terms is assumed to be as follows:

$$e_i = \rho_1 e_{i-1} + \rho_2 e_{i-2} + \cdots + \rho_p e_{i-p} + u_i, \qquad (11.16)$$

where the error term u_i satisfies the basic Assumptions 1, 2, and 3. An ordinary least-squares regression is run on the original variables. The estimated errors $e_1, e_2, \ldots, e_N$ are then used to estimate equation (11.16) to obtain estimates of $\rho_1, \rho_2, \ldots, \rho_p$. The original variables are then transformed as follows:

$$Y'_i = Y_i - \hat{\rho}_1 Y_{i-1} - \hat{\rho}_2 Y_{i-2} - \cdots$$
$$- \hat{\rho}_p Y_{i-p} \qquad \text{for } i = p + 1, \ldots, N \qquad (11.17)$$

and

$$X'_{1i} = X_{1i} - \hat{\rho}_1 X_{1,i-1} - \hat{\rho}_2 X_{1,i-2} - \cdots - \hat{\rho}_p X_{1,i-p}$$

$$\vdots \qquad (11.18)$$

$$X'_{Ki} = X_{Ki} - \hat{\rho}_1 X_{K,i-1} - \hat{\rho}_2 X_{K,i-2} - \cdots$$
$$- \hat{\rho}_p X_{K,i-p} \qquad \text{for } i = p + 1, \ldots, N.$$

The transformed variables are then used to estimate the equation

$$Y'_i = a_0(1 - \hat{\rho}_1 - \hat{\rho}_2 - \cdots - \hat{\rho}_p) + a_1X'_{1i} + \cdots + a_KX'_{Ki} + u_i. \quad (11.19)$$

Note, however, that there are p fewer observations on the transformed variables than on the original variables. The effect of fewer observations is to reduce the efficiency of the estimation procedure. Thus the elimination of autocorrelation by extending the order of the correlation must be balanced against the loss of observations.

The usual F and t tests should be applied to the parameter estimates of the transformed equation (11.12) or (11.19) and not to the ordinary least-squares estimates of the original regression equation (11.1). One must remember, however, that the number of observations of the transformed variables is $N - p$ and not N.

Heteroskedasticity

Assumption 3 concerning the error term states that the error terms $e_1, e_2, \ldots, e_N$ all have the same variance. If this is not the case, then the errors are said to be heteroskedastic. Heteroskedasticity arises, for example, when errors in the data are related to the size of the dependent variable. If one is estimating a consumption function of the form

$$C_i = a + b \cdot Y_i + e_i, \quad (11.20)$$

where C_i is consumption and Y_i is income, and the errors in the measurement of consumption are large in absolute value when consumption and income are large, and small in absolute value when consumption and income are small, the errors will be heteroskedastic. Heteroskedasticity is likely to arise particularly in studies based on cross-section data rather than time-series data. For example, in using household budget surveys to measure expenditure elasticities for various commodities, the errors in expenditures for small, low-income households are likely to be considerably smaller than for large, high-income households. Heteroskedasticity will be particularly prevalent if the data cover a large range of expenditures and incomes.

Heteroskedasticity alone does not introduce bias into ordinary least-squares estimates of coefficients, but the estimates of the variance of the least-squares coefficients $\hat{a}_0, \hat{a}_1, \ldots, \hat{a}_N$ may be biased downward. The usual t and F tests of significance will tend to give incorrect results; coefficients will be classed as significant when in fact they are not significant.

ESTIMATION PROCEDURES

If the variances $\sigma^2_{e_1}, \sigma^2_{e_2}, \ldots, \sigma^2_{e_N}$ are known, an efficient estimation procedure is to minimize the weighted sum of squares:

$$S = \sum_{i=1}^{N} \frac{e_i^2}{\sigma_{e_i}^2} = \sum_{i=1}^{N} \frac{(Y_i - a_0 - a_1 X_{1i} - \cdots - a_K X_{Ki})^2}{\sigma_{e_i}^2}. \tag{11.21}$$

It is easy to see from (11.21) that minimizing S is equivalent to transforming the original data as follows:

$$Y_i' = \frac{Y_i}{\sigma_{e_i}}, \quad X_{1i}' = \frac{X_{1i}}{\sigma_{e_i}}, \quad \ldots, \quad X_{Ki}' = \frac{X_{Ki}}{\sigma_{e_i}}, \tag{11.22}$$

and defining a new variable

$$X_{0i}' = \frac{1}{\sigma_{e_i}}. \tag{11.23}$$

We then find the ordinary least-squares estimates of the following equation in the transformed variables:

$$Y_i' = a_0 X_{0i}' + a_1 X_{1i}' + \cdots + a_K X_{Ki}' + \frac{e_i}{\sigma_{e_i}}. \tag{11.24}$$

The error term e_i/σ_{e_i} of this transformed equation is not heteroskedastic; that is, its variance is independent of the index i.

Note that the transformation (11.22) is equivalent to multiplying the original observations by a series of weights $1/\sigma_{e_1}, 1/\sigma_{e_2}, \ldots, 1/\sigma_{e_N}$. Hence, this kind of regression is called *weighted regression*. The problem, however, is how to find a reasonable method of determining the weights. Several different methods have been used by various authors. We shall describe two of them.

One possible approach is to assume that the variance of the error term is proportional to the square of one of the independent variables. For example,

$$\sigma_{e_i}^2 = \lambda X_{1i}^2. \tag{11.25}$$

The weights then are $1/X_{11}, 1/X_{12}, \ldots, 1/X_{1N}$. This is equivalent to regressing the independent variable Y_i/X_{1i} on $1/X_{1i}, X_{2i}/X_{1i}, \ldots, X_{Ki}/X_{1i}$ for $i = 1, \ldots, N$.

Example 11.4. Suppose we have the following data on average daily newspaper sales (Y), population (X_1), and Gross Domestic Product (X_2) for ten different countries:

Y_i (MILLIONS)	X_{1i} (MILLIONS)	X_{2i} (BILLIONS OF U.S. DOLLARS)
.2	2.3	.8
102.0	180.2	743.2
2.9	7.8	14.5
41.5	140.1	110.4
4.8	˙15.2	20.2
3.7	21.3	15.6
.8	5.7	3.0
4.2	86.8	15.3
1.8	14.9	5.7
45.0	543.2	115.2

Let us assume that we want to determine the relationship between newspaper sales, population, and income on the basis of the following linear regression equation:

$$Y_i = a_0 + a_1 X_{1i} + a_2 X_{2i} + e_i. \tag{11.26}$$

The error term e_i is likely to be quite heteroskedastic, since one would expect that absolute errors in measurement of newspaper sales would be much greater for a country with a population of 543.2 million as compared to a country with a population of 2.3 million. It would be reasonable to assume, however, that the variance of the error term is proportional to the square of the population (that is, the standard deviation is proportional to population). We could then transform all the variables by dividing each by population. The result is newspaper sales per capita (Y'), Gross Domestic Product per capita (X_2'), and a new variable, the inverse of population (X_0').

$Y_i' = Y_i/X_{1i}$ (NEWSPAPER SALES PER CAPITA)	$X_{0i} = 1/X_{1i}$ (INVERSE OF POPULATION)	$X_{2i} = X_{2i}/X_{1i}$ (GDP PER CAPITA IN U.S. DOLLARS)
.0870	.4348	348
.5660	.0055	4,124
.3718	.1282	1,859
.2962	.0071	788
.3158	.0658	1,329
.1737	.0469	732
.1404	.1754	526
.0484	.0115	176
.1208	.0671	383
.0828	.0018	212

These data may be used to estimate the coefficients of the regression equation

$$Y'_i = a_1 + a_0 X'_{0i} + a_2 X_{2i}. \tag{11.27}$$

The result is

$$\hat{a}_1 = .07992, \qquad \hat{a}_0 = .8412, \qquad \hat{a}_2 = .0001264$$

or

$$Y' = .07992 + .08412 X'_0 + .0001264 X'_2 \tag{11.28}$$

in terms of the transformed variables and

$$Y = .08412 + .07992 X_1 + .0001264 X_2 \tag{11.29}$$

in terms of the original variables.

Another way to handle heteroskedasticity is to divide the sample into groups. The estimated variance of the residuals within each of the groups can then be used in providing the weights for the data within each of the groups.

Example 11.5. Let us divide the data from Example 11.4 into two groups—countries with a population of less than 50 million and those with a population greater than 50 million. The overall regression line for the raw data is

$$Y = 1.8361 + .0589 X_1 + .1232 X_2.$$

The estimated variance of the residuals within the first group of data is calculated as follows:

Y_i	X_{1i}	X_{2i}	$\hat{Y}_i$	$\hat{e}_i = \hat{Y}_i - Y_i$	$\hat{e}_i^2$
.2	2.3	0.8	2.1	+1.9	3.61
2.9	7.8	14.5	4.1	+1.2	1.44
4.8	15.2	20.2	5.3	+.5	.25
3.7	21.3	15.6	5.1	+1.4	1.96
.8	5.7	3.0	2.6	+1.8	3.24
1.8	14.9	5.7	3.5	+1.7	2.89
					13.39

An estimate of the variance within the first group is

$$\hat{\sigma}_{e_1}^2 = \frac{\Sigma e_i^2}{6} = \frac{13.39}{6} = 2.23.$$

Within the second group, we also calculate an estimate of the variance.

Y_i	X_{1i}	X_{2i}	$\hat{Y}_i$	$\hat{e}_i = \hat{Y}_i - Y_i$	$\hat{e}_i^2$
102.0	180.2	743.2	102.7	$+.7$	.49
41.5	140.1	110.4	23.7	-17.8	316.84
4.2	86.8	15.3	8.9	$+4.7$	22.09
45.0	543.2	115.2	48.1	$+3.1$	9.61
					349.03

An estimate of the variance in the second group is

$$\hat{\sigma}_{e_2}^2 = \frac{\Sigma \hat{e}_i^2}{4} = \frac{349.03}{4} = 87.26.$$

The data in the first group then are multiplied by the weight

$$\frac{1}{\hat{\sigma}_{e_1}} = \frac{1}{1.49} = .6711,$$

while the data in the second group are multiplied by the weight

$$\frac{1}{\hat{\sigma}_{e_2}} = \frac{1}{9.34} = .1071.$$

The resulting transformed observations are:

	Y_i'	X_{1i}'	X_{2i}'
	.013	1.544	.537
	1.946	5.235	9.731
	3.221	10.201	13.556
Group 1	2.483	14.294	10.469
	.537	3.825	2.013
	1.208	9.999	3.825
	10.924	19.299	79.597
Group 2	4.445	15.005	11.824
	.450	9.296	1.639
	4.820	58.177	12.338

The regression equation using the transformed variables is

$$Y' = 1.2065 + .0171X_1' + .1063X_2'. \tag{11.30}$$

TESTING FOR HETEROSKEDASTICITY

A quick way to test for heteroskedasticity is to look at a scatter diagram of the residuals. In Fig. 11.5 the error terms seem to be heteroskedastic, since the residuals seem to become larger as the value of X increases. There

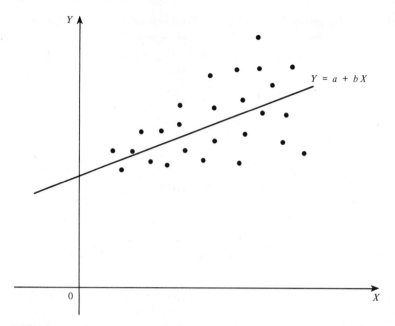

FIGURE 11.5   Heteroskedastic Residuals

are, however, a number of analytical tests for heteroskedasticity, all of which have various drawbacks in that arbitrary assumptions of one sort or another are required. We shall describe here Bartlett's test. This test involves breaking down the data into k groups and estimating the variance s_i^2 of the residuals within each group. Suppose the number of observations in each group is N_i for $i = 1, \ldots, K$. We calculate the ratio Q/l, where

$$Q = n \cdot \log_e \left(\sum_{i=1}^{K} \frac{n_i}{n} s_i^2 \right) - \sum_{i=1}^{K} n_i \cdot \log_e s_i^2, \tag{11.31}$$

$$l = 1 + \frac{1}{3(K-1)} \left(\sum_{i=1}^{K} \frac{1}{n_i} - \frac{1}{n} \right) \tag{11.32}$$

and where $n_i = N_i - K - 1$ is the number of degrees of freedom within each group and

$$n = \sum_{i=1}^{K} n_i.$$

It can be shown that when the error terms e_i are normally distributed, the ratio Q/l has a chi-square distribution with $K - 1$ degrees of freedom.†

---

† Q/l has a chi-square distribution only if each of the K groups has the same values of the X variables within the group. For most purposes, however, we can assume that Q/l is approximately distributed as chi-square even when this is not the case.

Example 11.6. In Example 11.5, we divided the data into two groups of countries—those with a population of less than 50 million and those with a population over 50 million. The estimated variance within the first group was

$$s_1^2 = \hat{\sigma}_{e_1}^2 = 2.23.$$

Within the second group the variance was estimated to be

$$s_2^2 = \hat{\sigma}_{e_2}^2 = 87.26.$$

The quantities Q and l are calculated as follows:

GROUP	n_i	$(n_i/n)s_i^2$	$n_i \cdot \log_e s_i^2$
1	$n_1 = 3$	$(3/4) \cdot (2.23) = 1.673$	$(3) \cdot (.8020) = 2.406$
2	$n_2 = 1$	$(1/4) \cdot (87.26) = 21.815$	$(1) \cdot (4.4689) = 4.469$
		23.488	6.875

$$Q = (4) \cdot \log_e (23.488) - 6.875 = 4(3.1564) - 6.875$$
$$= 12.6256 - 6.875 = 5.7506,$$
$$l = 1 + \frac{1}{3(2-1)}\left(\frac{1}{3} + \frac{1}{1} - \frac{1}{4}\right) = 1.5277,$$
$$\frac{Q}{l} = \frac{5.7506}{1.5277} = 3.764.$$

With one degree of freedom and a 5 percent critical region, the critical value of the chi-square variate is 3.841. Since 3.764 is less than the critical value, we conclude that there is not significant heteroskedasticity.

Errors in the Independent Variables

All of the analysis of stochastic linear models presented in this volume up to this point is based on the assumption that the independent variables $X_1, X_2, \ldots, X_K$ are not random. If these variables are in fact random, the same analysis applies if the error term e_i is uncorrelated with the independent variables. If the independent variables are random and are correlated with the error term e_i, however, ordinary least-squares produces biased and inconsistent estimates of the parameters of a linear function. The usual t and F tests of significance are not applicable. Errors in the independent variables occur often in economics, so that one must be very careful in applying ordinary least-squares techniques.

To keep matters simple, let us consider the case of a single independent variable. Let us denote the observed values of the variables Y and X by Y_i^* and X_i^* for $i = 1, \ldots, N$. The error in the dependent variable we denote by u_i and the error in the independent variable by v_i. The "true" or expected values are denoted by Y_i and X_i. Then

$$Y_i^* = Y_i + u_i, \qquad X_i^* = X_i + v_i. \tag{11.33}$$

The "true" values are presumed to be connected by a linear relationship

$$Y_i = a + b \cdot X_i. \tag{11.34}$$

The errors u_i and v_i are assumed to satisfy the usual Assumptions 1, 2, and 3.

WEIGHTED SUMS OF SQUARES

If the variance of the errors u_i in the dependent variable are proportional to the variance of the errors v_i in the independent variable, unbiased and efficient estimates can be obtained by minimizing a weighted sum of the squared residuals. This method, however, can only be used when one knows the coefficient of proportionality λ. (In general one would not know this.) The weighted sum of squared residuals is

$$\begin{aligned} S &= \lambda \sum_{i=1}^{N} u_i^2 + (1 - \lambda) \sum_{i=1}^{N} v_i^2 \\ &= \lambda \sum_{i=1}^{N} (Y_i^* - Y_i)^2 + (1 - \lambda) \sum_{i=1}^{N} (X_i^* - X_i)^2, \end{aligned} \tag{11.35}$$

where the variance of the errors u_i is related to the variance of the errors v_i as follows

$$\sigma_u^2 = \frac{1 - \lambda}{\lambda} \sigma_v^2. \tag{11.36}$$

Note that if $\lambda = 0$, the errors in the independent variable have zero variance, the sum S in (11.35) becomes the usual vertical sum of squares, and ordinary least-squares estimates minimize the sum S. If $\lambda = 0$, the errors in the dependent variable have zero variance.

Let us substitute the expression (11.34) for Y_i into the sum of squares (11.33). The result is

$$S = \lambda \sum_{i=1}^{N} (Y_i^* - a - b \cdot X_i)^2 + (1 - \lambda) \sum_{i=1}^{N} (X_i^* - X_i)^2. \tag{11.37}$$

The sum S becomes a function of the observed data Y_i^* and X_i^*, the parameters a and b, and the nonobserved "true" variables X_i for $i = 1, \ldots, N$.

We must find values for a, b, and the nonobserved X_i that minimize the sum S. The values for a and b that minimize the sum of squares are found by solving the following equations:†

$$a = \bar{Y}^* - b\bar{X}^*, \tag{11.38}$$

$$\alpha_1 b^2 + \alpha_2 b + \alpha_3 = 0, \tag{11.39}$$

where

$$\alpha_1 = \lambda \sum_{i=1}^{N} (Y_i^* - \bar{Y}^*)(X_i^* - \bar{X}^*),$$

$$\alpha_2 = (1 - \lambda) \sum_{i=1}^{N} (X_i^* - \bar{X}^*)^2 - \lambda \sum_{i=1}^{N} (Y_i^* - \bar{Y}^*)^2, \tag{11.40}$$

$$\alpha_3 = -(1 - \lambda) \sum_{i=1}^{N} (Y_i^* - \bar{Y}^*)(X_i^* - \bar{X}^*).$$

The equation (11.39) is a quadratic expression which, in general, has two solutions. They are

$$b = \frac{-\alpha_2 + \sqrt{\alpha_2^2 - 4\alpha_1\alpha_3}}{2\alpha_1} \tag{11.41}$$

and

$$b = \frac{-\alpha_2 - \sqrt{\alpha_2^2 - 4\alpha_1\alpha_3}}{2\alpha_1}. \tag{11.42}$$

The first of these solutions is always positive and the second always negative. The positive solution is used whenever

$$\sum_{1=i}^{N} (Y_i^* - \bar{Y}^*)(X_i^* - \bar{X}^*) > 0; \tag{11.43}$$

otherwise the negative solution is used.

Example 11.7.   Consider the following data:

Y_i^*	X_i^*	$(Y_i^*)^2$	$(X_i^*)^2$	$Y_i^* \cdot X_i^*$
3	4	9	16	12
2	1	4	1	2
5	7	25	49	35
6	8	36	64	48
4	5	16	25	20
20	25	90	155	117

† See the appendix to this chapter for a derivation.

$\bar{Y}^* = 4, \qquad \bar{X}^* = 5.$

$\Sigma (X_i^* - \bar{X}^*)^2 = \Sigma (X_i^*)^2 - N \cdot (\bar{X}^*)^2 = 155 - 5(25) = 30,$

$\Sigma (Y_i^* - \bar{Y}^*)^2 = \Sigma (Y_i^*)^2 - N \cdot (\bar{Y}^*)^2 = 90 - 5(16) = 10,$

$\Sigma (Y_i^* - \bar{Y}^*)(X_i^* - \bar{X}^*) = \Sigma Y_i^* X_i^* - N \cdot \bar{Y}^* \cdot \bar{X}^* = 117 - 5(20) = 17.$

Let us set $\lambda = 1/3$. Then from (11.40)

$$\alpha_1 = \frac{1}{3} \cdot 17 = \frac{17}{3},$$

$$\alpha_2 = \frac{2}{3} \cdot 30 - \frac{1}{3} \cdot 10 = \frac{50}{3},$$

$$\alpha_3 = -\frac{2}{3} \cdot 17 = -\frac{34}{3}.$$

We use the positive solution (11.41), since $\Sigma (Y_i^* - \bar{Y}^*)(X_i^* - \bar{X}^*) > 0$. Thus

$$b = \frac{-\dfrac{50}{3} + \sqrt{\left(\dfrac{50}{3}\right)^2 + 4\left(\dfrac{17}{3}\right)\left(\dfrac{34}{3}\right)}}{2\left(\dfrac{17}{3}\right)},$$

$$b = \frac{-50 + 69.37}{34} = .5697.$$

Note that if $\lambda = 0$, from (11.40) we have

$\alpha_1 = 0,$
$\alpha_2 = \Sigma (X_i^* - \bar{X}^*)^2,$
$\alpha_3 = -\Sigma (Y_i^* - \bar{Y}^*)(X_i^* - \bar{X}^*).$

If we substitute these values into (11.39) and solve for b, we obtain

$$b = \frac{\Sigma (Y_i^* - \bar{Y}^*)(X_i^* - \bar{X}^*)}{\Sigma (X_i^* - \bar{X}^*)^2},$$

which is the ordinary (vertical) least-squares estimate of b. Similarly if we set $\lambda = 1$, we obtain the least-squares estimate of b that minimizes the horizontal sum of squares. If $\lambda = 1/2$, the estimates minimize the perpendicular sum of squares (orthogonal regression as discussed in Chapter 3). In general, minimization of a weighted sum of squares is equivalent to minimizing an angular sum of squares as shown in Fig. 11.6. The angle θ is related to λ and the slope b of the regression line as follows:

$$\tan \theta = \frac{\lambda b + (1 - \lambda)}{\lambda + (1 - \lambda)b}. \tag{11.44}$$

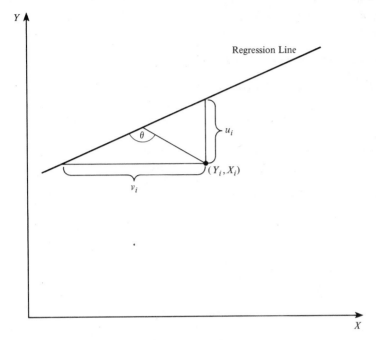

FIGURE 11.6   Angular Regression

The absolute value of the estimate of b as determined from the angular sum of squares lies between the two values that are produced by minimizing the horizontal or vertical sum of squares.

Example 11.8.   Using the data from Example 11.7, we have

$$b^* = \frac{\Sigma (Y_i^* - \bar{Y}^*)(X_i^* - \bar{X}^*)}{\Sigma (X_i^* - \bar{X}^*)^2} = \frac{17}{30} = .567$$

when minimizing the vertical sum of squares, and

$$b^{**} = \frac{\Sigma (Y_i - \bar{Y}^*)^2}{\Sigma (Y_i^* - \bar{Y}^*)(X_i^* - \bar{X}^*)} = \frac{10}{17} = .588$$

when minimizing the horizontal sum of squares. In Example 11.7, we estimated an angular sum of squares

$$b = .5697,$$

which lies between b^* and b^{**}.

The difficulty with the above method of estimation is that it assumes we know λ, the coefficient of proportionality between the variances of the two kinds of errors. One might be able to make a reasonable guess as to the magnitude of λ, but it is difficult to lay down general rules for its estimation.

INSTRUMENTAL VARIABLES

Another approach that gives consistent estimates of parameters when the independent variable is subject to errors is the method of instrumental variables. Suppose Z is a third variable which is independent of the error terms u and v. Consistent estimates of the parameters a and b are obtained by solving the following equations:

$$a = \bar{Y}^* - b \cdot \bar{X}^*, \tag{11.45}$$

$$b = \frac{\Sigma (Y_i^* - \bar{Y}^*)(Z_i - \bar{Z})}{\Sigma (X_i^* - \bar{X}^*)(Z_i - \bar{Z})}. \tag{11.46}$$

The variable Z is called an instrumental variable. While the estimate of b in (11.46) may be consistent, it may be very inefficient—that is, have a very high sampling variance. The variance is reduced if Z and X are highly correlated. Thus the choice of instrumental variables is not completely arbitrary. We should choose a variable that has a high correlation with the independent variable.

Example 11.9.   Suppose the variables X and Y represent Gross Domestic Product and government tax revenue, respectively. Suppose there are errors in both variables, however, so that we choose an instrumental variable that seems to have a high correlation with X. In this case, suppose Z is the number of automobiles registered, and that we have the following data:

Y_i^*	X_i^*	Z_i	$Y_i^* Z_i$	$X_i^* Z_i$
3	4	5	15	20
2	1	1	2	1
5	7	9	45	63
6	8	6	36	48
4	5	4	16	20
20	25	25	114	152

$$\Sigma (Y_i^* - \bar{Y}^*)(Z_i - \bar{Z}) = \Sigma Y_i^* Z_i - N \cdot \bar{Y}^* \cdot \bar{Z} = 114 - 5(4.5) = 14,$$
$$\Sigma (X_i^* - \bar{X}^*)(Z_i - \bar{Z}) = \Sigma X_i^* Z_i - N \cdot \bar{X}^* \cdot \bar{Z} = 152 - 5(5 \cdot 5) = 27,$$
$$b = \frac{14}{27} = .519.$$

Note that this estimate of b is less than both estimates obtained in Example 11.8 by minimizing the vertical and horizontal squares.

There are other techniques for handling errors in the independent varia-bles, and the weighted-sum-of-squares regression and instrumental-variable techniques can be extended to the case of more than one independent variable. None of these techniques is very adequate, however, and all involve the making of arbitrary assumptions that may be difficult to justify. In many cases, the best procedure is to assume that errors of measurement in the independent variable are relatively small compared to the errors in the dependent variable introduced by random shocks and the effects of nonincluded variables. If this is the case, then ordinary least-squares techniques are not seriously biased and the estimation pro-cedure is relatively efficient. Furthermore, errors in independent variables do not introduce serious bias whenever the correlation between the de-pendent and independent variables is high. The bias is most serious in weak fitting relationships. Finally, we ought to note that in certain cases, although the estimates of a and b are seriously biased, the use of the regres-sion equation for prediction purposes may produce predictions that are relatively unbiased or at any rate consistent.

Multicollinearity

Another problem that frequently arises in estimation of linear regression equations is that of *multicollinearity*. If there is more than one independent variable and these independent variables are correlated among them-selves, the sampling variances of the estimated parameters will tend to be quite large. Thus when two independent variables, say X_1 and X_2, are closely correlated, the parameters a_1 and a_2 are unlikely to be significant. Even though it may be quite clear that the *combined* effect of X_1 and X_2 on the dependent variable Y is significant, a high degree of correlation between them makes it difficult to distinguish the separate effects of these variables. For example, the regression equation

$$Y = a_0 + a_1 \cdot X_1 + a_2 \cdot X_2 \qquad (11.47)$$

may have a highly significant coefficient of determination R^2 but the two parameters a_1 and a_2 may not be significant.

The presence of multicollinearity has important effects on so-called "stepwise" regression procedures. Stepwise regression involves running the simple regression

$$Y = a_0 + a_1 \cdot X_1. \qquad (11.48)$$

If the coefficient a_1 proves to be significant, the variable X_1 is retained in the regression and a new regression is run by adding the second independent variable X_2. If the second variable does not add significant additional explanation of the variation in dependent variable (a_2 not significant), then this variable is dropped from the regression altogether. If X_1 and X_2 are highly correlated, a_1 may prove to be significant while a_2 is not significant. If the order of introduction is reversed, however, a_2 may prove to be significant while a_1 is not. That is, the variables retained in a stepwise regression procedure depend very crucially on the *order* in which they are introduced if there is a high degree of multicollinearity.

Example 11.10.   In order to illustrate some of the problems which occur with multicollinearity, let us consider the following hypothetical data.

Y	X_1	X_2
9	9	7
5	4	2
8	8	4
6	7	3
8	8	4
5	4	2
9	9	7
6	7	3

X_1 and X_2 are highly correlated. In fact their correlation coefficient is .857. If we regress Y on X_1 alone, we obtain the following regression equation:

$$Y = 1.56 + .79X_1.$$

The estimated standard deviation of the coefficient of X_1 is .31, so that the t ratio is

$$\frac{.79}{.31} = 2.548$$

with 6 degrees of freedom. This is significant at the 5 percent level. If, however, we use both X_1 and X_2 as regressors, we obtain the following regression equation:

$$Y = 2.35 + .42X_1 + .42X_2.$$

The estimated standard deviation for the coefficients of both X_1 and X_2 is .43. Thus neither coefficient is significant at the 5 percent level.

Note that the coefficient of X_1 changes markedly when X_2 is added to the regression. The coefficient is reduced almost by half. This is characteristic of multicollinearity. If X_1 and X_2 were uncorrelated, however, the coefficient of X_1 would not be changed by the addition of X_2 in the regression equation.

We have seen that multicollinearity makes the variance of estimates quite large and makes it difficult to determine the significance of variables. Multicollinearity, however, does *not* destroy the unbiasedness and efficiency of least-squares estimators. Furthermore, estimates of the sampling variance of the parameters are unbiased. Thus one can have the usual degree of confidence in tests of hypotheses.

It is frequently thought that if there is multicollinearity between two independent variables, one of the variables should be eliminated from the regression. This need not be the case. If one wishes to use a regression equation for purposes of prediction, the power of the predictions may be enhanced if both variables are included. This is especially true if the correlation between the variables is expected to continue in the future as it has in the past data on which the estimated equation is based.

There are a number of ways in which one might avoid the problems caused by multicollinearity. For example, multicollinearity can sometimes be avoided by changing the specification of the model.

Example 11.11.   In Example 11.4, the original data Y_i, X_{1i}, and X_{2i} are slightly collinear. The correlation coefficient between X_1 and X_2 is $R = .314$. The transformed data Y'_i, X'_{0i}, and X'_{2i} obtained by dividing through by X_{1i} do not indicate the same degree of multicollinearity. The correlation coefficient between X'_{0i} and X'_{2i} is $R = .229$. Thus the problems associated with multicollinearity are avoided somewhat by running the transformed regression equation (11.26) as opposed to the regression equation in the original variables.

Conclusion

This chapter has covered a number of problems associated with using ordinary least-squares estimates, testing procedures for determining whether significant difficulties exist, and some of the modified estimation procedures for dealing with these problems. These modified procedures have their weaknesses and frequently rely on somewhat arbitrary assumptions.

The estimators discussed in the section on autocorrelation and the weighted regression estimators discussed in connection with heteroskedasticity are special cases of what is called *generalized least-squares* estimators (also called Aitken estimators). A discussion of generalized least-squares requires a knowledge of matrix algebra. The interested reader may refer to the appendix of this chapter.

The discussion of multicollinearity in this chapter has assumed that the correlation among the independent variables arose through chance. If the independent variables are functionally related, however, an additional problem is introduced. The estimates may be biased and inconsistent. For example, we might postulate that expenditures on housing are a function of income and family size. Income and family size, however, may be related also, especially if there are several adults of working age in family groups covered by the data. Methods of estimating relationships of this sort are discussed in the next chapter.

APPENDIX   Further Discussion

Weighted Sums of Squares

Given the weighted sum of squares,

$$S = \lambda \sum_{i=1}^{N} u_i^2 + (1 - \lambda) \sum_{i=1}^{N} v_i^2$$

$$= \lambda \sum_{i=1}^{N} (Y_i^* - a - b - X_i)^2 + (1 - \lambda) \sum_{i=1}^{N} (X_i^* - X_i)^2, \qquad (11.49)$$

determine the parameters a and b and the unknown X_i so as to minimize S
We differentiate S with respect to a, b, and X_i for $i = 1, \ldots, N$ and set the
derivatives equal to zero. Thus

$$\frac{\partial S}{\partial a} = -2\lambda \sum_{i=1}^{N} (Y_i^* - a - b \cdot X_i) = 0, \qquad (11.50)$$

$$\frac{\partial S}{\partial b} = -2\lambda \sum_{i=1}^{N} (Y_i^* - a - b \cdot X_i)(X_i) = 0, \qquad (11.51)$$

$$\frac{\partial S}{\partial X_i} = -2\lambda(Y_i^* - a - b \cdot X_i) \cdot b - 2(1 - \lambda)(X_i^* - X_i) = 0, \qquad (11.52)$$

for $i = 1, \ldots, N$. Let us solve (11.52) for X_i.

$$X_i = \frac{\lambda b \cdot Y_i^* - \lambda a \cdot b + (1 - \lambda)X_i^*}{1 - \lambda + \lambda b^2}. \qquad (11.53)$$

If we substitute (11.53) into (11.52) and (11.51), we get

$$\frac{\partial S}{\partial a} = \frac{-2\lambda(1 - \lambda)}{1 - \lambda + \lambda b^2} \sum_{i=1}^{N} (Y_i^* - a - b \cdot X_i^*) = 0, \qquad (11.54)$$

$$\frac{\partial S}{\partial b} = \frac{-2\lambda(1 - \lambda)}{(1 - \lambda + \lambda b^2)^2} \sum_{i=1}^{N} (Y_i^* - a - b \cdot X_i^*)[\lambda b \cdot Y_i^* - \lambda a \cdot b + (1 - \lambda)X_i^*] = 0,$$

$$\qquad (11.55)$$

which gives $\partial S/\partial a$ in terms of the unknown parameters a and b and the
observed values Y_i^* and X_i^*. We can divide both sides of these equations by
$2\lambda(1 - \lambda)/(1 - \lambda + \lambda b^2)$ and collect terms to obtain

$$-\sum_{i=1}^{N} (Y_i^* - a - b \cdot X_i^*) = 0, \tag{11.56}$$

$$-\lambda b \sum_{i=1}^{N} (Y_i^* - a - b \cdot X_i^*)Y_i^* - \lambda b \cdot a \sum_{i=1}^{N} (Y_i^* - a - b \cdot X_i^*)$$

$$- (1 - \lambda) \sum_{i=1}^{N} (Y_i^* - a - b \cdot X_i^*)X_i^* = 0. \tag{11.57}$$

From (11.56) we see that the second term in (11.57) goes to zero, so that we may write

$$-\lambda b \sum_{i=1}^{N} (Y_i^* - a - b \cdot X_i^*)Y_i^* - (1 - \lambda) \sum_{i=1}^{N} (Y_i^* - a - b \cdot X_i^*)X_i^* = 0. \tag{11.58}$$

Now (11.56) may be solved for a.

$$N \cdot a = \Sigma Y_i^* - b \Sigma X_i^* \quad \text{or} \quad a = \bar{Y}^* - b \cdot \bar{X}^*. \tag{11.59}$$

Substitute (11.59) into (11.58) to obtain

$$-\lambda b[\Sigma (Y_i^* - \bar{Y}^*)Y_i^* - b \Sigma (X_i^* - \bar{X}^*)Y_i^*]$$
$$- (1 - \lambda)[\Sigma (Y_i^* - \bar{Y}^*)X_i^* - b \Sigma (X_i^* - \bar{X}^*)X_i^*] = 0$$

or

$$b^2\lambda \Sigma (X_i^* - \bar{X}^*)Y_i^* + b[(1 - \lambda) \Sigma (X_i^* - \bar{X}^*)X_i^*$$
$$- \lambda \Sigma (Y_i^* - \bar{Y}^*)Y_i^*] - (1 - \lambda) \Sigma (Y_i^* - \bar{Y}^*)X_i^* = 0. \tag{11.60}$$

Thus

$$\alpha_1 b^2 + \alpha_2 b + \alpha_3 = 0, \tag{11.61}$$

where

$$\alpha_1 = \lambda \sum_{i=1}^{N} (X_i^* - \bar{X}^*)Y_i^*,$$

$$\alpha_2 = (1 - \lambda) \sum_{i=1}^{N} (X_i^* - \bar{X}^*)X_i^* - \lambda \sum_{i=1}^{N} (Y_i^* - \bar{Y}^*)Y_i^*, \tag{11.62}$$

$$\alpha_3 = -(1 - \lambda) \sum_{i=1}^{N} (Y_i^* - \bar{Y}^*)X_i^*.$$

Since, however,

$$\Sigma \ (X_i^* - \bar{X}^*)\bar{Y}^* = \Sigma \ (Y_i^* - \bar{Y})\bar{Y}^*$$
$$= \Sigma \ (X_i^* - \bar{X}^*)\bar{X}^* = \Sigma^*(Y_i^* - \bar{Y}^*)\bar{X}^* = 0, \qquad (11.63)$$

equation (11.62) may be written the same as equations (11.40) in the text of this chapter.

Generalized Least-Squares

The methods suggested for estimating parameters with autocorrelated error terms and weighted regression, as suggested when errors are heteroskedastic, are special cases of what is known as generalized least-squares estimates (sometimes called Aitken estimates). Let σ_{ii}^2 for $i = 1, \ldots , N$ represent the variance of the error term e_i and let σ_{ij}^2 represent the co-variance between the error term e_i and the term e_j, where $j \neq i$. By defini-tion the covariance $\sigma_{ij}^2 = \sigma_{ji}^2$. These variances and covariances can be arranged in the form of a matrix V, where

$$V = \begin{bmatrix} \sigma_{11}^2 & \sigma_{12}^2 & \cdots & \sigma_{1N}^2 \\ \sigma_{21}^2 & \sigma_{22}^2 & \cdots & \sigma_{2N}^2 \\ \cdot & & & \\ \cdot & & & \\ \cdot & & & \\ \sigma_{N1}^2 & \sigma_{N2}^2 & \cdots & \sigma_{NN}^2 \end{bmatrix} \qquad (11.64)$$

If there is no autocorrelation (that is, the successive error terms are dis-tributed independently), then $\sigma_{ij}^2 = 0$ for all $i \neq j$. Then V becomes

$$V = \begin{bmatrix} \sigma_{11}^2 & 0 & \cdots & 0 \\ 0 & \sigma_{22}^2 & \cdots & 0 \\ \cdot & & & \\ \cdot & & & \\ \cdot & & & \\ 0 & 0 & \cdots & \sigma_{NN}^2 \end{bmatrix} \qquad (11.65)$$

If there is no autocorrelation and no heteroskedasticity, then $\sigma_{ij}^2 = 0$ for all $i \neq j$, and σ_{ii}^2 are equal for all $i = 1, \ldots , N$. That is, $\sigma_{ii}^2 = \sigma^2$ for $i = 1, \ldots , N$, and V becomes

$$V = \begin{bmatrix} \sigma^2 & 0 & \cdots & 0 \\ 0 & \sigma^2 & \cdots & 0 \\ \cdot & & & \\ \cdot & & & \\ \cdot & & & \\ 0 & 0 & & \sigma^2 \end{bmatrix} \qquad (11.66)$$

Now the general regression equation is

$$Y_i = a_0 + a_1 X_{1i} + \cdots + a_K X_{Ki} + e_i \tag{11.67}$$

or

$$e_i = Y_i - a_0 - (a_1 X_{1i} + \cdots + a_K X_{Ki}). \tag{11.68}$$

The generalized least-squares estimates are obtained by minimizing the generalized sum of squares,

$$S = (e_1, e_2, \ldots, e_N) V^{-1} (e_1, e_2, \ldots, e_N)', \tag{11.69}$$

where $(e_1, e_2, \ldots, e_N)$ is a row vector of the error terms, $(e_1, e_2, \ldots, e_N)'$ is a column vector of the error terms, and V^{-1} is the inverse of the variance-covariance matrix V. In the case of no heteroskedasticity and no autocorrelation, the variance-covariance matrix V in (11.64) has the following inverse:

$$V^{-1} = \begin{bmatrix} \frac{1}{\sigma^{2N}} & 0 & \cdots & 0 \\ 0 & \frac{1}{\sigma^{2N}} & \cdots & 0 \\ \cdot & & & \\ \cdot & & & \\ \cdot & & & \\ 0 & 0 & \cdots & \frac{1}{\sigma^{2N}} \end{bmatrix} \tag{11.70}$$

The sum of squares S in (11.69) then becomes

$$S = \frac{1}{\sigma^{2N}} \sum_{i=1}^{N} e_i^2. \tag{11.71}$$

Since $1/\sigma^{2N}$ is a constant, minimizing S is equivalent to minimizing $\Sigma\, e_i^2$, which is the ordinary sum of squared residuals.

In the case of heteroskedasticity alone, the inverse of V in (11.65) is

$$V^{-1} = \begin{bmatrix} \frac{1}{\sigma_{11}^2} & 0 & \cdots & 0 \\ 0 & \frac{1}{\sigma_{22}^2} & \cdots & 0 \\ \cdot & & & \\ \cdot & & & \\ \cdot & & & \\ 0 & 0 & \cdots & \frac{1}{\sigma_{NN}^2} \end{bmatrix} \tag{11.72}$$

and the sum of squares is

$$S = \sum_{i=1}^{N} \left(\frac{e_i}{\sigma_{ii}}\right)^2,$$

(11.73)

which is a sum of squares of weighted residuals where the weights are the inverse of the square of the variance.

Finally, if there is linear first-order autocorrelation, the variance-covariance matrix V is

$$V = \begin{bmatrix} \sigma^2 & \rho\sigma^2 & \rho^2\sigma^2 & \cdots & \rho^{N-1}\sigma^2 \\ \rho\sigma^2 & \sigma^2 & \rho\sigma^2 & \cdots & \rho^{N-2}\sigma^2 \\ \rho^2\sigma^2 & \rho\sigma^2 & \sigma^2 & \cdots & \rho^{N-3}\sigma^2 \\ \cdot & & & & \\ \cdot & & & & \\ \cdot & & & & \\ \rho^{N-1}\sigma^2 & \rho^{N-2}\sigma^2 & \rho^{N-3}\sigma^2 & \cdots & \sigma^2 \end{bmatrix}$$

(11.74)

where ρ is the coefficient of autocorrelation. That is, the error terms e_i satisfy the following relationship:

$$e_i = \rho e_{i-1} + u_i,$$

(11.75)

where u_i is an error term that is neither heteroskedastic nor autocorrelated. If σ_u^2 is the variance of the error term u_i, the inverse of V is

$$V^{-1} = \frac{1}{\sigma_u^2} \begin{bmatrix} 1 & -\rho & 0 & \cdots & 0 \\ -\rho & 1+\rho^2 & -\rho & \cdots & 0 \\ 0 & -\rho & 1+\rho^2 & \cdots & 0 \\ \cdot & & & & \\ \cdot & & & & \\ \cdot & & & & \\ 0 & 0 & 0 & \cdots & 1 \end{bmatrix}$$

(11.76)

The sum of error terms in (11.69) becomes

$$S = \left[(1-\rho^2)e_1^2 + \sum_{i=2}^{N}(e_i - \rho e_{i-1})^2\right] \bigg/ \sigma_u^2 = \left[(1-\rho^2)e_1^2 + \sum_{i=2}^{N} u_i^2\right] \bigg/ \sigma_u^2.$$

(11.77)

Since σ_u^2 is a constant term, minimization of the sum of squares S in (11.77) is approximately equivalent to minimization of the sum of squares (11.7) in the text.

PROBLEMS

1. Consider the following quarterly data (seasonally adjusted) on consumption of fine paper (Y) in the United States and real Gross National Product (X).

YEAR	QUARTER	Y	X
1956	1	385	443.6
	2	390	445.6
	3	398	444.5
	4	396	450.3
1957	1	368	453.4
	2	365	453.2
	3	364	455.2
	4	374	448.2
1958	1	372	437.5
	2	371	439.5
	3	387	450.7
	4	407	461.6
1959	1	406	468.6
	2	432	479.9
	3	431	475.0
	4	425	480.4
1960	1	445	490.2
	2	437	489.7
	3	427	487.3
	4	417	483.7
1961	1	430	482.6
	2	453	497.8
	3	470	501.5
	4	489	511.7

The estimated regression line is

$$Y = -284.692 + 1.485X.$$

Using a 5 percent critical region, determine whether there is significant linear first-order autocorrelation.

2. (a) Estimate the first-order autocorrelation coefficient ρ in problem 1. (b) Using the transformation (11.1) in the text, calculate a new regression equation that takes into account a linear first-order autocorrelation structure.

3. The following data give cotton acreage in an African country, the deviation of cotton acreage from a linear trend, and the price paid to growers of seed cotton.

YEAR	COTTON ACREAGE (MILLIONS OF ACRES)	DEVIATION FROM TREND (Y)	PRICES PAID TO GROWERS (X)
1955	1.59	+.05	61
1956	1.57	−.02	55
1957	1.62	−.03	56
1958	2.01	+.30	58
1959	1.57	−.20	47
1960	1.52	−.30	48
1961	2.07	+.17	55
1962	1.80	−.14	57
1963	2.01	+.02	57
1964	2.14	+.09	51
1965	2.24	+.13	56
1966	2.17	.00	60
1967	2.15	−.17	40

(a) Determine the regression line between deviation from trend Y and prices paid to growers (X). (b) Use Bartlett's test to determine whether there is any heteroskedasticity. Divide the sample into two groups consisting of the first seven observations (1955-1961) and the last six observations (1962-1967). Use a 5 percent critical region.

4. Recalculate the regression line from problem 3 using the estimated standard deviations as weights for the variables in the two groups.

5. Suppose the X variable in problem 1 is subject to error and that the variance of that error is assumed to be the same as the variance of the error in the dependent variable Y. Calculate a weighted least-squares regression line.

6. (a) Calculate a regression line from the data of problem 1 using housing construction as an instrumental variable Z. The data on Z are as follows:

YEAR	QUARTER	Z	YEAR	QUARTER	Z
1956	1	20.7	1959	1	23.6
	2	20.2		2	22.1
	3	19.9		3	21.1
	4	20.0		4	20.7
1957	1	19.8	1960	1	20.8
	2	19.6		2	21.1
	3	20.8		3	21.7
	4	22.8		4	22.6
1958	1	24.9	1961	1	23.1
	2	25.4		2	23.9
	3	24.7		3	24.2
	4	23.8		4	23.9

(b) Is housing construction a particularly good choice as an instrumental variable?

7. Suppose there are two independent variables X_1 and X_2. X_2 is always three times X_1. This is a case of perfect multicollinearity. Show that in a case of perfect multicollinearity it is impossible to determine the coefficients of the regression equation. [*Hint:* Set up the normal equations and solve for a_0, a_1, and a_2.]

Chapter **12**

Identification and Simultaneous Estimation

This chapter deals with econometric models in which more than one equation is being estimated. For example, a model of a market might have a demand equation and a supply equation. An aggregate model of an economy might have hundreds of equations. When the structure of an economic model involves more than one equation, the methods of estimation discussed in the preceding chapters are not necessarily applicable. Methods of estimating the parameters of a multiequation model are called *simultaneous estimation procedures*.

Bias in Multiequation Models

Let us consider a simple model of supply and demand. The demand equation is

$$D_t = a + bP_t + e_{dt}, \tag{12.1}$$

which indicates that quantity demanded (D_t) in period t is a function of price (P_t) in period t and an error term e_{dt}. The supply equation may be represented by

$$S_t = c + dP_t + e_{st}, \tag{12.2}$$

which indicates that quantity supplied (S_t) in period t is also a function of price (P_t) in period t and an error term e_{st}. Equilibrium requires that quantity demanded equal quantity supplied or

$$D_t = S_t. \tag{12.3}$$

Suppose the demand curve (12.1) shifts from period to period as the result of fluctuations in the error term. This is shown in Fig. 12.1 as a shift from demand curve D_1 to D_2 to D_3 in periods 1, 2, and 3. If the supply curve also shifts from S_1 to S_2 to S_3, the series of equilibrium prices and quantities are represented by the points E_1, E_2, and E_3. If we fit a least-squares line to these three equilibrium points, we obtain the dotted regression line AB. The line AB is neither a demand curve nor a supply curve; rather it is something of a cross between demand and supply. If we interpret AB as a demand curve, our estimates of the parameters are seriously biased downward. The slope of AB is much smaller than the slope of the true demand curve. If we interpret AB as a supply curve, the estimates of the parameters are seriously biased upward.

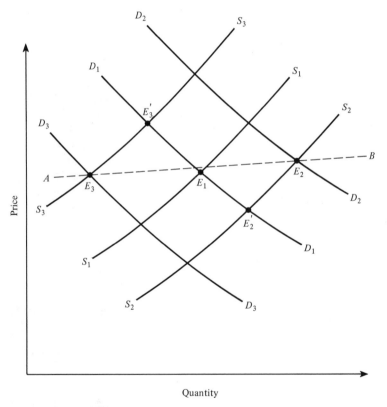

FIGURE 12.1   Shifting Demand and Supply Functions

If the demand curve were to remain stable, however, while the supply curve shifted owing to fluctuations in the error term, the three equilibrium points would be E_1, E_2', and E_3'. A least-squares line through these points would provide an unbiased estimate of the demand curve. Thus the ordinary least-squares estimation procedures can produce relatively unbiased estimates in special circumstances. In general, however, ordinary least-squares is inappropriate in situations where the values of economic variables are determined simultaneously by more than one relationship.

Note that the demand function (12.1) and the supply function (12.2) "look alike" in that, since demand equals supply ($D_t = S_t$) in equilibrium, both equations (12.1) and (12.2) contain the same two variables—quantity demanded (equals quantity supplied) and price. Whenever the same variables are contained in two different equations, it is impossible to estimate the parameters of either equation. As Fig. 12.1 illustrated, if we regress price on quantity, we get a relationship that is a cross between a demand curve and a supply curve. In such a case, we say the equations cannot be statistically identified. Suppose, however, that the demand and supply equations do not "look alike" and we can statistically identify either equation. For example, assume that the demand equation contains an income variable (Y_t) and that the supply equation has a weather variable W_t. The demand and supply equations are

$$D_t = a_0 + b_0 P_t + c_0 Y_t + e_{dt}, \tag{12.4}$$

$$S_t = a_1 + b_1 P_t + c_1 W_t + e_{st}, \tag{12.5}$$

where as before $D_t = S_t$ in equilibrium.

Although the demand and supply equations are identified, least-squares estimates of the parameters of both equations are still biased. Least-squares estimates of the parameters of the demand equation are biased because the price variable P_t is correlated with the error terms e_{dt}.[†] This dependence can be seen if we set quantity demanded equal to quantity supplied and solve for P_t.

$$a_0 + b_0 P_t + c_0 Y_t + e_{dt} = a_1 + b_1 P_t + c_1 W_t + e_{st}$$

or

$$P_t = \frac{(a_1 - a_0) + c_1 W_t - c_0 Y_t + e_{st} - e_{dt}}{b_0 - b_1}. \tag{12.6}$$

Note that P_t depends on the error term e_{dt}. We cannot get around this dependence by switching P_t to the left-hand side of (12.4) and treating D_t as an independent variable, since D_t also depends on the error term e_{dt}. This can be shown by substituting the expression (12.6) into (12.4). The result is

$$D_t = \frac{a_1 b_0 - a_0 b_1 - b_0 c_1 W_t - b_1 c_0 Y_t + b_0 e_{dt} - b_1 e_{st}}{b_0 - b_1} \tag{12.7}$$

[†] In the Appendix to Chapter 8 we showed that the coefficient of a regression equation is unbiased if the independent variable is uncorrelated with the error term.

The bias in the ordinary least-squares estimates of the parameters remains no matter how large a sample we use. In other words, the estimates are inconsistent as well as biased.

Another well-known example of bias in ordinary least-squares concerns the estimation of a consumption function. Let the consumption function be represented by

$$C_t = a + b \cdot Y_t + e_t, \tag{12.8}$$

where C_t is consumption in period t, Y_t is income in period t, and e_t is the error term in period t. The parameter b is the marginal propensity to consume. In addition to equation (12.4) we have the identity

$$C_t + I_t = Y_t, \tag{12.9}$$

where I_t is investment in period t. The consumption function (12.4) is represented by the line A_1B_1 in Fig. 12.2. The consumption function shifts from period to period (owing to fluctuations in the error term e_t) from A_1B_1 to A_2B_2 to A_3B_3. If OC is a 45-degree line and if investment I is constant from period to period, the equilibrium consumption and income are represented by the points E_1, E_2, and E_3. A least-squares regression line fitted through these points is represented by the dotted line GH. Note that in this example the dotted line is parallel to the 45-degree line; that is, it has a slope equal to unity. The marginal propensity to consume is less than unity. The marginal propensity is represented by the slope of the consumption function A_1B_1. The regression line GH produces an overestimate of the marginal propensity to consume. Ordinary least-squares estimates of a consumption function give upwardly biased estimates of the marginal propensity to consume. Furthermore, ordinary least-squares estimates are inconsistent. No matter how large the sample, the estimates do not tend to lose their bias.

The bias arises in the consumption-function case again because the "independent" variable Y_t on the right-hand side of (12.8) is correlated with the error term e_t. To see this, substitute the expression (12.8) for C_t into (12.9) and solve for Y_t. We get

$$Y_t = \frac{a + I_t + e_t}{1 - b}. \tag{12.10}$$

Indirect Least-Squares

One way to produce better estimates of the parameters of a multiequation model is to use the method of indirect least-squares. This method involves estimating not the original *structural equations* of the model but estimating the *reduced-form equations*. Indirect least-squares produces estimates that are biased but consistent. The method is applicable only to structural

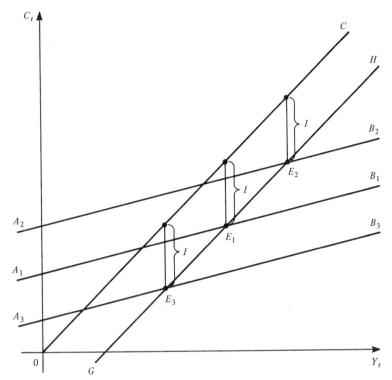

FIGURE 12.2   Estimating a Consumption Function

equations that are *exactly identified*. The meaning of identification is discussed below. For the moment, however, let us illustrate the method with the consumption-function model consisting of equations (12.8) and (12.9). These two equations are called the structural equations.

The reduced form of a model involves expressing the *endogenous* variables of the model in terms of the *exogenous* variables. Endogenous variables are variables whose values are determined simultaneously in equilibrium as the solution to the equations of the model. There are usually as many endogenous variables as there are equations in the model. Exogenous variables are variables whose values are determined independently or "outside" of the model. Exogenous variables are analogous to independent variables in the case of a single-equation model, while an endogenous variable is analogous to a dependent variable.

Which variables are endogenous and which is exogenous is a question of economic theory. There is no way a priori to determine which variables are endogenous and which exogenous merely by looking at the equations of a model. Specifying the exogenous variables of a system is equivalent to making an assumption about the way in which an economic system works.

Example 12.1.   In the Keynesian model represented by equations (12.8) and (12.9), investment I_t is usually assumed to be exogenous while income Y_t and consumption C_t are assumed to be jointly determined by the solution to (12.8) and (12.9). Suppose, for example, the exact specification of the consumption function is

$$C_t = 1.0 + .90Y_t, \tag{12.11}$$

where we assume for the sake of argument that the error term $e_t = 0$ for the particular period under consideration. Suppose investment $I_t = 5.0$. Then (12.5) becomes

$$C_t + 5.0 = Y_t. \tag{12.12}$$

The equilibrium values of consumption and income are determined by solving (12.11) and (12.12) for Y_t and C_t. Substituting the expression (12.12) for Y_t into (12.11), we obtain

$$C_t = 1.0 + .90(C_t + 5.0) \quad \text{or} \quad C_t = \frac{5.5}{.10} = 55.0.$$

Substituting this value for C_t into (12.12), we obtain

$$Y_t = 60.0.$$

If investment assumes any other value, we can determine C_t and Y_t in a similar fashion.

The reduced form of a model is obtained by solving the structural equations for the endogenous variables in terms of the exogenous variables. For (12.8) and (12.9) we may do this by first substituting the expression (12.9) for Y_t into (12.8). We obtain

$$C_t = a + b(C_t + I_t) + e_t$$

or

$$C_t = \frac{a}{1 - b} + \frac{b}{1 - b} I_t + \frac{e_t}{1 - b}. \tag{12.13}$$

If this expression for C_t is substituted into (12.9), we obtain

$$Y_t = \frac{a}{1 - b} + \frac{1}{1 - b} I_t + \frac{e_t}{1 - b}. \tag{12.14}$$

Equations (12.13) and (12.14) constitute the reduced-form equations of the structural equations (12.8) and (12.9).

For the demand and supply model (12.4) and (12.5) we normally think of quantity demanded (equals quantity supplied in equilibrium) and price as the two endogenous variables, while income Y_t and weather W_t are regarded as exogenous variables. The reduced form is given by the two equations (12.6) and (12.7), of which the first expresses price in terms of the exogenous variables W_t and Y_t and the second expresses quantity in terms of the exogenous variables.

The method of indirect least-squares involves estimating the reduced-form equations by regressing each endogenous variable on *all* the exogenous variables. This results in a set of estimated parameters called the reduced-form parameters. These reduced-form parameters are unbiased if the exogenous variables are uncorrelated with the error terms. The parameters of the structural equations are then deduced from the estimates of the parameters of the reduced form. The estimated structural parameters will generally be biased but, unlike ordinary least-squares estimators, they will be consistent.

The method of indirect least-squares may be illustrated with the consumption-function example. The reduced-form equations are (12.13) and (12.14). Only one of these need be estimated, since both parameters a and b of the structural equation can be deduced from either of the equations. In this case there is only one exogenous variable—investment I_t. Suppose we regress consumption C_t on investment. The result may be written

$$C_t = \hat{\pi}_0 + \hat{\pi}_1 I_t, \tag{12.15}$$

where $\hat{\pi}_0$ and $\hat{\pi}_1$ are the estimated parameters of the reduced-form equation (12.13). The estimated values of a and b may be deduced by setting the estimated parameters of (12.15) equal to the parameters of (12.13). Thus

$$\hat{\pi}_0 = \frac{\hat{a}}{1 - \hat{b}}, \qquad \hat{\pi}_1 = \frac{\hat{b}}{1 - \hat{b}}, \tag{12.16}$$

where $\hat{a}$ and $\hat{b}$ are the estimated values of a and b. We solve (12.16) for $\hat{a}$ and $\hat{b}$ in terms of $\hat{\pi}_0$ and $\hat{\pi}_1$.

$$\hat{a} = \frac{\hat{\pi}_0}{1 + {}_1\hat{\pi}}, \qquad \hat{b} = \frac{\hat{\pi}_1}{1 + \hat{\pi}_1}, \tag{12.17}$$

Example 12.2. Consider the following data on disposable personal income, personal consumption, and personal savings for the United States (all in billions of current dollars):

YEAR	DISPOSABLE INCOME Y	CONSUMPTION C	SAVINGS S
1959	337.3	318.3	19.1
1960	350.0	333.0	17.0
1961	364.4	343.3	21.2
1962	385.3	363.7	21.6
1963	404.6	384.7	19.9
1964	438.1	411.9	26.2
1965	473.2	444.8	28.4
1966	511.9	479.3	32.5
1967	546.5	506.2	40.4
1968	590.0	551.6	38.4

If we estimate the consumption function by the method of direct least-squares, we obtain the following relationship:

$$C = a + b \cdot Y = 14.26 + .9075Y.$$

Since $C + S = Y$, we have the reduced-form equation

$$C = \frac{a}{1 - b} + \frac{b}{1 - b} S.$$

If we estimate the reduced-form equation, we obtain

$$C = 170.051 + 9.2040S.$$

The estimated reduced-form equation can be translated into an estimated consumption function as follows:

$$170.051 = \frac{a}{1 - b}, \qquad 9.2040 = \frac{b}{1 - b}.$$

The second of these equations may be solved for b.

$$(1 - b)9.2040 = b, \qquad 9.2040 = 10.2040b$$

or

$$b = \frac{9.2040}{10.2040} = .9020.$$

Solving the first equation for a, we obtain

$$
\begin{aligned}
a &= 170.051(1 - b) \\
&= 170.051(0.0980) \\
&= 16.66
\end{aligned}
$$

Thus the consumption function as estimated by indirect least-squares is

$C = 16.66 + .9020X$.

The estimated marginal propensity to consume in the direct least-squares case was .9075. We know that this estimate is upwardly biased. The indirect least-squares estimate is .9020, which is less than the direct least-squares estimate.

Identification

The method of indirect least-squares can be applied to estimation of the parameters of a structural equation only if the parameters of the equation are *exactly identified*. The parameters of a structural equation are exactly identified if the structural parameters can be deduced unambiguously from almost any set of reduced-form parameters.

Example 12.3.   Suppose the quantity demanded of wheat in period t (Y_{1t}) is a function of price (Y_{2t}) and the level of national income (X_{1t}). Thus we have the following demand equation:

$$Y_{1t} = a_{10} + b_{12} \cdot Y_{2t} + a_{11} \cdot X_{1t} + e_{1t}, \tag{12.18}$$

where e_{1t} is an error term. Suppose the quantity supplied of wheat is a function of price Y_{2t} alone. Since, in equilibrium, the quantity demanded equals the quantity supplied, we can simplify matters by letting Y_{1t} stand for both quantity demanded and quantity supplied. Thus

$$Y_{1t} = a_{20} + b_{22} \cdot Y_{2t} + e_{2t}. \tag{12.19}$$

The endogenous variables are quantity Y_{1t} and price Y_{2t}. There is only one exogenous variable: income X_{1t}. Let us solve (12.18) and (12.19) for the endogenous variables in terms of the exogenous variables. First substitute the expression (12.19) for Y_{1t} into (12.18).

$$a_{20} + b_{22} \cdot Y_{2t} + e_{2t} = a_{10} + b_{12} \cdot Y_{2t} + a_{11} \cdot X_{1t} + e_{1t},$$

or by rearranging terms we can solve for Y_{2t}:

$$Y_{2t} = \frac{a_{10} - a_{20}}{b_{22} - b_{12}} + \frac{a_{11}}{b_{22} - b_{12}} X_{1t} + \frac{e_{1t} - e_{2t}}{b_{22} - b_{12}}. \tag{12.20}$$

This expression for Y_{2t} may be substituted into (12.19) to give a solution for Y_{1t}.

$$Y_{1t} = \frac{b_{22} \cdot a_{10} - b_{12} \cdot a_{20}}{b_{22} - b_{12}} + \frac{b_{22} \cdot a_{11}}{b_{22} - b_{12}} X_{1t} + \frac{b_{22}e_{1t} - b_{12}e_{2t}}{b_{22} - b_{12}}. \tag{12.21}$$

Equations (12.20) and (12.21) then constitute the reduced-form equations. They are estimated by regressing each of the endogenous variables on all the exogenous variables. The estimated equations can be represented by

$$Y_{1t} = \hat{\pi}_{10} + \hat{\pi}_{11}X_{1t}, \tag{12.22}$$

$$Y_{2t} = \hat{\pi}_{20} + \hat{\pi}_{21}X_{1t}, \tag{12.23}$$

where $\hat{\pi}_{10}$, $\hat{\pi}_{11}$, $\hat{\pi}_{20}$, and $\hat{\pi}_{21}$ represent the estimated reduced-form parameters. Since (12.22) is the estimated form of (12.21), and (12.23) is the estimated form of (12.20), the a and b parameters of the structural equations must be deduced from the following equations:

$$\hat{\pi}_{10} = \frac{\hat{b}_{22} \cdot \hat{a}_{10} - \hat{b}_{12} \cdot \hat{a}_{20}}{\hat{b}_{22} - \hat{b}_{12}}, \tag{12.24}$$

$$\hat{\pi}_{11} = \frac{\hat{b}_{22} \cdot \hat{a}_{11}}{\hat{b}_{22} - \hat{b}_{12}}, \tag{12.25}$$

$$\hat{\pi}_{20} = \frac{\hat{a}_{10} - \hat{a}_{20}}{\hat{b}_{22} - \hat{b}_{12}}, \tag{12.26}$$

$$\hat{\pi}_{21} = \frac{\hat{a}_{11}}{\hat{b}_{22} - \hat{b}_{12}}. \tag{12.27}$$

Now from (12.27), the equation (12.25) can be written

$$\hat{\pi}_{11} = \hat{b}_{22} \cdot \frac{\hat{a}_{11}}{\hat{b}_{22} - \hat{b}_{12}} = \hat{b}_{22} \cdot \hat{\pi}_{21}. \tag{12.28}$$

We can solve (12.28) for $\hat{b}_{22}$.

$$\hat{b}_{22} = \frac{\hat{\pi}_{11}}{\hat{\pi}_{21}}. \tag{12.29}$$

Equation (12.24) can be written

$$\hat{\pi}_{10} = \frac{\hat{b}_{22}\hat{a}_{10} - \hat{b}_{12}\hat{a}_{20} + \hat{b}_{22}\hat{a}_{20} - \hat{b}_{22}\hat{a}_{20}}{\hat{b}_{22} - \hat{b}_{12}}$$

or

$$\hat{\pi}_{10} = \frac{\hat{b}_{22}(\hat{a}_{10} - \hat{a}_{20}) + (\hat{b}_{22} - \hat{b}_{12})\hat{a}_{20}}{\hat{b}_{22} - \hat{b}_{12}}$$

or

$$\hat{\pi}_{10} = \hat{b}_{22} \cdot \frac{\hat{a}_{10} - \hat{a}_{20}}{\hat{b}_{22} - \hat{b}_{12}} + \hat{a}_{20},$$

or from (12.26) we have

$$\hat{\pi}_{10} = \hat{b}_{22} \cdot \hat{\pi}_{20} + \hat{a}_{20}. \tag{12.30}$$

Substitute (12.29) into (12.30) and obtain

$$\hat{a}_{20} = \hat{\pi}_{10} - \frac{\hat{\pi}_{11}}{\hat{\pi}_{21}} \hat{\pi}_{20}. \tag{12.31}$$

Thus equations (12.29) and (12.31) enable us to determine the two parameters $\hat{b}_{22}$ and $\hat{a}_{20}$ from the reduced-form parameters. The structural equation (12.19) or supply equation is said to be exactly identified, since both its parameters can be deduced unambiguously from the reduced form. The structural equation (12.18), however, is not identified. The reason is that there are a total of five structural parameters, a_{10}, b_{12}, a_{11}, a_{20}, and b_{22}, while only four equations (12.24) through (12.27) are available to determine these five parameters. Although the last two of these parameters, a_{20} and b_{22}, can be deduced unambiguously from these four equations, it is impossible to determine the other three parameters.

Note that we said that a structural equation is exactly identified if the structural parameters can be deduced unambiguously from *almost any* set of reduced-form parameters. The phrase *almost any* was used to distinguish certain exceptional cases. Suppose, for example, that the estimated value of $\hat{\pi}_{21}$ in Example 12.3 turned out to be zero. Then (12.29) and (12.31) could not be used to solve for b_{22} and a_{20}. Similarly, in the estimation of the consumption function discussed above, if the parameter $\hat{\pi}_1$ happens to be -1, the equations (12.17) cannot be used to solve for $\hat{a}$ and $\hat{b}$.

One way to determine whether an equation is exactly identified is to determine the reduced form in terms of the structural parameters and set the reduced-form estimates equal to the proper expressions in terms of the reduced-form parameters, as was done in equations (12.24) to (12.27) in Example 12.3. If the structural parameters can be deduced unambiguously from the reduced-form parameters, the equation is exactly identified. This operation can be quite tedious, however, and there is a simple rule that is easy to apply and is satisfied only if the equation is identified. This rule is called the *counting rule* for exact identification.

Counting Rule (Exact Identification).   A structural equation is exactly identified only if the number of variables (both endogenous and exogenous) excluded from the equation is one less than the number of structural equations.†

Example 12.4.   Using the structural equations (12.18) and (12.19) from Example 12.3, let us apply the counting rule. There are two endogenous variables Y_{1t} and Y_{2t} and one exogenous variable X_{1t} in the model as a whole. Equation (12.18), the demand equation, contains all three of these variables and therefore is not exactly identified. The supply equation (12.19) contains only Y_{1t} and Y_{2t}. One variable, X_{1t}, is excluded. Since

---

† An equivalent rule is that an equation is exactly identified only if the number of excluded exogenous variables is one less than the number of included endogenous variables.

there are two equations in the system, the counting rule is satisfied and the supply equation may be identified.

Example 12.5.   Let us consider the following three structural equations:

$$
\begin{aligned}
Y_{1t} + \beta_{12}Y_{2t} &= \gamma_{10} + \gamma_{11}X_{1t} + \gamma_{12}X_{2t} \\
Y_{2t} + \beta_{23}Y_{3t} &= \gamma_{20} \qquad\qquad + \gamma_{22}X_{2t} \\
\beta_{32}Y_{2t} + \quad Y_{3t} &= \gamma_{30}
\end{aligned}
\tag{12.32}
$$

The endogenous variables are Y_{1t}, Y_{2t}, and Y_{3t}. The exogenous variables are X_{1t} and X_{2t}.

The first structural equation is missing just one variable. The number of equations is three, and three less one is two. Since the number of missing variables is less than two, the first equation is not exactly identified.

The number of variables missing from the second equation is two. The counting rule is satisfied, and the second equation may be exactly identified.

The number of variables missing from the third equation is three. The counting rule is not satisfied, and the third equation is not exactly identified.

The counting rule above entails a number of problems. The rule in this form is applicable only when the structural parameters are restricted to be zero or nonzero. That is, a variable is either excluded or not excluded from an equation. It does not take into account other types of restrictions on the parameters of the structural equations. In particular, we might wish to restrict the parameters to conform to certain linear constraints.

Example 12.6.   Suppose we are attempting to estimate the following Cobb-Douglas production function:

$$
Y = \alpha \cdot K^{\beta} \cdot L^{(1-\beta)},
\tag{12.33}
$$

where Y is output, K is the capital stock, L is labor input, and α and β are constants. We know that the share of labor in total output is equal to $(1 - \beta)$, the elasticity of output with respect to labor.

$$
\frac{w \cdot L}{Y} = 1 - \beta,
\tag{12.34}
$$

where w is the wage rate. We may rewrite (12.33) and (12.34) in terms of natural logarithms:

$$
\log_e Y = \log_e \alpha + \beta \log_e K + (1 - \beta) \log_e L
\tag{12.35}
$$

and

$$
\log_e w + \log_e L - \log_e Y = \log_e (1 - \beta).
\tag{12.36}
$$

The two equations contain four variables, $\log_e Y$, $\log_e K$, $\log_e L$, and $\log_e w$. We may rewrite these equations in a more general form as follows:

$$
\begin{aligned}
\log_e Y + a_{12} \log_e L &= b_{10} + b_{11} \log_e w + b_{12} \log_e K, \\
a_{21} \log_e Y + \log_e L &= b_{20} + b_{21} \log_e w + b_{22} \log_e K.
\end{aligned}
\tag{12.37}
$$

Since the first equation (12.35) does not contain $\log_e w$ and the second equation (12.36) does not contain $\log_e K$, we may write

$$b_{11} = 0, \qquad\qquad (12.38)$$
$$b_{22} = 0. \qquad\qquad (12.39)$$

In the first equation, the coefficient of $\log_e K$ is β and of $\log_e L$ is $(1 - \beta)$. Thus

$$a_{12} = 1 - b_{12}. \qquad\qquad (12.40)$$

In equation (12.36), the coefficients of all variables are equal to unity. Thus

$$a_{21} = 1, \qquad\qquad (12.41)$$
$$b_{21} = 1. \qquad\qquad (12.42)$$

Thus the parameters of the model expressed in the form (12.37) must satisfy the five linear restrictions (12.38) through (12.42). [Note that the constant term in (12.36) is the logarithm of the coefficient of $\log L$ in (12.35). This is a nonlinear restriction, however, and we shall ignore it for the sake of illustration.]

The counting rule may be modified to take into account general linear restrictions quite easily. First, however, let us adopt some rather general notation. Let the structural equations be represented by

$$Y_{1t} + \beta_{12}Y_{2t} + \cdots + \beta_{1P}Y_{Pt} = \gamma_{10} + \gamma_{11}X_{1t} + \cdots + \gamma_{1Q}X_{Qt} + e_{1t}$$
$$\beta_{21}Y_{1t} + Y_{2t} + \cdots + \beta_{2P}Y_{Pt} = \gamma_{20} + \gamma_{21}X_{1t} + \cdots + \gamma_{2Q}X_{Qt} + e_{2t}$$

$$\cdot$$
$$\cdot \qquad\qquad (12.43)$$
$$\cdot$$

$$\beta_{G1}Y_{1t} + \beta_{G2}Y_{2t} + \cdots + Y_{Pt} = \gamma_{G0} + \gamma_{G1}X_{1t} + \cdots + \gamma_{GQ}X_{Gt} + e_{Gt}$$

where the endogenous variables are $Y_1, Y_2, \cdots, Y_P$; the exogenous variables are $X_1, X_2, \cdots, X_Q$; and $e_{1t}, e_{2t}, \cdots, e_{Gt}$ are the error terms. Note that the coefficient of Y_1 in equation 1 is unity, the coefficient of Y_2 in equation 2 is unity, and the coefficient of Y_P in equation P is unity. This is an arbitrary convention, which is permissible, since if the ith equation is not in this form, it may be made so by dividing both sides by β_{ii}. This form (12.43) is called a *normalized* form of the structural equations.

The linear restrictions on the parameters of the ith equation can be represented by

$$\phi_{i11}\beta_{i1} + \cdots + \phi_{i1P}\beta_{iP} + \mu_{i10}\gamma_{i0} + \mu_{i11}\gamma_{i1} + \cdots + \mu_{i1Q}\gamma_{iQ} = r_{i1}$$

$$\cdot$$
$$\cdot \qquad\qquad (12.44)$$
$$\cdot$$

$$\phi_{iR1}\beta_{i1} + \cdots + \phi_{iRP}\beta_{iP} + \mu_{iR0}\gamma_{i0} + \mu_{iR1}\gamma_{i1} + \cdots + \mu_{iRQ}\gamma_{iQ} = r_{iR},$$

where the ϕ's, μ's, and r's are all constants and there are R linear restrictions on the parameters of the ith equation.†

Counting Rule (Exact Identification). The ith structural equation is exactly identified only if the number R of linear restrictions on the parameters of the ith equation is equal to $G - 1$, one less than the number of structural equations.

Both versions of the counting rule provide only *necessary* conditions for exact identification. That is, exact identification occurs *only if* the counting rule is satisfied, but even this does not guarantee identification. A sufficient condition for identification can be formulated in terms of ranks of matrices. A discussion of the sufficient condition is contained in an appendix to this chapter.

The method of indirect least-squares is applicable if an equation is exactly identified. If an equation is not exactly identified, it may be either *overidentified* or *underidentified*. If an equation is underidentified, there is no way in which the parameters can be estimated. It is impossible to deduce any estimates of all the structural parameters from the reduced-form parameters. A simple counting rule provides a sufficient condition for underidentification.

Counting Rule (Underidentification.) The ith structural equation is underidentified if the number of variables excluded from the ith equation (alternatively, the number of linear restrictions on the parameters of the ith equation) is less than $G - 1$ or one less than the number of equations.

Example 12.7. In Example 12.3 the two structural equations are (12.18) and (12.19). The first of these equations is underidentified, since no variables are excluded from it. In Example 12.5 the structural equations are given by (12.32). The first of these equations is underidentified. One variable is excluded from this equation. There are $G = 3$ equations, so that $G - 1 = 2$.

If an equation is overidentified, there are a number of estimation procedures that can be used. An equation is overidentified if in some sense there are too many a priori restrictions on the structural parameters. The a priori restrictions on the structural parameters may (or may not) imply certain restrictions on the reduced-form parameters. A structural equation is overidentified if and only if the structural parameters can be deduced from a restricted set of reduced-form parameter estimates.

---

† We assume that the restrictions are linearly independent.

Example 12.8. Suppose we have the following structural equations:

$$Y_1 = \gamma_{10} + \gamma_{11}X_1, \tag{12.45}$$
$$\beta_{21}Y_1 + Y_2 = \gamma_{20} \qquad\qquad + \gamma_{22}X_2.$$

If we substitute the first of these equations into the second, we may obtain the following reduced form:

$$Y_1 = \gamma_{10} + \gamma_{11}X_1, \tag{12.46}$$
$$Y_2 = (\gamma_{20} - \beta_{21}\gamma_{10}) - \beta_{21}\gamma_{11}X_1 + \gamma_{22}X_2.$$

If we regress each endogenous variable on all the exogenous variables, we obtain the following estimates of the reduced-form equations:

$$Y_1 = \hat{\pi}_{10} + \hat{\pi}_{11}X_1 + \hat{\pi}_{21}X_2, \tag{12.47}$$
$$Y_2 = \hat{\pi}_{20} + \hat{\pi}_{21}X_1 + \hat{\pi}_{22}X_2.$$

Comparing the first equation in (12.47) with the first equation in (12.46), we have

$$\gamma_{10} = \hat{\pi}_{10}, \tag{12.48}$$
$$\gamma_{11} = \hat{\pi}_{11}, \tag{12.49}$$
$$0 = \hat{\pi}_{21}. \tag{12.50}$$

The last of these equalities will not be true generally. The estimated value $\hat{\pi}_{21}$ will equal zero only accidentally. If, however, we were to estimate the reduced-form parameters subject to the condition that $\hat{\pi}_{21} = 0$, we could in fact deduce estimates of the two structural parameters γ_{10} and γ_{11} from (12.48) and (12.49). The first equation, therefore, is overidentified. The reader may wish to verify that the second equation is exactly identified.

Example 12.9. Consider the structural equations below:

$$Y_1 = \gamma_{11}X_1 + \gamma_{12}X_2, \tag{12.51}$$
$$\beta_{21}Y_1 + Y_2 = 0.$$

The reduced form is

$$Y_1 = \gamma_{11}X_1 + \gamma_{12}X_2, \tag{12.52}$$
$$Y_2 = -\beta_{21}\gamma_{11}X_1 - \beta_{21}\gamma_{12}X_2.$$

The estimated reduced form is

$$Y_1 = \hat{\pi}_{11}X_1 + \hat{\pi}_{12}X_2, \tag{12.53}$$
$$Y_2 = \hat{\pi}_{21}X_1 + \hat{\pi}_{22}X_2.$$

Comparing (12.52) and (12.53), we have

$$\hat{\pi}_{11} = \gamma_{11}, \qquad \hat{\pi}_{12} = \gamma_{12}, \tag{12.54}$$
$$\hat{\pi}_{21} = -\beta_{21}\gamma_{11}, \qquad \hat{\pi}_{22} = -\beta_{21}\gamma_{12}. \tag{12.55}$$

Substitute (12.54) into (12.55),

$$\hat{\pi}_{21} = -\beta_{21}\hat{\pi}_{11}, \qquad \hat{\pi}_{22} = -\beta_{21}\hat{\pi}_{12}, \qquad\qquad (12.56)$$

or

$$\beta_{21} = \frac{-\hat{\pi}_{21}}{\hat{\pi}_{11}}, \qquad \beta_{21} = \frac{-\hat{\pi}_{22}}{\hat{\pi}_{12}}. \qquad\qquad (12.57)$$

Thus two different estimates of the structural parameter β_{21} can be deduced from the reduced-form estimates. These two separate estimates generally will not be the same, and we have no real way of choosing between them.

If, however, we were to restrict the parameters of the second reduced-form equation to satisfy the constraints (12.56), the two estimates (12.57) of β_{21} would be the same. Thus the second equation is overidentified.

A counting rule for overidentification may be formulated as follows:

Counting Rule (Overidentification). If the number of variables excluded from the ith equation (alternatively the number of linear restrictions on the ith equation) is greater than $G - 1$ or one less than the number of equations, then the ith equation is overidentified.†

Example 12.10.   In Example 12.8 the first structural equation of (12.45) is overidentified, since there are two variables excluded and two equations.

Example 12.11.   The second structural equation of (12.51) of Example 12.9 is overidentified. There are two variables excluded and two equations.

To summarize, then, an equation is (1) underidentified if estimates of the structural parameters cannot be deduced from the reduced-form estimators; (2) exactly identified if estimates of the structural parameters can be deduced from almost any set of reduced-form parameters; and (3) overidentified if estimates of the structural parameters can be deduced from a restricted set of reduced-form estimators.

The counting rules tell us that if the number of linear restrictions on the structural parameters of the ith equation is: (1) equal to $G - 1$, the ith equation *may* be exactly identified; (2) less than $G - 1$, the ith equation is underidentified; and (3) greater than $G - 1$, the ith equation is overidentified.

---

† The validity of this counting rule requires the assumption that there is no linear combination of the remaining $G - 1$ equations that satisfies the linear restrictions on the structural parameters of the ith equation. In particular, there can be no other equation or equations that have the same variables excluded as the ith equation.

Example 12.12.   Consider the following structural equations:

$$\begin{aligned}
Y_1 &= \gamma_{11}X_1 &&+ \gamma_{13}X_3 \\
\beta_{21}Y_1 + \quad Y_2 &= \gamma_{21}X_1 &&+ \gamma_{23}X_3 \\
\beta_{31}Y_1 + \beta_{32}Y_2 + Y_3 &= \quad \gamma_{32}X_2 &&
\end{aligned} \tag{12.58}$$

The first equation is overidentified. Three variables are excluded and there are three equations. The second equation conceivably could be exactly identified. There are two missing variables. (In fact it is underidentified.) The third equation could also be exactly identified. (In fact it is.)

Simultaneous Estimation of Overidentified Equations

All of the methods of estimating the structural parameters of an overidentified equation produce biased estimates, but they are consistent. That is, as the number of observations on the data increases indefinitely, the bias is eliminated.

SIMULTANEOUS LEAST-SQUARES

The structural parameters of an overidentified equation can be deduced from reduced-form estimators only if the reduced-form estimators satisfy certain linear restrictions. The method of simultaneous least-squares involves minimizing the total sum of squared residuals from all of the reduced-form equations subject to the linear restrictions that must be imposed on the reduced-form parameters to obtain unambiguous estimates of the structural parameters.

Example 12.13.   In Example 12.8, we examined the structural equations (12.45) and the reduced form (12.46). The first structural equation is overidentified. Note, however, that the first reduced-form equation must satisfy the restriction that the coefficient of the variable X_2 must be zero. Thus to obtain estimates for the first reduced-form equation we eliminate the variable X_2 and regress Y_1 on X_1 alone. To estimate the second reduced-form equation, we regress Y_1 on X_1 and X_2. We obtain

$$\begin{aligned}
Y_1 &= \hat{\pi}_{10} + \hat{\pi}_{11}X_1, \\
Y_2 &= \hat{\pi}_{20} + \hat{\pi}_{21}X_1 + \hat{\pi}_{22}X_2.
\end{aligned} \tag{12.59}$$

Comparing (12.59) with the reduced form (12.46), we obtain

$$\begin{aligned}
\hat{\pi}_{10} &= \gamma_{10}, & \hat{\pi}_{11} &= \gamma_{11}, \\
\hat{\pi}_{20} &= \gamma_{20} - \beta_{21}\gamma_{10}, & \hat{\pi}_{21} &= -\beta_{21}\gamma_{11}, & \hat{\pi}_{22} &= \gamma_{22}.
\end{aligned} \tag{12.60}$$

There are five equations and five structural parameters. We can reduce the equations (12.60) as follows:

$$\gamma_{10} = \hat{\pi}_{10},$$
$$\gamma_{11} = \hat{\pi}_{11},$$
$$\beta_{21} = -\frac{\hat{\pi}_{21}}{\hat{\pi}_{11}}, \tag{12.61}$$
$$\gamma_{20} = \hat{\pi}_{20} - \frac{\hat{\pi}_{21}}{\hat{\pi}_{11}}\hat{\pi}_{10},$$
$$\gamma_{22} = \hat{\pi}_{22}.$$

to obtain estimates of the parameters of both structural equations.

Example 12.14. In Example 12.9 we considered the structural equations (12.51) and the reduced form (12.52). The method of simultaneous least-squares may be applied to estimate the parameter of the second structural equation, which is overidentified. First we estimate the first reduced-form equation of (12.53), and we obtain reduced-form estimates $\hat{\pi}_{11}$ and $\hat{\pi}_{12}$. The estimated parameters $\hat{\pi}_{21}$ and $\hat{\pi}_{22}$ of the second reduced-form equation must satisfy the constraints (12.56) if we are to obtain unambiguous estimates of the parameter β_{21} of the second structural equation. Substituting the constraints (12.56) into the second equation of (12.53), we obtain

$$Y_2 = -\beta_{21}\hat{\pi}_{11}X_1 - \beta_{21}\hat{\pi}_{12}X_2. \tag{12.62}$$

Since $\hat{\pi}_{11}$ and $\hat{\pi}_{12}$ have already been estimated from the first reduced-form equation, we may estimate β_{21} directly from (12.62). We can rewrite this equation as follows:

$$Y_2 = -\beta_{21}(\hat{\pi}_{11}X_1 + \hat{\pi}_{12}X_2). \tag{12.63}$$

Thus we can estimate β_{21} in (12.63) by regressing the endogenous variable Y_2 on a "composite" exogenous variable $\hat{\pi}_{11}X_1 + \hat{\pi}_{12}X_2$.

Simultaneous least-squares may be applied to estimate the parameters of any overidentified structural equation by constraining the reduced-form parameter estimates so that the structural parameters may be inferred unambiguously. The restrictions on the reduced-form parameters that permit estimation of the parameters of one overidentified equation, however, may not be the same restrictions that permit estimation of another overidentified equation in the structural model. It is theoretically possible to impose simultaneously all the necessary constraints on the reduced form to estimate all overidentified equations from the same reduced-form estimates. The computation involved, however, may be quite difficult.†

---

† The normal equations are nonlinear and may not be particularly easy to solve. If, however, only the restrictions are imposed that are required to estimate one particular overidentified equation, simultaneous least squares is equivalent to two-stage least squares discussed below. See also the appendix to this chapter.

Example 12.15. Consider structural equations of the following form:

$$Y_1 + \beta_{12}Y_2 = \gamma_{11}X_1 + \gamma_{12}X_2 \qquad (12.64)$$
$$\beta_{21}Y_1 + \quad Y_2 = \qquad\qquad \gamma_{23}X_3 + \gamma_{24}X_4$$

Both equations are overidentified. By solving (12.64) for Y_1 and Y_2 in terms of the exogenous X variables and comparing the resulting reduced-form expressions with the estimated reduced-form parameters, we can show that the first structural equation can be estimated if the following restrictions on the reduced-form estimates are satisfied:

$$\hat{\pi}_{13} = -\beta_{12}\hat{\pi}_{23}, \qquad \hat{\pi}_{14} = -\beta_{12}\hat{\pi}_{24}. \qquad (12.65)$$

If we substitute these expressions for $\hat{\pi}_{13}$ and $\hat{\pi}_{14}$ into the general reduced-form expressions, we obtain

$$Y_1 = \hat{\pi}_{11}X_1 + \hat{\pi}_{12}X_2 + (-\beta_{12}\hat{\pi}_{23})X_3 + (-\beta_{12}\hat{\pi}_{24})X_4, \qquad (12.66)$$
$$Y_2 = \hat{\pi}_{21}X_1 + \hat{\pi}_{22}X_2 + \hat{\pi}_{23}X_3 + \hat{\pi}_{24}X_4.$$

We can first find the reduced-form estimates for the second equation, $\hat{\pi}_{21}$, $\hat{\pi}_{22}$, $\hat{\pi}_{23}$, and $\hat{\pi}_{24}$. Then we may form the composite variable $\hat{\pi}_{23}X_3 + \hat{\pi}_{24}X_4$ and estimate the first reduced-form equation,

$$Y_1 = \hat{\pi}_{11}X_1 + \hat{\pi}_{12}X_2 - \beta_{12}(\hat{\pi}_{23}X_3 + \hat{\pi}_{24}X_4), \qquad (12.67)$$

by regressing Y_1 on X_1, X_2, and the composite variable. The estimated value of β_{12} comes directly from the regression equation (12.67), while the other structural parameters γ_{11} and γ_{12} of the first equation can be determined by

$$\gamma_{11} = \hat{\pi}_{11} + \beta_{12}\hat{\pi}_{21}, \qquad \gamma_{12} = \hat{\pi}_{12} + \beta_{12}\hat{\pi}_{22}. \qquad (12.68)$$

Similarly, we can estimate the parameters of the second structural equation of (12.64) by imposing the restrictions

$$\hat{\pi}_{21} = -\beta_{21}\hat{\pi}_{11}, \qquad \hat{\pi}_{22} = -\beta_{21}\hat{\pi}_{12} \qquad (12.69)$$

on the reduced-form parameters of the second equation. If, however, we attempt to impose both the restrictions (12.65) and (12.69), the general reduced-form expressions become

$$Y_1 = \hat{\pi}_{11}X_1 + \hat{\pi}_{12}X_2 + (-\beta_{12}\hat{\pi}_{23})X_3 + (-\beta_{12}\hat{\pi}_{24})X_4 \qquad (12.70)$$
$$Y_2 = (-\beta_{21}\hat{\pi}_{11})X_1 + (-\beta_{21}\hat{\pi}_{12})X_2 + \hat{\pi}_{23}X_3 + \hat{\pi}_{24}X_4.$$

Now the first of these reduced-form equations can be estimated easily if we first know $\hat{\pi}_{23}$ and $\hat{\pi}_{24}$ from the second equation. The second equation can be estimated easily if we first know $\hat{\pi}_{11}$ and $\hat{\pi}_{12}$ from the first equation. But then we would have to estimate the second equation before the first equation and the first equation before the second equation, which clearly is impossible. In theory, we can estimate both equations simultaneously, but in general this is quite difficult to do.†

TWO-STAGE LEAST-SQUARES

A very simple method of simultaneous estimation of an overidentified equation that gives consistent estimates is called two-stage least-squares. Two-stage least-squares does not require a complete knowledge of all the structural equations. Only the equation being estimated must be completely specified. If, however, the structural equation being estimated is the only overidentified equation in the system, then two-stage least-squares and simultaneous least-squares are equivalent. (See the appendix to this chapter.)

Suppose we are trying to estimate the first structural equation.

$$Y_1 + \beta_{12}Y_2 + \cdots + \beta_{1G*}Y_{G*} = \gamma_{10} + \gamma_{11}X_1 + \cdots + \gamma_{1P*}X_{P*}. \quad (12.71)$$

There are G^* endogenous variables in the first equation and P^* exogenous variables in the first equation. Suppose there are a total of P different exogenous variables in the structural model. The exogenous variables $X_{P*+1}, \ldots, X_P$, however, do not appear in the first equation.

Two-stage least-squares involves estimating all the reduced-form equations corresponding to the endogenous variables included in the first equation except for Y_1.

$$Y_2 = \hat{\pi}_{20} + \hat{\pi}_{21}X_1 + \cdots + \hat{\pi}_{2P}X_P$$
$$Y_3 = \hat{\pi}_{30} + \hat{\pi}_{31}X_1 + \cdots + \hat{\pi}_{3P}X_P$$
$$\cdot$$
$$\cdot \quad (12.72)$$
$$\cdot$$
$$Y_{G*} = \hat{\pi}_{G0} + \hat{\pi}_{G1}X_1 + \cdots + \hat{\pi}_{GP}X_P$$

Note that the reduced-form estimates involve regressing each of the included endogenous variables on *all* of the exogenous variables. The first structural equation (12.71) is then estimated by regressing Y_1 on the *predicted* endogenous variables $\hat{Y}_2, \cdots, \hat{Y}_{G*}$ from the reduced form *and* the exogenous variables $X_1, \cdots, X_{P*}$, which are included in the first equation. That is, we estimate the following equation

$$Y_1 = -\beta_{12}\hat{Y}_2 - \cdots - \beta_{1G}*\hat{Y}_{G*} + \gamma_{10} + \gamma_{11}X_1 + \cdots + \gamma_{1P*}X_{P*} \quad (12.73)$$

to obtain estimates of the structural parameters $\beta_{12}, \ldots, \beta_{1G*}$ and $\gamma_{11}, \ldots, \gamma_{1P*}$.

There is a degree of arbitrariness in the use of two-stage least-squares. Rather than estimate the reduced-form equations corresponding to Y_2,

---

† The estimation procedure is quite simple, however, when the matrix of β coefficients is triangular. See the appendix to this chapter.

$Y_3, \ldots, Y_{G*}$, we could estimate the reduced-form equations corresponding to *any* $G^* - 1$ of the included endogenous variables. For example, we might regress $Y_1, Y_3, Y_4, \ldots, Y_{G*}$ on all the exogenous variables and then regress Y_2 on the predicted values, $\hat{Y}_1, \hat{Y}_3, \hat{Y}_4, \ldots, \hat{Y}_{G*}$ and the exogenous variables $X_1, X_2, \ldots, X_{P*}$ to obtain estimates of the structural parameters of the first equation. Sometimes, however, it seems natural in terms of the economic theory to associate a particular endogenous variable to a particular structural equation. In such a case, it seems sensible to regress this particular variable on the predicted values of the other endogenous variables and the exogenous variables.

Example 12.16.   Let us consider a simple Keynesian income-determination model. Consumption (C) is a function of income

$$C = \alpha + \beta \cdot Y. \tag{12.74}$$

Investment (I) is determined by the rate of interest (r) and the profit rate P.

$$I = \gamma + \delta \cdot r + \mu \cdot P. \tag{12.75}$$

Income is the sum of consumption and investment

$$C + I = Y. \tag{12.76}$$

The endogenous variables in this model are consumption (C), income (Y), and investment (I). The rate of interest (r) and the profit rate (P) are the exogenous variables. For sake of illustration, let us solve (12.76) for Y and substitute into (12.74). This eliminates the variable Y and the identity (12.76) from the model. The model then consists of

$$C = \frac{\alpha}{1 - \beta} + \frac{\beta}{1 - \beta} I, \tag{12.77}$$
$$I = \gamma + \delta \cdot r + \mu \cdot P.$$

The first equation is overidentified. There are two variables excluded (r and P) and only two equations. We think of the first equation as the one that determines consumption. Therefore, in using two-stage least-squares we would estimate the parameters of this equation by regressing the other endogenous variable I on the exogenous variables r and P, and then regressing C on the predicted value of I alone, where

$$\hat{I} = \hat{\gamma} + \hat{\delta} \cdot r + \hat{\mu} \cdot P, \tag{12.78}$$

$\hat{I}$ is the predicted value of I, and $\hat{\gamma}$, $\hat{\delta}$, and $\hat{\mu}$ are the least-squares estimates of γ, δ, and μ.

We could also obtain consistent two-stage least-squares estimates of the first equation in (12.77) by regressing C on r and P and then regressing I

on C rather than C on I. This procedure does not seem sensible in terms of economic theory, and in fact may tend to produce less efficient estimates in a statistical sense.

MAXIMUM-LIKELIHOOD ESTIMATES

In addition to the methods above, which are fairly simple to apply, estimates of structural parameters may be obtained by maximum-likelihood methods. The reader may remember our earlier discussion (in the fifth section of Chapter 7) of maximum-likelihood estimators, which were defined as estimators that maximize the probability (likelihood) that a particular sample of observed values of the variables could have been generated.

Maximum-likelihood methods involve a considerable computational effort and for this reason are rarely used in practice. Two-stage least-squares, indirect least-squares, and simultaneous least-squares are much simpler to use and produce consistent estimates of parameters. The advantage of maximum-likelihood estimators is their asymptotic efficiency. That is, as the sample becomes very large, maximum-likelihood estimators have less sampling variance than other kinds of estimators. Like other methods of simultaneous estimation, however, maximum-likelihood estimators are biased.

Maximum-likelihood estimates involve minimizing sums of squared residuals from each of the reduced-form equations *and* cross-products of residuals from different equations. The minimization of the sums of squares and cross-products is done subject to the a priori restrictions on the parameters of the reduced form. The main difference between simultaneous least-squares and maximum-likelihood estimators is the presence of cross-products of residuals from different equations in the quantity to be minimized in maximum-likelihood estimation. Like simultaneous least-squares, only the a priori restrictions on the reduced-form parameters necessary to estimate a particular overidentified equation may be imposed. In this case the maximum-likelihood estimates are called *limited-information maximum-likelihood estimators*. If all of the a priori restrictions on the reduced-form equations are imposed simultaneously, we obtain *full-information maximum-likelihood estimators*.

Concluding Remarks

None of the methods of estimation discussed above apply when a structural equation is underidentified. Identification can often be achieved, however, by a judicious reformulation of the model. Consider, for example, a simple supply-and-demand model:

$$Q_t = \alpha + \beta \cdot P_t, \qquad Q_t = \gamma + \delta \cdot P_t, \tag{12.79}$$

where Q_t is quantity in period t and P_t is price in period t. The first equation is a demand equation and the second is a supply equation. Neither of the two equations is identified. If, however, we add an income variable to the demand equation, the supply equation becomes identified. If we add a weather variable, say rainfall, to the supply equation, the demand equation becomes identified. Thus we have the model

$$Q_t = \alpha + \beta \cdot P_t + \mu \cdot Y_t, \tag{12.80}$$
$$Q_t = \gamma + \delta \cdot P_t + \rho \cdot R_t,$$

where Y_t is income and R_t is rainfall. Both equations are exactly identified, and we can use indirect least-squares to obtain consistent estimates of the structural parameters.

The use of simultaneous estimation procedures is clearly desirable when one is dealing with large samples. These procedures are more efficient than ordinary least-squares and produce "nearly" unbiased estimates when samples are large. Furthermore, for large samples, we can rank the various estimation procedures according to efficiency. Full-information maximum-likelihood estimators are more efficient than simultaneous, indirect, or two-stage least-squares. Full-information maximum-likelihood methods are more efficient than limited-information methods. Finally, for large samples, one may use the usual t and F tests of significance to the parameters.

The small-sample properties of simultaneous estimators are not very well known, however. One is not necessarily justified in performing the usual significance tests, and the various estimators cannot be ranked according to efficiency. Fairly extensive experimental tests of the desirability of various methods have been conducted. While the results of these tests are not conclusive, they do not contradict the large-sample conclusions.

APPENDIX Further Discussion

In this appendix we shall develop the rank condition for identifiability and develop a bit more systematically the indirect and simultaneous least-squares estimation procedures using matrix algebra.

Matrix and Vector Multiplication

A matrix A with n rows and m columns ($n \times m$) can premultiply a matrix B with m rows and p columns ($m \times p$). The result is a matrix $A \cdot B = C$ with n rows and p columns ($n \times p$), where the element of the ith row and jth column of C is

$$c_{ij} = \sum_{p=1}^{m} a_{ip} \cdot b_{pj}, \tag{12.81}$$

where a_{ip} is the element of the ith row and pth column of A and b_{pj} is the element of the pth row and jth column of B.

Matrix multiplication is not commutative. That is, if we can multiply $A \cdot B$ and $B \cdot A$, it is not necessarily true that $A \cdot B = B \cdot A$. Note also that while it may be possible to premultiply B by A ($A \cdot B$), it is not always possible to postmultiply B by A ($B \cdot A$). The two matrices A and B must be square (that is, have the same number of rows and columns) for both kinds of multiplication to be possible.

A vector is a matrix with one row or one column. A matrix with one row is called a row vector. A matrix with one column is a column vector. An m-dimensional row vector X may premultiply an $m \times n$ matrix A to obtain the product $X \cdot A$, which is an n-dimensional row vector. An n-dimensional column vector Y may postmultiply an $m \times n$ matrix A to obtain an m-dimensional column vector $A \cdot Y$.

Matrix Representation of the Structure and Reduced Form

The structural equations of an econometric model may be written in terms of matrix notation. Let B be a $G \times G$ matrix of β coefficients, Γ be a $G \times P$ matrix of γ coefficients, Y be a column vector of the endogenous variables, $Y_1, Y_2, \ldots, Y_G$, X be a P-dimensional column vector of exogenous variables $X_1, X_2, \ldots, X_P$ and e be a G-dimensional column vector of error terms. Then we may write

$$B \cdot Y = \Gamma \cdot X + e. \tag{12.82}$$

For example, if we have the structure

$$Y_1 + \beta_{12}Y_2 + \beta_{13}Y_3 = \gamma_{11}X_1 + \gamma_{12}X_2 + e_1,$$
$$\beta_{21}Y_1 + Y_2 + \beta_{23}Y_3 = \gamma_{21}X_1 + \gamma_{22}X_2 + e_2, \qquad (12.83)$$
$$\beta_{31}Y_1 + \beta_{32}Y_2 + Y_3 = \gamma_{31}X_1 + \gamma_{32}X_2 + e_3,$$

then the matrices B and Γ and the vectors Y, X, and e are

$$B = \begin{bmatrix} 1 & \beta_{12} & \beta_{13} \\ \beta_{21} & 1 & \beta_{23} \\ \beta_{31} & \beta_{32} & 1 \end{bmatrix}$$

$$\Gamma = \begin{bmatrix} \gamma_{11} & \gamma_{12} \\ \gamma_{21} & \gamma_{22} \\ \gamma_{31} & \gamma_{32} \end{bmatrix} \qquad (12.84)$$

$$Y = \begin{bmatrix} Y_1 \\ Y_2 \\ Y_3 \end{bmatrix}, \qquad X = \begin{bmatrix} X_1 \\ X_2 \\ X_3 \end{bmatrix}, \qquad e = \begin{bmatrix} e_1 \\ e_2 \\ e_3 \end{bmatrix} \qquad (12.85)$$

The reduced-form equations in matrix form may be obtained by premultiplying both sides of (12.82) by B^{-1}, the inverse of the matrix B.

$$Y = B^{-1} \cdot \Gamma \cdot X + B^{-1} \cdot e, \qquad (12.86)$$

since $B^{-1} \cdot B \cdot Y = Y$. We may also write the reduced form as

$$Y = \Pi \cdot X + \mu, \qquad (12.87)$$

where Π is a $G \times P$ matrix of reduced-form coefficients and where

$$\Pi = B^{-1} \cdot \Gamma. \qquad (12.88)$$

Rank of a Matrix

A *square* matrix is a matrix with an equal number of rows and columns. The *order* of a square matrix is the number of rows or columns.

A square matrix whose determinant is zero is called a singular matrix. A square matrix with a nonzero determinant is nonsingular.

A submatrix of an $m \times n$ matrix A is any matrix formed by eliminating rows and columns of A. For example, take the matrix

$$A = \begin{bmatrix} a_{11} & a_{12} & a_{13} & a_{14} & a_{15} \\ a_{21} & a_{22} & a_{23} & a_{24} & a_{25} \\ a_{31} & a_{32} & a_{33} & a_{34} & a_{35} \end{bmatrix} \qquad (12.89)$$

By eliminating the third and fifth columns of A, we obtain a square submatrix of order three.

$$\begin{bmatrix} a_{11} & a_{12} & a_{14} \\ a_{21} & a_{22} & a_{24} \\ a_{31} & a_{32} & a_{34} \end{bmatrix}$$

By eliminating the second, third, and fourth columns and the second row, we obtain a square matrix of order two.

$$\begin{bmatrix} a_{11} & a_{15} \\ a_{31} & a_{35} \end{bmatrix}$$

The *rank* of any $m \times n$ matrix A is the largest order of a nonsingular square submatrix of A. For example, consider the matrix

$$\begin{bmatrix} 3 & 2 & 4 & 1 \\ 5 & 4 & 3 & 6 \\ 6 & 4 & 8 & 2 \end{bmatrix}$$

There are four square submatrices of order three:

$$\begin{bmatrix} 2 & 4 & 1 \\ 4 & 3 & 6 \\ 4 & 8 & 2 \end{bmatrix} \quad \begin{bmatrix} 3 & 4 & 1 \\ 5 & 3 & 6 \\ 6 & 8 & 2 \end{bmatrix} \quad \begin{bmatrix} 3 & 2 & 1 \\ 5 & 4 & 6 \\ 6 & 4 & 2 \end{bmatrix} \quad \begin{bmatrix} 3 & 2 & 4 \\ 5 & 4 & 3 \\ 6 & 4 & 8 \end{bmatrix}$$

The determinants of all of these matrices are zero; that is, all the square submatrices of order three are singular. However, the determinant of the submatrix

$$\begin{bmatrix} 3 & 2 \\ 5 & 4 \end{bmatrix}$$

is 2. Thus the largest order of a nonsingular square submatrix is 2 and the rank is 2.

The Rank Criterion

Let us combine the matrices B and Γ to obtain a system matrix $[B, \Gamma]$, where

$$[B, \Gamma] = \begin{bmatrix} 1 & \beta_{12} & \cdots & \beta_{1G} & \gamma_{11} & \gamma_{12} & \cdots & \gamma_{1P} \\ \beta_{21} & 1 & \cdots & \beta_{2G} & \gamma_{21} & \gamma_{22} & \cdots & \gamma_{2G} \\ \cdot & \cdot & & \cdot & \cdot & \cdot & & \cdot \\ \cdot & \cdot & & \cdot & \cdot & \cdot & & \cdot \\ \cdot & \cdot & & \cdot & \cdot & \cdot & & \cdot \\ \beta_{G1} & \beta_{G2} & \cdots & 1 & \gamma_{G1} & \gamma_{G2} & \cdots & \gamma_{GP} \end{bmatrix} \qquad (12.90)$$

In order to determine whether the ith equation is identified, we must know the rank of a particular submatrix of $[B, \Gamma]$ that is obtained by striking out the ith row of $[B, \Gamma]$ and all columns of $[B, \Gamma]$ corresponding to variables that are *included* in the ith equation. Let us denote the rank of this submatrix by ρ_i.

Rank Criterion.   The ith structural equation is exactly identified or overidentified if and only if $\rho_i = G - 1$, where G is the number of structural equations.

For example, consider the following structural equations:

$$\begin{aligned}
Y_1 + \beta_{12}Y_2 \quad\quad &= \gamma_{11}X_1 \\
\beta_{21}Y_1 + \quad Y_2 + \beta_{23}Y_3 &= \quad\quad \gamma_{21}X_1 + \gamma_{22}X_2 \\
\beta_{32}Y_2 + \quad Y_3 &= \gamma_{31}X_1 \quad\quad + \gamma_{32}X_2
\end{aligned} \tag{12.91}$$

The system matrix is

$$[B, \Gamma] = \begin{bmatrix}
1 & \beta_{12} & 0 & \gamma_{11} & 0 & 0 \\
\beta_{21} & 1 & \beta_{23} & 0 & \gamma_{21} & \gamma_{22} \\
0 & \beta_{32} & 1 & \gamma_{31} & 0 & \gamma_{32}
\end{bmatrix} \tag{12.92}$$

In order to determine whether the first equation is exactly identified or overidentified, we eliminate the first row and the first, second, and fourth column of $[B, \Gamma]$. The submatrix is

$$\begin{bmatrix}
\beta_{23} & \gamma_{21} & \gamma_{22} \\
1 & 0 & \gamma_{32}
\end{bmatrix} \tag{12.93}$$

The rank of this matrix is $2 = G - 1$. Thus the first equation is either exactly identified or overidentified. In fact, it is overidentified according to the counting rule for overidentification.

Indirect Least-Squares

For convenience, let us consider the method of indirect least-squares for estimating the first structural equation:

$$Y_1 + \beta_{12}Y_2 + \cdots + \beta_{1G*}Y_{G*} = \gamma_{11}X_1 + \gamma_{12}X_2 + \cdots + \gamma_{1P*}X_{P*}, \tag{12.94}$$

where G^* is the number of endogenous variables included in the first equation and P^* is the number of exogenous variables included in the first equation.

If Π is the matrix of reduced-form coefficients, then it must satisfy (12.88). We can premultiply both sides of (12.88) by B to obtain

$$B \cdot \Pi = \Gamma. \tag{12.95}$$

Since we are interested only in estimating the first equation, let us consider only the first row of the B and Γ matrices. From (12.94) and (12.95) we have

$$[1, \beta_{12}, \ldots, \beta_{1G*}, 0, \ldots, 0] \cdot \Pi = [\gamma_{11}, \gamma_{12}, \ldots, \gamma_{1P*}, 0, \ldots, 0]. \tag{12.96}$$

Performing the multiplication in (12.96), we obtain the following P equations:

$$\begin{aligned}
\pi_{11} + \beta_{12}\pi_{21} + \cdots + \beta_{1G*}\pi_{G*1} &= \gamma_{11} \\
\pi_{12} + \beta_{12}\pi_{22} + \cdots + \beta_{1G*}\pi_{G*2} &= \gamma_{12} \\
& \cdot \\
& \cdot \\
& \cdot \\
\pi_{1P*} + \beta_{12}\pi_{2P*} + \cdots + \beta_{1G*}\pi_{G*P*} &= \gamma_{1P*}
\end{aligned} \tag{12.97}$$

$$\pi_{1,P*+1} + \beta_{12}\pi_{2P*+1} + \cdots + \beta_{1G*}\pi_{G*P*+1} = 0$$

$$\vdots \tag{12.98}$$

$$\pi_{1P} + \beta_{12}\pi_{2P} + \cdots + \beta_{1G*}\pi_{G*P} = 0$$

The last $P - P^*$ equations or (12.98) can usually be solved for the $G^* - 1$ parameters, $\beta_{12}, \beta_{13}, \ldots, \beta_{1G*}$ of the first structural equation if there are $G^* - 1$ equations—that is, if

$$P - P^* = G^* - 1. \tag{12.99}$$

The equality (12.99) implies that the number of exogenous variables excluded $(P - P^*)$ is one less than the number of endogenous variables included in the first equation. We may add G to both sides of (12.99) to obtain

$$P - P^* + G = G^* - 1 + G$$

or

$$(P - P^*) + (G - G^*) = G - 1, \tag{12.100}$$

which states that the total number of variables excluded from the first equation must be one less than the total number of equations. Thus we have derived the counting rule discussed in the main text of Chapter 12.

Indirect least-squares then involves (1) estimating the reduced-form π parameters by ordinary least-squares, (2) solving (12.98) for the β coefficients of the first structural equation, and (3) substituting the resulting β values into (12.97) to obtain estimates of the γ parameters of the first equation.

Simultaneous Least-Squares

If in (12.98) there are more equations than there are coefficients $\beta_{12}, \ldots, \beta_{1G*}$, then the first equation is overidentified. This is the case if

$$P - P^* > G^* - 1 \tag{12.101}$$

or, adding G to both sides of (12.101),

$$P - P^* + G > G^* - 1 + G$$

or

$$(P - P^*) + (G - G^*) > G - 1. \tag{12.102}$$

The total number of variables excluded is greater than one less than the number of equations.

Any subset of $G^* - 1$ of the $P - P^*$ equations can be used to solve for $\beta_{12}, \ldots, \beta_{1G*}$. There is no way to choose the subset that should be used, so we resort to the method of simultaneous least-squares. This method involves solving (12.98) for $\pi_{1,P*+1}, \ldots, \pi_{1P}$ or

$$\pi_{1,P*+1} = -(\beta_{12}\pi_{2,P*+1} + \cdots + \beta_{1G*}\pi_{G*,P*+1})$$

$$\cdot$$
$$\cdot$$
$$\cdot \tag{12.103}$$

$$\pi_{1P} = -(\beta_{12}\pi_{2P} + \cdots + \beta_{1G*}\pi_{G*P})$$

The first reduced-form equation may be written as

$$Y_1 = \pi_{11}X_1 + \cdots + \pi_{1,P*}X_{P*} + \pi_{1,P*+1}X_{P*+1} + \cdots + \pi_{1P}X_P. \tag{12.104}$$

We can substitute (12.103) into (12.104) to obtain

$$Y_1 = \pi_{11}X_1 + \cdots + \pi_{1P*}X_{P*} - \beta_{12}(\pi_{2,P*+1}X_{P*+1} + \cdots + \pi_{2P}X_P)$$
$$- \cdots - \beta_{1G*}(\pi_{G*,P*+1}X_{P*+1} + \cdots + \pi_{G*P}X_P). \tag{12.105}$$

The method of simultaneous least-squares requires us to first estimate all of the reduced-form equations expressed in the same form as the first reduced-form equation is expressed in (12.105). The estimated values of $\beta_{12}, \ldots, \beta_{1G*}$ can be obtained directly from (12.105), while $\gamma_{11}, \ldots, \gamma_{1G*}$ can be determined from (12.97).

If we impose on the reduced-form parameters only the restrictions necessary to estimate the structural parameters of the first equation, simultaneous least-squares and two-stage least-squares are equivalent. To see this, solve (12.97) for $\Pi_{11}, \ldots, \Pi_{1P*}$ and substitute into (12.105). We obtain

$$Y_1 = \gamma_{11}X_1 + \cdots + \gamma_{1P*}X_{P*} - \beta_{12}(\pi_{21}X_1 + \cdots + \pi_{2P}X_P)$$
$$- \cdots - \beta_{1G*}(\pi_{G*1}X_1 + \cdots + \pi_{G*P}X_P). \tag{12.106}$$

We estimate all the reduced-form equations except the first and determine the composite variables $(\hat{\pi}_{21}X_1 + \cdots + \hat{\pi}_{2P}X_P), \ldots, (\hat{\pi}_{G*1}X_1 + \cdots + \hat{\pi}_{G*P}X_P)$, which are, in fact, the estimated values of the endogenous variables $Y_2, Y_3, \ldots, Y_{G*}$. Thus we can estimate (12.106) as in the two-stage least-squares procedure in which we regress Y_1 on the included exogenous variables and the predicted values of the last $G^* - 1$ included endogenous variables.

If the matrix B of structural parameters is triangular, then the method of simultaneous least-squares is simple to apply. Suppose, for example, we have

$$\begin{aligned} Y_1 &= \gamma_{11}X_1 + \gamma_{12}X_2 \\ \beta_{21}Y_1 + Y_2 &= \gamma_{22}X_2 + \gamma_{23}X_3 \\ \beta_{31}Y_1 + \beta_{32}Y_2 + Y_3 &= 0 \end{aligned} \tag{12.107}$$

The first and third equations are overidentified and the second equation is exactly identified. The first equation can be estimated directly by regressing Y_1 on X_1 and X_2. The predicted value of Y_1 then is

$$\hat{Y}_1 = \hat{\gamma}_{11}X_1 + \hat{\gamma}_{12}X_2, \tag{12.108}$$

where $\hat{\gamma}_{11}$ and $\hat{\gamma}_{12}$ are ordinary least-squares estimates of γ_{11} and γ_{12}. Substitute this predicted value for Y_1 in the second equation.

$$Y_2 = -\beta_{21}\hat{\gamma}_{11}X_1 + (\gamma_{22} - \beta_{21}\hat{\gamma}_{12})X_2 + \gamma_{23}X_3 \qquad (12.109)$$

or

$$Y_2 = -\beta_{21}(\hat{\gamma}_{11}X_1 + \hat{\gamma}_{12}X_2) + \gamma_{22}X_2 + \gamma_{23}X_3. \qquad (12.110)$$

We may estimate β_{21}, γ_{22}, and γ_{23} by regressing Y_2 on $(\hat{\gamma}_{11}X_1 + \hat{\gamma}_{12}X_2)$, X_2, and X_3. The predicted value of Y_2 then is

$$\hat{Y}_2 = -\hat{\beta}_{21}\hat{\gamma}_{11}X_1 + (\hat{\gamma}_{22} - \hat{\beta}_{21}\hat{\gamma}_{12})X_2 + \hat{\gamma}_{23}X_3. \qquad (12.111)$$

We substitute (12.108) and (12.111) for Y_1 and Y_2, respectively, in the third structural equation of (12.107). We obtain

$$Y_3 = -\beta_{31}(\hat{\gamma}_{11}X_1 + \hat{\gamma}_{12}X_2) - \beta_{32}[-\hat{\beta}_{21}\hat{\gamma}_{11}X_1 + (\hat{\gamma}_{22} - \hat{\beta}_{21}\hat{\gamma}_{12})X_2 + \hat{\gamma}_{23}X_3]. \qquad (12.112)$$

We may estimate β_{31} and β_{32} from (12.112) by regressing Y_3 on the two composite variables $(\hat{\gamma}_{11}X_1 + \hat{\gamma}_{12}X_2)$ and $[-\hat{\beta}_{21}\hat{\gamma}_{11}X_1 + (\hat{\gamma}_{22} - \hat{\beta}_{21}\hat{\gamma}_{12})X_2 + \hat{\gamma}_{23}X_3]$.

PROBLEMS

1. Suppose the demand for tea is given by the following equation:

   $$Q_t = \alpha_0 + \alpha_1 P_t + \alpha_2 Y_t + \alpha_3 S_t + e_{dt},$$

   where Q_t is quantity demanded of tea, P_t is the price of tea, Y_t is per capita income, S_t is the price of sugar, and e_{dt} is an error term. The supply curve is

   $$Q_t = \beta_0 + \beta_1 P_t + \beta_4 R_t + e_{st},$$

   where R_t is an index of rainfall and e_{st} is an error term. Solve these structural equations to obtain the reduced form.
2. Consider the following data on consumption, disposable income, and savings for France for the years 1958 through 1967. Estimate the consumption function using the method of indirect least-squares.

YEAR	DISPOSABLE INCOME Y	CONSUMPTION C	SAVINGS S
1958	174	161	13
1959	186	174	12
1960	208	189	19
1961	224	206	18
1962	256	226	30
1963	287	255	28
1964	311	278	33
1965	336	298	38
1966	363	321	42
1967	394	346	48

3. Given the structural equations in problem 1, suppose the estimated reduced form is as follows:

$$P_t = .321 + 4.742Y_t - 36.302S_t + 1.242R_t,$$
$$Q_t = .013 + .314Y_t + 1.479S_t - .060R_t.$$

(a) What are the estimated parameters of the first structural equation, namely α_0, α_1, α_2, and α_3? (b) Calculate at least two separate estimates of the parameter β_1 of the second structural equation that are consistent with the reduced-form estimates.

4. Consider the following econometric model of an economy

$$I_t = \alpha_0 + \alpha_1 r_t + \alpha_2 P_t + \alpha_3 C_t + \alpha_4 Y_{t-1} \qquad\qquad + \alpha_6 E_t + e_{1t}$$
$$C_t = \beta_0 + \beta_1 r_t + \beta_2 P_t + \qquad\qquad\qquad + \beta_5 I_t \qquad\quad + e_{2t}$$
$$M_t = \gamma_0 + \gamma_1 r_t \qquad\qquad + \gamma_3 C_t \qquad\qquad + \gamma_5 I_t \qquad\quad + e_{3t}$$

where I_t = investment in period t,
$\quad r_t$ = the interest rate in period t,
$\quad P_t$ = the rate of profit in period t,
$\quad Y_{t-1}$ = income in period $t - 1$,
$\quad C_t$ = consumption in period t,
$\quad M_t$ = money supply in period t,
$\quad E_t$ = excess capacity in period t,
$\quad e_{it}$ = error term.

The exogenous variables are P_t, M_t, and Y_{t-1}. Which equations are overidentified and which are underidentified?

5. Consider the following simple model of an economy

$$C = \alpha_0 + \alpha_1 Y + e_{1t},$$
$$I = \beta_0 + \beta_1 P + e_{2t},$$
$$Y = C + I + Z,$$

where C = consumption,
$\quad Y$ = national income,
$\quad I$ = investment,
$\quad P$ = corporate profits before taxes,
$\quad Z$ = government expenditures plus net exports,
$\quad e_{it}$ = error term.

The first equation is a consumption function, the second an investment function, and the third the well-known national income identity. The consumption function is overidentified in this model. The endogenous variables are consumption C, income Y, and investment I. The exogenous variables are corporate profits P and government expenditures plus net exports Z. (a) Show that in order to solve uniquely for all the structural parameters in terms of the reduced-form parameters we must have $\pi_{12} = 0$ and $\pi_{21} = \pi_{22} \cdot \pi_{11}$, where the π's are the reduced-form parameters—that is,

$$I = \pi_{10} + \pi_{11}P + \pi_{12}Z,$$
$$C = \pi_{20} + \pi_{21}P + \pi_{22}Z.$$

[*Hint:* Work only with the reduced-form expressions for C and I. The expression for Y is extraneous.] (b) Using the data below, estimate the parameters of the structural equations by the method of simultaneous least-squares.

YEAR	Y	C	I	P	Z
1959	484	311	75	29	97
1960	504	325	75	27	104
1961	520	335	72	27	113
1962	560	355	83	31	122
1963	591	375	87	33	128
1964	632	401	94	38	137
1965	685	433	108	47	144
1966	750	466	121	50	162
1967	794	492	116	47	185
1968	866	537	126	50	203

6. Estimate the consumption function in problem 5 using two-stage least-squares.

Chapter 13

Further Topics

In the previous chapters we developed in considerable detail some widely used statistical and econometric techniques. In this chapter we introduce a variety of topics that are not developed very fully. In some cases a fuller development would require considerably more background in mathematics and statistical theory than is provided in previous chapters. For those who wish to explore further the topics presented here and other useful subjects, a list of more advanced works in statistics and econometrics is presented at the end of this chapter.

Autoregressive Models

Frequently in econometric research one encounters linear models that contain lagged values of the dependent variable as independent variables. This kind of model is called autoregressive. For example, suppose the acreage planted of a certain crop is a function not only of the previous season's price of the crop but also the acreage planted in the previous season.

$$A_t = a_0 + a_1 A_{t-1} + a_2 P_{t-1} + e_t, \tag{13.1}$$

where A_t is acreage planted in period t, A_{t-1} is acreage planted in the previous period, and P_{t-1} is price in the previous period.

There are a number of reasons why one might want to use such a model. The two best-known approaches are the "lagged-adjustment" model and the "adaptive-price-expectations" model. The lagged-adjustment model assumes that farmers adjust their acreage to the desired level only gradually after there has been a change in price. Let A_t^* be the acreage planted in period t after all adjustments have taken place, if the price in the previous period P_{t-1} remains the same in the future. We call A_t^* the "desired" acreage planted as opposed to the actual acreage planted. It is a function of the previous period's price, or

$$A_t^* = \alpha_0 + \alpha_1 P_{t-1} + u_t, \tag{13.2}$$

where u_t is an error term. Actual acreage planted A_t in period t, however, is equal to the acreage planted last period A_{t-1} plus a proportion λ $(0 < \lambda < 1)$ of the difference between desired acreage in period t and acreage planted last period. That is,

$$A_t = A_{t-1} + \lambda(A_t^* - A_{t-1}) + v_t, \tag{13.3}$$

where v_t is an error term. λ is called the adjustment coefficient. We can solve this last equation for A_t^* and substitute the resulting expression into (13.2). Thus

$$A_t^* = \frac{1}{\lambda} A_t - \frac{1-\lambda}{\lambda} A_{t-1} - \frac{1}{\lambda} v_t. \tag{13.4}$$

Substituting this expression into (13.2), we obtain

$$\frac{1}{\lambda} A_t - \frac{1-\lambda}{\lambda} A_{t-1} - \frac{1}{\lambda} v_t = \alpha_0 + \alpha_1 P_{t-1} + u_t. \tag{13.5}$$

Solve (13.5) for A_t.

$$A_t = \lambda\alpha_0 + (1 - \lambda)A_{t-1} + \lambda\alpha_1 P_{t-1} + \lambda u_t + v_t. \tag{13.6}$$

This last expression is the same as the autoregressive form (13.1), in which $a_0 = \lambda\alpha_0$, $a_1 = 1 - \lambda$, $a_2 = \lambda\alpha_1$, and $e_t = \lambda u_t + v_t$.

In the adaptive-expectations model, acreage planted in period t is assumed to be a function not of actual price but expected price P_t^* in period t. That is,

$$A_t = \alpha_0 + \alpha_1 P_t^* + u_t, \tag{13.7}$$

where u_t is an error term. Expected price P_t^* in period t is assumed to depend on expected price in the previous period plus a proportion δ $(0 < \delta < 1)$ of the difference between the actual price in the previous period and expected price in the previous period. Thus

$$P_t^* = P_{t-1}^* + \delta(P_{t-1} - P_{t-1}^*). \tag{13.8}$$

That is, expected price is "adapted," at least partially, to conform to actual price behavior. The coefficient δ is called the coefficient of adaptation. If δ is close to zero, expected prices are adapted very little with respect to price performance. If δ is close to unity, adaptation is nearly complete; that is, expected price in period t is very nearly equal to actual price in period $t - 1$.

If we substitute (13.8) into (13.7), we obtain

$$A_t = \alpha_0 + (1 - \delta)\alpha_1 P_{t-1}^* + \delta\alpha_1 P_{t-1} + u_t. \tag{13.9}$$

Now P_{t-1}^* as an "expected price" is not an observable quantity. In order to transform (13.9) into a form that may be estimated empirically we must express P_{t-1}^* in terms of observable variables. This can be done by expressing equation (13.7) in terms of $t - 1$ rather than t and solving for P_{t-1}^*. Thus

$$A_{t-1} = \alpha_0 + \alpha_1 P_{t-1}^* + u_{t-1}. \tag{13.10}$$

Solve for P_{t-1}^*.

$$P_{t-1}^* = \frac{1}{\alpha_1} A_{t-1} - \frac{\alpha_0}{\alpha_1} - \frac{1}{\alpha_1} u_{t-1}. \tag{13.11}$$

Substitute (13.11) into (13.9).

$$A_t = \delta\alpha_0 + (1 - \delta)A_{t-1} + \delta\alpha_1 P_{t-1} + u_t - (1 - \delta)u_{t-1}. \tag{13.12}$$

Thus we have a regression equation of the autoregressive type (13.1) in which $a_0 = \delta\alpha_0$, $a_1 = (1 - \delta)$, $a_2 = \delta\alpha_1$, and $e_t = u_t - (1 - \delta)u_{t-1}$.

Although we have discussed autoregressive models in terms of acreage planted to a crop, such models arise frequently in econometric research. For example, the well-known Friedman permanent-income hypothesis can be formulated in terms of an autoregressive model. In general, an autoregressive model may have a number of lagged dependent variables. The general form is

$$Y_t = a_0 + a_1 X_1 + \cdots + a_K X_K + b_1 Y_{t-1} + b_2 Y_{t-2}$$
$$+ \cdots + b_T Y_{t-T} + e_t, \tag{13.13}$$

where T is called the order of the autoregression.

In an autoregressive model, the assumptions of Chapter 9 concerning the error term cannot all be satisfied. In particular, the independent variables Y_{t-1}, Y_{t-2}, ... , Y_{t-T} are random and they are correlated with the lagged error terms e_{t-1}, e_{t-2}, ... , e_{t-T}, respectively. Thus the estimated values of the a and b coefficients of (13.13) will be biased. They are, however, consistent and thus asymptotically unbiased if all the other assumptions concerning the error term are satisfied.† A word of caution in this regard: in the model of adaptive expectations, the error terms will not in general satisfy the other assumptions. To see this, let us examine equation (13.12). The variable A_{t-1} will depend on the error term u_{t-1}, as indicated by (13.10). The error term $e_t = u_t - (1 - \delta)u_{t-1}$ in (13.12) also depends on u_{t-1}. Thus the independent variable A_{t-1} is correlated with the contemporaneous error term e_t as well as the lagged error terms e_{t-1}, e_{t-2}, and so on.

---

† This result has been generally proved only in stable systems, although the same result has been proved when there are types of instability. See E. Malinvaud, *Statistical Methods of Econometrics* (Skokie, Ill.: Rand McNally, 1966), 449–451.

In this situation, the least-squares estimates of the coefficients of the regression equation will be inconsistent as well as biased. Thus, in general, the adaptive expectations model gives biased and inconsistent estimates of the autoregressive form. In the lagged-adjustment model, however, least-squares estimates are biased but consistent.

In Chapter 11 we noted that serial correlation of the error terms alone does not bias least-squares estimates, although they may be inefficient. If the error terms are serially correlated *and* there is a lagged dependent variable in the equation, least-squares estimates are both biased and inconsistent. This is true for both the adaptive-expectations and lagged-adjustment models.

Distributed Lags

In the previous section we discussed models in which lagged values of the *dependent* variable were in the equations. Another kind of model often encountered in econometric research contains lagged values of the *independent* variables. For example, we might have a consumption function of the form

$$C_t = a_0 + a_1 Y_t + a_2 Y_{t-1} + a_3 Y_{t-2}, \tag{13.14}$$

where C_t is consumption, Y_t is income in period t, Y_{t-1} is income in the previous period, and Y_{t-2} is income two periods ago. This kind of model is called a distributed-lag model.

Distributed-lag models often arise in contexts similar to those in which one might wish to work with an autoregressive model—that is, when adjustments do not take place immediately or when expectations in the current period are conditioned by past values of the independent variables. In fact, an autoregressive model can always be cast in a distributed-lag form. For example, take the lagged-adjustment model of the form (13.6). In terms of $t - 1$, we may write

$$A_{t-1} = \lambda \alpha_0 + (1 - \lambda)A_{t-2} + \lambda \alpha_1 P_{t-2} + \lambda u_{t-1} + v_{t-1}. \tag{13.15}$$

Substitute this expression into (13.6).

$$
\begin{aligned}
A_t = {}& [1 + (1 - \lambda)]\lambda \alpha_0 + \lambda \alpha_1 P_{t-1} + \lambda \alpha_1 (1 - \lambda)P_{t-2} \\
& + (1 - \lambda)^2 A_{t-2} + (\lambda u_t + v_t) + (1 - \lambda)(\lambda u_{t-1} + v_{t-1}).
\end{aligned} \tag{13.16}
$$

We can repeat this process by expressing (13.6) in terms of $t - 2$ and substituting the expression for A_{t-2} into (13.16). If the process is repeated say T times, we end up with the following expression:

$$
\begin{aligned}
A_t = {}& [1 + (1 - \lambda) + (1 - \lambda)^2 + \cdots + (1 - \lambda)^T]\lambda \alpha_0 \\
& + \lambda \alpha_1 P_{t-1} + \lambda \alpha_1 (1 - \lambda)P_{t-2} + \cdots + \lambda \alpha_1 (1 - \lambda)^T P_{t-T-1} \\
& + (\lambda u_t + v_t) + (1 - \lambda)(\lambda u_{t-1} + v_{t-1}) + \cdots \\
& + (1 - \lambda)^T (\lambda u_{t-T} + v_{t-T}) + (1 - \lambda)^T A_{t-T}.
\end{aligned} \tag{13.17}
$$

Since $(1 - \lambda)$ is less than unity, the last term in the expression (13.17), $(1 - \lambda)^T A_{t-T}$, will tend to be quite small. If we ignore this term, (13.17) becomes a pure distributed-lag model.

The same process may be used to convert the autoregressive form of the adaptive-expectations model (13.12). The resulting expression is

$$
\begin{aligned}
A_t = {} & [1 + (1 - \delta) + (1 - \delta)^2 + \cdots + (1 - \delta)^T] \delta \alpha_0 \\
& \delta \alpha_1 P_{t-1} + \delta \alpha_1 (1 - \delta) P_{t-2} + \cdots + \delta \alpha_1 (1 - \delta)^T P_{t-T-1} \\
& + u_t - (1 - \delta)^T u_{t-T} + (1 - \delta)^T A_{t-T}.
\end{aligned} \tag{13.18}
$$

Again, since $(1 - \delta)$ is less than unity, the last *two* terms in (13.18) will tend to be quite small. If we ignore these terms, (13.18) takes on a pure distributed-lag form.

There is an important difference between the distributed-lag form of the adaptive-expectations model (13.18) and the lagged-adjustment model (13.17). In the lagged-adjustment model the error terms will be serially correlated even if the error terms u_t and v_t are not serially correlated. In the adaptive-expectations model, however, the error term in (13.18) is not serially correlated if the error term u_t is not serially correlated.

The distributed-lag forms (13.17) and (13.18) are both cases of a special kind of distributed-lag model called a geometrically declining distributed-lag model. The coefficient of the lagged price variable in both cases declines geometrically by a factor $(1 - \lambda)$ as the lag increases. The more general specification of a distributed-lag model is

$$
Y_t = a_0 + a_1 X_t + a_2 X_{t-1} + \cdots + a_T X_{t-T-1} + e_t, \tag{13.19}
$$

where $a_1, a_2, \cdots, a_T$ are called the distributed-lag coefficients. The lag coefficients may be assumed to follow a variety of particular schemes. The geometrically declining scheme is one particular example. A rather general approach is to assume that the distributed-lag coefficients are in the form of a polynomial in τ, where τ is the degree of the lag. That is,

$$
a_\tau = \alpha_0 + \alpha_1 \tau + \alpha_2 \tau^2 + \cdots + \alpha_h \tau^h, \tag{13.20}
$$

where a_τ is the distributed-lag coefficient and h is the degree of the polynomial. Almon (see the references for this chapter) has developed a computational scheme for estimating the distributed lags in which the degree h of the polynomial may be determined arbitrarily. In the Almon scheme one may restrict the first lag coefficient a_1 to be zero or the last lag coefficient a_h to be zero, or both may be restricted to be zero.

Limited Dependent Variables

Another problem frequently encountered in econometric research is a limited dependent variable. For example, suppose we are trying to explain the pattern of home ownership in a particular community. Households are classified into two groups, those owning homes and those not owning homes. Home ownership is assumed to be a function of household income, age of the household members, number of children in the household, and so on. The model then may be formulated as a linear regression model of the form

$$Y_i = a_0 X_{1i} + a_2 X_{2i} + \cdots + a_K X_{Ki} + e_i, \tag{13.21}$$

where Y_i, the dependent variable, is 0 for nonhomeowning households and 1 for homeowning households, the X's are the independent variables such as income, family size, and so on, and e_i is an error term.

This approach, however, is not very satisfactory. If the dependent variable can only assume either the value 0 or the value 1, the error term must follow a rather strange pattern and must be correlated with the X variables. Figure 13.1 illustrates this dependence. Since Y is only 0 or 1, all the observation points must lie on either the horizontal line $Y = 0$ or at $Y = 1$. If AB is the true regression line, the errors must just offset the difference between the regression line and either the horizontal line $Y = 0$ or $Y = 1$. Thus the errors will tend to be positive for very small X and negative for large X. Furthermore, the expected value of the error terms e_i cannot be zero. The errors will always be just enough positive or negative to make $Y = 0$ or $Y = 1$. Finally, the errors will be heteroskedastic; their variance will not be constant but depend on the value of the X variables. Thus very few of the usual assumptions are satisfied if a linear regression model is used when the dependent variable is limited to 0 or 1.

In order to avoid these difficulties, a different approach must be used. One approach is called *probit analysis*. We assume that there is an index I that measures the likelihood or probability that a household owns a home. If I is large, the likelihood is high, and if I is small, the likelihood is small. The index I is a function of the X variables—that is,

$$I = a_0 + a_1 X_1 + a_2 X_2 + \cdots + a_K X_K. \tag{13.22}$$

Each individual household is assumed to have a critical value of the index, denoted by I_i^*, where i refers to the particular household. If the value of the index I for a particular household exceeds the critical value I_i^*, the household owns a home; if it is less than the critical value, the household does not own a home. That is,

$$Y_i = \begin{cases} 1 & \text{if } I_i \geq I_i^*, \\ 0 & \text{if } I_i < I_i^*. \end{cases} \tag{13.23}$$

The critical value I_i^* is assumed to be a stochastic variable. Thus the probability that a household is a homeowner is

$$\Pr(Y_i = 1 \mid I_i) = \Pr(I_i \geq I_i^*). \tag{13.24}$$

Similarly, the probability that a household is not a homeowner is

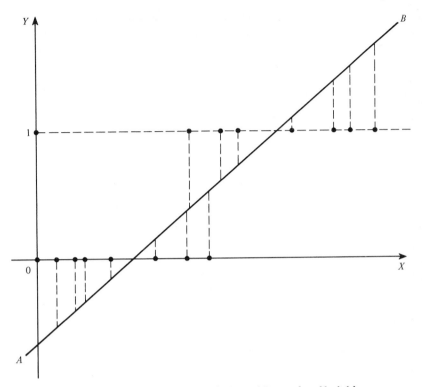

FIGURE 13.1 Pattern of Error Terms with Limited Dependent Variable

$$\text{Pr}\,(Y_i = 0 \mid I_i) = \text{Pr}\,(I_i < I_i^*). \tag{13.25}$$

The probabilities (13.24) and (13.25) may be combined to form a likelihood function. Suppose that there are N_1 homeowning households and N_2 nonhomeowning households in a particular sample. Then the likelihood of the sample is

$$L = \text{Pr}\,(I_1^* \geq I_1) \cdot \ldots \cdot \text{Pr}\,(I_{N_1}^* \geq I_{N_1}) \cdot \text{Pr}\,(I_{N_1+1}^* < I_{N_1+1}) \cdot \ldots \cdot$$
$$\text{Pr}\,(I_{N_1+N_2}^* < I_{N_1+N_2}). \tag{13.26}$$

In Chapter 6 we defined the probability $\text{Pr}\,(I_i \geq I_i^*)$ in (13.24) as the cumulative distribution function for the random variable I_i^*. We denote the cumulative distribution function by $F(I_i)$. The probability (13.25), moreover, is equal to one minus the cumulative distribution function. Thus the likelihood function (13.26) can be written as

$$L = F(I_1) \cdot \ldots \cdot F(I_{N_1}) \cdot [1 - F(I_{N_1+1}) \cdot \ldots \cdot 1 - F(_{N_1+N_2})]. \tag{13.27}$$

Since the index I is a function of the independent X variables as in (13.22), it is clear that the likelihood function (13.27) is a function of the a's and the observed X's for each individual household in the sample. The a's are chosen so that they maximize the likelihood function. The normal equations that result from this maximization procedure are generally nonlinear and quite difficult to solve. In practice, one uses numerical approximation techniques to determine values for the a's that come close to maximizing the likelihood function.

Another approach to limited dependent variables is *discriminant analysis*. With this technique one assumes that there are two different theoretical populations—a homeowning population and a nonhomeowning population. The homeowning and nonhomeowning populations are assumed to have certain characteristics in terms of the X variables. The vector $(X_1, X_2, \ldots, X_K)$ of X variables is assumed to be randomly distributed with one probability distribution for the homeowning population and another distribution for the nonhomeowning population. The problem then is to find an index I that is a function of the X's and a critical value of the index I^* such that one minimizes the probability of misclassification for a given set of X values. That is, the index $I = f(X)$ is such that if $I_i \geqq I_i^*$, then the ith household is classified as homeowning, and if $I_i < I_i^*$, then the ith household is classified as nonhomeowning.

Now if the variances of the X variables and the covariances between different X variables are the same in each of the populations and if the vector $(X_1, X_2, \ldots, X_K)$ has a joint normal distribution in both populations, then the index I is a linear function of the X variables. That is, the functional form of the index I is given by (13.22). The a's can be estimated by regressing a dummy variable Y_i on the X's. The dummy variable takes the value of unity for homeowning households and the value of zero for nonhomeowning households.

The critical value of the index I may be estimated in a number of different ways. If one knows an a priori probability that a household is homeowning, then the critical value of I is

$$I^* = q_1 \bar{I}_1 + (1 - q_1)\bar{I}_2,$$

where q_1 is the a priori probability that a household is homeowning, $\bar{I}_1$ is the sample mean of the estimated index among the homeowning households, and $\bar{I}_2$ is the sample mean among nonhomeowning households. If the a priori probability is not known, then a minimax solution is to set the critical value of the index I equal to

$$I^* = \tfrac{1}{2}\bar{I}_1 + \tfrac{1}{2}\bar{I}_2. \tag{13.28}$$

If the variances and covariances of the X variables are different in the homeowning and nonhomeowning groups, then the index $I = f(X)$ is a quadratic function of the X variables. The critical value of the index I is determined in a manner similar to that in the equal variance-covariance case.

Principal Components

Sometimes one may be working with data on a large number of variables. In order to reduce the number of variables one may wish to take a few linear combinations of the variables. For example, suppose one has data on exports of a wide range of commodities over a period of time. Rather than work with all commodities, one may wish to combine various groups of commodities and deal only with these combined groups.

One method of combining variables is called the analysis of principal components. Suppose there are K variables $X_1, \ldots, X_K$ and N observations on these variables. For simplicity let us assume that the X variables are expressed in terms of deviations from the mean. The *first* principal component of these K variables is a linear combination

$$P_1 = \alpha_{11}X_1 + \alpha_{21}X_2 + \cdots + \alpha_{K1}X_K, \qquad (13.29)$$

which has the maximum variance, given that the squared α's sum up to unity—that is,

$$\alpha_{11}^2 + \alpha_{21}^2 + \cdots + \alpha_{K1}^2 = 1. \qquad (13.30)$$

If we did not place this restriction on the α's, there would be no limit to the variance of the first principal component. We could always increase the variance by making the α's larger. Since the α's are restricted to satisfy (13.30), we call P_1 a *normalized* linear combination of the X's.

In all there are K principal components. The *second* principal component is a normalized linear combination of the X's that is uncorrelated with the first principal component and has the maximum variance of all uncorrelated, normalized linear combinations. Similarly, the pth principal component is a normalized linear combination

$$P_p = \alpha_{1p}X_1 + \alpha_{2p}X_2 + \cdots + \alpha_{Kp}X_K, \qquad (13.31)$$

where

$$\alpha_{1p}^2 + \alpha_{2p}^2 + \cdots + \alpha_{Kp}^2 = 1, \qquad (13.32)$$

which is uncorrelated with the first $p - 1$ principal components and has the maximum variance of all uncorrelated, normalized linear combinations.

If there is a high degree of intercorrelation among the X's, the variance of the first principal component will be quite large, and the other $K - 1$ components will have very little variance. Thus the first principal component will "catch" most of the variation in the X's, and one could work with only the first component or possibly the first few components.

The sum of the variances of all K principal components is equal to the sum of the variances of the X variables. Thus if the first component catches, say, 90 percent of the sum of the variances of X_1 through X_K, the other 10 percent of the variance of the X variables will be spread out over the remaining $K - 1$ principal components. Thus the first component will catch far more variance than any of the remaining principal components.

The method of calculating the α's is relatively straightforward in terms of matrices and vectors. If one differentiates the expression for the variance of P_1 subject to the restriction (13.30), it is easy to show that variance of P_1 is the largest characteristic root of the variance-covariance matrix of the X variables. The variance of the second principal component equals the second largest characteristic root, and so forth. The α's are found by solving a system of linear equations that depend on the characteristic roots.

Analysis of principal components is most suitable when all of the X variables are measured in the same units. If the variables are not measured in the same units, the rationale of the analysis is questionable. In fact the principal components can be shown to depend on the units of measurement. For example, if one X variable is measured in terms of dollars and the other in tons, the components will be different from those where one is measured in thousands of dollars and the other in pounds.

Factor Analysis

Related to analysis of principal components is a technique called factor analysis. The idea is to break down a large number of measurements into a few factors. There are several different ways to estimate the factors; the method of principal components is one. In addition to estimating the factors, however, one also tries to estimate "factor loadings" for each of the variables. That is, if $X_1, \ldots, X_K$ are the variables and $f_1, \ldots, f_P$ are the factors $(P < K)$, then the variables may be expressed as a linear combination of the factors as follows:

$$
\begin{aligned}
X_1 &= \lambda_{11} f_1 + \lambda_{12} f_2 + \cdots + \lambda_{1P} f_P + e_1 \\
X_2 &= \lambda_{21} f_1 + \lambda_{22} f_2 + \cdots + \lambda_{2P} f_P + e_2 \\
&\ \vdots \\
X_K &= \lambda_{K1} f_1 + \lambda_{K2} f_2 + \cdots + \lambda_{KP} f_P + e_K
\end{aligned}
\tag{13.33}
$$

where the λ's are called factor loadings, and the e's are error terms. The estimated factor loadings can be determined in a number of different ways.

Canonical Correlation

A problem related to factor analysis and principal components is that of finding a linear combination of one set of variables that has the maximum correlation with a linear combination of another set of variables. Suppose there are P variables $Y_1, Y_2, \ldots, Y_P$ and K variables $X_1, X_2, \ldots, X_K$.

Again we can simplify matters if we assume that both the Y's and the X's are measured as deviations from their means. Let

$$U_1 = \alpha_{11}Y_1 + \alpha_{12}Y_2 + \cdots + \alpha_{1P}Y_P \tag{13.34}$$

and

$$V_1 = \beta_{11}X_1 + \beta_{12}X_2 + \cdots + \beta_{1K}X_K. \tag{13.35}$$

That is, U_1 is a linear combination of the Y's, and V_1 is a linear combination of the X's. U_1 and V_1 are called the first canonical variates. The α and β coefficients are normalized—that is,

$$\alpha_{11}^2 + \alpha_{12}^2 + \cdots + \alpha_{1P}^2 = 1 \tag{13.36}$$

and

$$\beta_{11}^2 + \beta_{12}^2 + \cdots + \beta_{1K}^2 = 1. \tag{13.37}$$

The α's and β's are chosen to maximize the correlation between U_1 and V_1.

The second canonical variates U_2 and V_2 are normalized linear combinations of the Y's and X's, respectively, that are uncorrelated with the first canonical variates and that maximize the correlation between U_2 and V_2. Similarly the Tth canonical variates U_T and V_T are uncorrelated with the first $T - 1$ variates and are normalized linear combinations of the Y's and X's that maximize the correlation between U_T and V_T. In general there are P such canonical variates (assuming $P \leqq K$).

Another way of viewing canonical variates is to think of the variables $Y_1, Y_2, \ldots, Y_P$ as dependent variables with respect to the independent variables $X_1, X_2, \ldots, X_K$. We wish to find a normalized linear combination of the dependent variables that is a linear function of the independent variables

$$\alpha_1Y_{1i} + \alpha_2Y_{2i} + \cdots + \alpha_PY_{Pi} = \beta_1X_{1i} + \beta_2X_{2i} + \cdots + \beta_KX_{Ki} + e_i \tag{13.38}$$

such that the sum of the squared error terms

$$S = \sum_{i=1}^{N} (e_i^2) \tag{13.39}$$

is minimized. The α's are normalized so that they must satisfy (13.36). It can be shown that linear combination of the dependent variables that minimizes the sum of squares is the first canonical variate, U_1—that is,

$$\alpha_1Y_{1i} + \alpha_2Y_{2i} + \cdots + \alpha_PY_{Pi} = \alpha_{11}Y_{1i} + \alpha_{12}Y_{2i} + \cdots + \alpha_{1P}Y_{Pi}, \tag{13.40}$$

and the right-hand side of (13.38) is the first canonical variate V_1 multiplied by a constant λ:

$$\beta_1 X_1 + \beta_2 X_2 + \cdots + \beta_K X_K = \lambda(\beta_{11} X_1 + \beta_{12} X_2 + \cdots + \beta_{1K} X_K).$$

(13.41)

The value of λ is given by

$$\lambda = \rho \frac{\hat{\sigma}_u}{\hat{\sigma}_v},$$

(13.42)

where ρ is the first canonical correlation, $\hat{\sigma}_u$ is the estimated variance of the first canonical variate U_1, and $\hat{\sigma}_v$ is the estimated variance of the first canonical variate V_1.

The method of estimating the α's and β's of the canonical variates involves solving a polynomial characteristic equation of degree P. There are generally P such roots, the largest of which is the first canonical correlation. The second largest is the second canonical correlation, and so forth. The α's and β's are determined by solving a system of linear equations that depend on the characteristic root.

Spectral Analysis

In dealing with time-series data, one may wish to decompose a series into several periodic components. For example, one may wish to decompose a time series into a seasonal component and one or more business-cycle components of various lengths—say a 2-year cycle, a 5-year cycle, and a 20-year cycle. Or one might wish to determine whether a price series for a tree crop such as cocoa or coffee exhibits a cycle that is related to the gestation period for the crop. One method of decomposing a time series into one or more periodic components is called spectral analysis.

Spectral analysis is usually applied to time series that are stationary. Roughly speaking, a stationary time series exhibits no trend. If trend is present in a time series, this is sometimes removed by fitting a trend line to the series by least-squares and working with deviations from the trend. The deviations from the trend are assumed to be stationary. Another method is to work with first differences of the time series to remove trend; the first differences are assumed to be stationary.

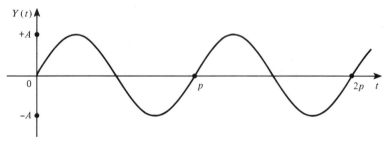

FIGURE 13.2   Sine Function

Any stationary time series can be expressed as an infinite sum of periodic components. The usual approach is to express the time series as a sum of sine or cosine wave components (see Fig. 13.2). The sine wave depicted in Fig. 13.2 has a period p; that is, a cycle is completed after p time periods. The frequency of the sine wave is $f = 1/p$; that is, the frequency is the number of cycles completed per time period. For example, if $p = 5$, then $f = 1/5$, or 1/5 of a cycle is completed for every unit of time.

The formula for a sine wave $X(t)$ with period p or frequency f and amplitude A is

$$X(t) = A \sin (2\pi f \cdot t). \tag{13.43}$$

The amplitude A is the maximum height of the sine wave above or below zero. The time series Y_t can be viewed as an infinite series, where t runs from $-\infty$ to $+\infty$. The time-series element Y_t can be expressed as the sum

$$Y_t = \sum_{i=1}^{\infty} A_i \sin (2\pi f_i \cdot t), \tag{13.44}$$

where A_i is the amplitude of the sine wave with frequency f_i. The frequency can be anything between 0 and 1. A graph of the relationship between amplitude and frequency is shown in Fig. 13.3. This is called a frequency power spectrum. Notice there are peaks at frequencies of 1/4, 1/2, and 7/8. The spectrum is said to exhibit "power" at frequencies of 1/4, 1/2, and 7/8.

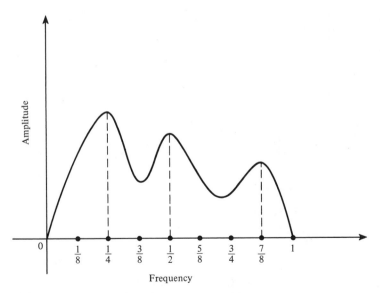

FIGURE 13.3 Power Spectrum

Spectral analysis is a method of determining the amplitude A_i at each of the frequencies f_i. More accurately, spectral analysis is a method for estimating the *average* amplitude A within a frequency *band* where the frequency domain between 0 and 1 is divided into a number of bands. A typical estimated power spectrum is shown in Fig. 13.4. The frequency domain between 0 and 1 is divided into ten bands of equal width. The estimated average amplitude shows a peak in the first spectral band between $f = 0$ and $f = 1/10$ and in the fourth spectral band between $f = 3/10$ and $f = 4/10$.

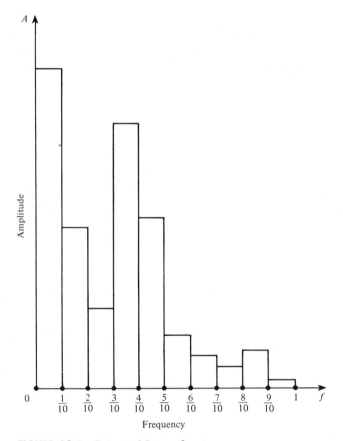

FIGURE 13.4   Estimated Power Spectrum

Concluding Remarks

This chapter's discussion of various topics is far from complete. The reader interested in doing further reading should consult the references.

Much of the analysis presented in this book is based on rather rigid sets of assumptions about the statistical nature of the data with which one is working. Rarely do actual data satisfy all these assumptions completely, and the econometrician must constantly use his judgment in determining whether the violations of the assumptions render useless the standard statistical techniques, or in deciding the degree of caution to be used in interpreting his results. The only way to acquire this judgment is by extensive reading of descriptions of actual econometric research and by experience in doing one's own econometric analysis. No textbook can be a substitute for practical experience.

Outline of Further Reading for Part V

A. Autocorrelated Errors
   1. Christ [1969], pp. 481–488, 522–530
   2. Cochran and Orcutt [1949]
   3. Durbin and Watson [1950], [1951]
   4. Goldberger [1964], pp. 231–247
   5. Johnston [1960], pp. 177–200
   6. Kane [1968], 351–380
   7. Malinvaud [1966], pp. 420–448
   8. Sargan [1961]
   9. Wonnacott and Wonnacott [1970], pp. 132–148
   10. Zellner [1961]

Durbin and Watson provided the well-known test for autocorrelation of the error terms. Zellner is a survey of problems associated with autocorrelated errors, and Sargan develops maximum-likelihood estimation techniques when autocorrelated errors are present. The other references are basic texts which cover the problems associated with autocorrelated errors.

B. Heteroskedasticity and Generalized Least-Squares
   1. Aitken [1934]
   2. Basmann [1960]
   3. Christ [1969], pp. 390–395
   4. Cochran [1947]
   5. David and Johnson [1951]
   6. Goldberger [1964], pp. 231–248
   7. Horsnell [1953]
   8. Johnston [1960], pp. 179–191
   9. Kane [1968], pp. 351–380
   10. Malinvaud [1966], pp. 254–258

Cochran, David and Johnson, and Horsnell are early attempts to modify estimation and testing procedures in analysis of variance when heteroskedasticity is present. As early as 1934, however, Aitken had formulated the generalized least-squares estimation procedure which is applicable when either heteroskedasticity or autocorrelated errors are present. Basmann derived the properties of asymptotic distributions of generalized least-squares estimates. The other references are basic textbook treatments of generalized least-squares and heteroskedasticity.

C. Multicollinearity
   1. Christ [1969], pp. 387–390
   2. Goldberger [1964], pp. 192–194
   3. Johnston [1960], pp. 201–207
   4. Kane [1968], 277–280
   5. Klein and Mitsugu [1962]

D. Errors in Variables
   1. Christ [1969], pp. 251–252
   2. Goldberger [1964], pp. 282–284
   3. Halperin [1961]
   4. Johnston [1960], pp. 148–176
   5. Kendall and Stuart [1967], pp. 375–418
   6. Madansky [1959]
   7. Malinvaud [1966], pp. 326–366
   8. Wonnacott and Wonnacott [1970], pp. 149–171, 337–342

E. Simultaneous Estimation and Identification
   1. Bronfenbrenner [1953]
   2. Chernoff and Divinsky [1953]
   3. Chernoff and Rubin [1953]
   4. Christ [1960]
   5. Christ [1969], pp. 298–346, 395–494
   6. Fisher [1966], *passim*
   7. Goldberger [1964], pp. 288–388
   8. Haavelmo [1943]
   9. Hooper [1959]
   10. Johnston [1960], pp. 231–296
   11. Kane [1968], pp. 303–350
   12. Klein [1953], pp. 64–143
   13. Koopmans [1950], pp. 238–265
   14. Koopmans [1953]
   15. Madansky [1964]
   16. Malinvaud [1966], pp. 497–613
   17. Nagar [1969]
   18. Orcutt [1950]
   19. Sargan [1958]
   20. Summers [1965]

21. Theil [1965], pp. 204–240, 334–354
22. Tintner [1952], pp. 154–188
23. Valavanis [1959]
24. Wonnacott and Wonnacott [1970], pp. 172–195, 343–400
25. Working [1927]
26. Zellner and Theil [1962]

In the twenties Working recognized the problem of identification in the context of the analysis of demand. Haavelmo was concerned with the simultaneous equation bias in least-squares estimates of the consumption function and proposed the method of indirect least-squares. In the late forties and early fifties considerable work was done on problems of identification and estimation of simultaneous systems at the Cowles Commission, first at the University of Chicago and later at Yale. Koopmans, Bronfenbrenner, Chernoff and Divinsky, and Chernoff and Rubin are some of the works under Cowles Foundation sponsorship. Theil generalized the various estimation procedures which arose from the Cowles research and called them k-class estimators. Zellner and Theil later developed another estimation technique called three-stage least-squares. Fisher has extended the work done originally on identification of systems of linear equations to the case of nonlinear equations. Since the early fifties most textbooks on econometrics—such as Tintner, Klein, Johnston, Goldberger, Malinvaud, Kane, Christ, and Wonnacott and Wonnacott—have devoted substantial space to the problems of identification and simultaneous estimation. Valavanis attempts to give an intuitive feel for these subjects in a nontechnical text.

One of the problems associated with simultaneous estimation techniques is that little is known about the sampling properties of simultaneous estimators. Summers has attempted to learn something about the sampling properties of these estimators through Monte Carlo techniques (that is, computer simulation of experiments). Summers' work, and that of others, is summarized in Johnston and later in Christ. Some analytical work on the properties of the sampling distributions has been done by Basmann, Madansky, and Nagar among others.

F. Matrix Theory
1. Graybill [1969], *passim*
2. Hadley [1961], pp. 60–131
3. Perlis [1952], *passim*
4. Stein [1967], *passim*

The understanding of matrix theory is essential for much of the literature in econometrics. Stein and Graybill are a good pair of books; the first deals with elementary matrix theory, and the second develops theorems useful in the analysis of linear statistical models.

G. Distributed Lags and Autoregressive Models
 1. Almon [1965]
 2. Christ [1966], pp. 204–208, 487–488
 3. Eisner [1960]
 4. Friedman [1957], *passim*
 5. Goldberger [1964], pp. 272–278
 6. Griliches [1961]
 7. Griliches [1967]
 8. Hannan [1965]
 9. Jorgenson [1966]
 10. Klein [1958]
 11. Koyck [1954]
 12. Malinvaud [1961]
 13. Malinvaud [1966], pp. 449–496
 14. Nerlove [1958a], [1958b]
 15. Solow [1960]
 16. Theil and Stern [1960]

Solow, Theil and Stern, Klein, Malinvaud [1961], Koyck, Jorgenson, Hannan, and Almon are all articles proposing different forms of the lag distribution function. One of the most powerful and frequently used forms is the polynomial lag distribution function proposed by Almon. Griliches [1967] is a survey of the various lag distribution functions which have been proposed by various authors. Griliches [1961] shows how least-squares estimation procedures produce inconsistent estimates in lagged models with autocorrelated errors. Nerlove shows how distributed lags and auto-regressive models arise from theoretical consideration of farmers' supply response based on lagged adjustment and expectational models. Malinvaud and Christ are textbook treatments of distributed lag and autoregressive models. Eisner and Koyck apply distributed lags to estimation of invest-ment functions. Friedman's permanent income hypothesis concerning the consumption function may be formulated as a distributed lag model.

H. Limited Dependent Variables
 1. Anderson [1958], pp. 126–153
 2. Anderson and Bahadur [1961]
 3. Frank [1970]
 4. Goldberger [1964], pp. 248–254
 5. Hope [1968], pp. 102–124
 6. Rosett [1959]
 7. Tintner [1952], pp. 96–101
 8. Tobin [1955]
 9. Tobin [1958]
 10. Warner [1962]

Hope is an elementary treatment of discriminant analysis. Anderson provides a comprehensive treatment of discriminant analysis, while Warner applies discriminant analysis to choice of mode of travel. Tobin [1955] applies probit analysis to estimation of consumer durable expenditure functions. The 1958 article of Tobin develops probit analysis and applies it to expenditure data. Rosett applies probit analysis to a general model of friction in economics. Anderson and Bahadur develop discriminant analysis under the assumption that the covariance functions of the two populations are unequal. Frank further develops this theory and applies it to measurement of debt servicing capacity.

I. Principal Components, Factor Analysis, and Canonical Correlation
   1. Anderson [1958], pp. 272–306, 323–324, 329–330, 326–328
   2. Anderson and Rubin [1956]
   3. Dempster [1969], pp. 136–140, 176–185
   4. Harman [1967], *passim*
   5. Hope [1968], pp. 33–67, 86–101
   6. Horst [1965], *passim*
   7. Hotelling [1936]
   8. Kloek and Mennes [1960]
   9. Lawley and Maxwell [1963], *passim*
   10. Rummel [1970], *passim*
   11. Tintner [1962], pp. 102–121
   12. Wilks [1962], pp. 564–592

Harman, Hope, and Horst are easy to read and elementary treatments of factor analysis and principal components. Harman and Horst are exceptionally comprehensive. Rummel is also an elementary book on factor analysis, with emphasis on applications in the social sciences. Lawley and Maxwell, Dempster, Wilks, Anderson, and Anderson and Rubin are advanced treatments of factor analysis and principal components for the experienced mathematical statistician. Anderson, Dempster, and Wilks cover canonical correlation at an advanced level, Tintner at an intermediate level, and Hope at an elementary level. Hotelling provides the seminal article on canonical correlation. Kloek and Mennes apply the analysis of principal components to the theory of multiple regression.

J. Spectral Analysis
   1. Adelman [1965]
   2. Fishman [1968], *passim*
   3. Godfrey [1967]
   4. Godfrey and Granger [1964]
   5. Granger [1966]
   6. Granger and Hatanaka [1961], *passim*
   7. Granger and Rees [1968]

   8. Howrey [1968]
   9. Jenkins and Watts [1968], *passim*
  10. Nerlove [1964]

Fishman, Granger and Hatanaka, and Jenkins and Watts are basic texts on spectral analysis. The other references are applications of spectral analysis to various problems in economics.

Appendix of
Statistical
Tables

TABLE A   Areas under normal distribution

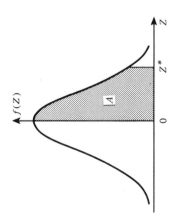

Z^*	AREA A	Z^*	AREA A	Z^*	AREA A	Z^*	AREA A	Z^*	AREA A	Z^*	AREA A	Z^*	AREA A
.00	.000	.30	.118	.60	.226	1.20	.385	1.80	.464	2.40	.4918	3.30	.49952
.01	.004	.31	.122	.62	.232	1.22	.389	1.82	.466	2.43	.4925	3.33	.49957
.02	.008	.32	.126	.64	.239	1.24	.393	1.84	.467	2.46	.4931	3.36	.49961
.03	.012	.33	.129	.66	.245	1.26	.396	1.86	.469	2.49	.4936	3.39	.49965
.04	.016	.34	.133	.68	.252	1.28	.400	1.88	.470	2.52	.4941	3.42	.49969
.05	.020	.35	.137	.70	.258	1.30	.403	1.90	.471	2.55	.4946	3.45	.49972
.06	.024	.36	.141	.72	.264	1.32	.407	1.92	.473	2.58	.4951	3.48	.49975
.07	.028	.37	.144	.74	.270	1.34	.410	1.94	.474	2.61	.4955	3.51	.49978
.08	.032	.38	.148	.76	.276	1.36	.413	1.96	.475	2.64	.4959	3.54	.49980
.09	.036	.39	.152	.78	.282	1.38	.416	1.98	.476	2.67	.4962	3.57	.49982
.10	.040	.40	.155	.80	.288	1.40	.419	2.00	.477	2.70	.4965	3.60	.49984
.11	.044	.41	.159	.82	.294	1.42	.422	2.02	.478	2.73	.4968	3.63	.49986

x	A	x	A	x	A	x	A	x	A	x	A	x	A
.12	.048	.42	.163	.84	.300	1.44	.425	2.04	.479	2.76	.4971	3.66	.49987
.13	.052	.43	.166	.86	.305	1.46	.428	2.06	.480	2.79	.4974	3.69	.49989
.14	.056	.44	.170	.88	.311	1.48	.431	2.08	.481	2.82	.4976	3.72	.49990
.15	.060	.45	.174	.90	.316	1.50	.433	2.10	.482	2.85	.4978	3.75	.49991
.16	.064	.46	.177	.92	.321	1.52	.436	2.12	.483	2.88	.4980	3.78	.49992
.17	.067	.47	.181	.94	.326	1.54	.438	2.14	.484	2.91	.4982	3.81	.49993
.18	.071	.48	.184	.96	.331	1.56	.441	2.16	.485	2.94	.4984	3.84	.49994
.19	.075	.49	.188	.98	.336	1.58	.443	2.18	.4854	2.97	.4985	3.87	.49995
.20	.079	.50	.191	1.00	.341	1.60	.445	2.20	.4861	3.00	.4987	3.90	.49995
.21	.083	.51	.195	1.02	.346	1.62	.447	2.22	.4868	3.03	.4988	3.93	.49996
.22	.087	.52	.198	1.04	.351	1.64	.450	2.24	.4875	3.06	.4989	3.96	.49996
.23	.091	.53	.202	1.06	.355	1.66	.451	2.26	.4881	3.09	.4990	3.99	.49997
.24	.095	.54	.205	1.08	.360	1.68	.454	2.28	.4887	3.12	.4991		
.25	.099	.55	.209	1.10	.364	1.70	.455	2.30	.4893	3.15	.4992		
.26	.103	.56	.212	1.12	.369	1.72	.457	2.32	.4898	3.18	.4993		
.27	.106	.57	.216	1.14	.373	1.74	.459	2.34	.4904	3.21	.49934		
.28	.110	.58	.219	1.16	.377	1.76	.461	2.36	.4909	3.24	.49940		
.29	.114	.59	.222	1.18	.381	1.78	.462	2.38	.4913	3.27	.49946		

Table A is reprinted from *Mathematics of Statistics* by J. F. Kenney and E. S. Keeping © 1939, 1951, by Litton Educational Publishing, Inc., by permission of Van Nostrand Reinhold Company (New York).

TABLE B   Chi-square distribution*

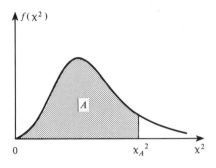

DEGREES OF FREEDOM	AREA A					
	.995	.99	.975	.95	.9	.75
	1 − AREA A					
	.005	.01	.025	.05	.1	.25
1	7.879	6.635	5.024	3.842	2.706	1.323
2	10.597	9.210	7.378	5.992	4.605	2.773
3	12.838	11.345	9.348	7.815	6.251	4.108
4	14.860	13.277	11.143	9.488	7.779	5.385
5	16.750	15.086	12.832	11.070	9.236	6.626
6	18.548	16.812	14.449	12.592	10.645	7.841
7	20.278	18.475	16.013	14.067	12.017	9.037
8	21.955	20.090	17.535	15.507	13.362	10.219
9	23.589	21.666	19.023	16.919	14.684	11.389
10	25.188	23.209	20.483	18.307	15.987	12.549
12	28.300	26.217	23.337	21.026	18.549	14.845
15	32.801	30.578	27.488	24.996	22.307	18.245
20	39.997	37.566	34.170	31.410	28.412	23.828
24	45.558	42.980	39.364	36.415	33.196	28.241
30	53.672	50.892	46.979	43.773	40.256	34.800
40	66.766	63.691	59.342	55.758	51.805	45.616
60	91.952	88.379	83.298	79.082	74.397	66.981
120	163.64	158.95	152.21	146.57	140.23	130.06

* Table gives χ_A^2.
Table B is reprinted from Lewis E. Vogler and Kenneth A. Norton, *Graphs and Tables of the Significance Levels F(ν_1, ν_2, p) for the Fisher-Snedecor Variance Ratio*, from NBS Report 5069, National Bureau of Standards, Boulder Laboratories, by the kind permission of the authors and the Boulder Laboratory.

AREA A						
.5	.25	.1	.05	.025	.01	.005

1 − AREA A						
.5	.75	.9	.95	.975	.99	.995

.4549	.1015	.01579	.00393	.00098	.00016	.00004
1.386	.5754	.2107	.1026	.05064	.02010	.01002
2.366	1.213	.5844	.3519	.2158	.1148	.07172
2.357	1.923	1.064	.7107	.4844	.2971	.2070
4.352	2.675	1.610	1.146	.8312	.5543	.4117
5.348	3.455	2.204	1.635	1.237	.8721	.6757
6.346	4.255	2.833	2.167	1.690	1.239	.9893
7.344	5.071	3.490	2.733	2.180	1.647	1.344
8.343	5.899	4.168	3.325	2.700	2.088	1.735
9.342	6.737	4.865	3.940	3.247	2.558	2.156
11.340	8.438	6.304	5.226	4.404	3.571	3.074
14.339	11.036	8.547	7.261	6.262	5.229	4.601
19.337	15.452	12.443	10.851	9.591	8.260	7.434
23.337	19.037	15.659	13.848	12.401	10.856	9.886
29.336	24.478	20.599	18.493	16.791	14.954	13.787
39.335	33.660	29.050	26.509	24.433	22.164	20.706
59.335	52.294	46.459	43.188	40.482	37.485	35.535
119.33	109.22	100.62	95.701	91.576	86.926	83.851

TABLE C   Student's t distribution*

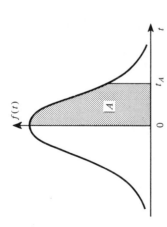

DEGREES OF FREEDOM	AREA A											
	.495	.490	.475	.450	.400	.350	.300	.250	.200	.150	.100	.005
1	63.657	31.821	12.706	6.314	3.078	1.963	1.376	1.000	.727	.510	.325	.158
2	9.925	6.965	4.303	2.920	1.886	1.386	1.061	.816	.617	.445	.289	.142
3	5.841	4.541	3.182	2.353	1.638	1.250	.978	.765	.584	.424	.277	.137
4	4.604	3.747	2.776	2.132	1.533	1.190	.941	.741	.569	.414	.271	.134
5	4.032	3.365	2.571	2.015	1.476	1.156	.920	.727	.559	.408	.267	.132
6	3.707	3.143	2.447	1.943	1.440	1.134	.906	.718	.553	.404	.265	.131
7	3.499	2.998	2.365	1.895	1.415	1.119	.896	.711	.549	.402	.263	.130
8	3.355	2.896	2.306	1.860	1.397	1.108	.889	.706	.546	.399	.262	.130
9	3.250	2.821	2.262	1.833	1.383	1.100	.883	.703	.543	.398	.261	.129
10	3.169	2.764	2.228	1.812	1.372	1.093	.879	.700	.542	.397	.260	.129
11	3.106	2.718	2.201	1.796	1.363	1.088	.876	.697	.540	.396	.260	.129

12	3.055	2.681	2.179	1.782	1.356	1.083	.873	.695	.539	.395	.259	.128
13	3.012	2.650	2.160	1.771	1.350	1.079	.870	.694	.538	.394	.259	.128
14	2.977	2.624	2.145	1.761	1.345	1.076	.868	.692	.537	.393	.258	.128
15	2.947	2.602	2.131	1.753	1.341	1.074	.866	.691	.536	.393	.258	.128
16	2.921	2.583	2.120	1.746	1.337	1.071	.865	.690	.535	.392	.258	.128
17	2.898	2.567	2.110	1.740	1.333	1.069	.863	.689	.534	.392	.257	.128
18	2.878	2.552	2.101	1.734	1.330	1.067	.862	.688	.534	.392	.257	.127
19	2.861	2.539	2.093	1.729	1.328	1.066	.861	.688	.533	.391	.257	.127
20	2.845	2.528	2.086	1.725	1.325	1.064	.860	.687	.533	.391	.257	.127
22	2.819	2.508	2.074	1.717	1.321	1.061	.858	.686	.532	.390	.256	.127
24	2.797	2.492	2.064	1.711	1.318	1.059	.857	.685	.531	.390	.256	.127
26	2.779	2.479	2.056	1.706	1.315	1.058	.856	.684	.531	.390	.256	.127
28	2.763	2.467	2.048	1.701	1.313	1.056	.855	.683	.530	.389	.256	.127
30	2.750	2.457	2.042	1.697	1.310	1.055	.854	.683	.530	.389	.256	.127
∞	2.576	2.326	1.960	1.645	1.282	1.036	.842	.675	.524	.385	.253	.126

* Table gives value of t_A.

Table C is taken from the Tables of R. A. Fisher: *Statistical Methods for Research Workers*, 13th ed., rev., published by Oliver & Boyd, Edinburgh, 1958, p. 174, and by permission of the author and publishers.

TABLE D Critical values of F distribution

D.1. One-Tailed Test, 5 Percent Critical Region

$k_2 =$ DE-GREES OF FREEDOM IN DE-NOMINATOR	$k_1 =$ DEGREES OF								
	1	2	3	4	5	6	7	8	9
1	161.45	199.50	215.71	224.58	230.16	233.99	236.77	238.88	240.54
2	18.513	19.000	19.164	19.247	19.296	19.330	19.353	19.371	19.385
3	10.128	9.552	9.277	9.117	9.014	8.941	8.887	8.845	8.812
4	7.709	6.944	6.591	6.388	6.256	6.163	6.094	6.041	5.999
5	6.608	5.786	5.410	5.192	5.050	4.950	4.876	4.818	4.773
6	5.987	5.143	4.757	4.534	4.388	4.284	4.207	4.147	4.099
7	5.591	4.737	4.347	4.120	3.972	3.866	3.787	3.726	3.677
8	5.318	4.459	4.066	3.838	3.688	3.581	3.501	3.438	3.388
9	5.117	4.257	3.863	3.633	3.482	3.374	3.293	3.230	3.179
10	4.965	4.103	3.708	3.478	3.326	3.217	3.136	3.072	3.020
12	4.747	3.885	3.490	3.259	3.106	2.996	2.913	2.849	2.796
15	4.543	3.682	3.287	3.056	2.901	2.791	2.707	2.641	2.588
20	4.351	3.493	3.098	2.866	2.711	2.599	2.514	2.447	2.393
24	4.260	3.403	3.009	2.776	2.621	2.508	2.423	2.355	2.300
30	4.171	3.316	2.922	2.690	2.534	2.421	2.334	2.266	2.211
40	4.085	3.232	2.839	2.606	2.450	2.336	2.249	2.180	2.124
60	4.001	3.150	2.758	2.525	2.368	2.254	2.167	2.097	2.040
120	3.920	3.072	2.680	2.447	2.290	2.175	2.087	2.016	1.959
∞	3.842	2.996	2.605	2.372	2.214	2.099	2.010	1.938	1.880

Table D is based on tables from Lewis E. Vogler and Kenneth A. Norton, *Graphs and Tables of the Significance Levels F(ν_1, ν_2, p) for the Fisher-Snedecor Variance Ratio*, from NBS Report 5069, National Bureau of Standards, Boulder Laboratories, by the kind permission of the authors and the Boulder Laboratory.

FREEDOM IN NUMERATOR

10	12	15	20	24	30	40	60	120	∞
241.88	243.91	245.95	248.01	249.05	250.09	251.14	252.20	253.25	254.32
19.396	19.413	19.429	19.446	19.454	19.462	19.471	19.479	19.487	19.496
8.786	8.745	8.703	8.660	8.6385	8.617	8.594	8.572	8.549	8.527
5.964	5.912	5.858	5.803	5.774	5.746	5.717	5.688	5.658	5.628
4.735	4.678	4.619	4.558	4.527	4.496	4.464	4.431	4.398	4.365
4.060	4.000	3.938	3.874	3.842	3.808	3.774	3.740	3.705	3.669
3.637	3.575	3.511	3.445	3.411	3.376	3.340	3.304	3.267	3.230
3.347	3.284	3.218	3.150	3.115	3.079	3˙043	3.005	2.967	2.928
3.137	3.073	3.006	2.937	2.901	2.864	2.826	2.787	2.748	2.707
2.978	2.913	2.845	2.774	2.737	2.700	2.661	2.621	2.580	2.538
2.753	2.687	2.617	2.544	2.506	2.466	2.426	2.384	2.341	2.296
2.544	2.475	2.404	2.328	2.288	2.247	2.204	2.160	2.114	2.066
2.348	2.278	2.203	2.124	2.083	2.039	1.994	1.946	1.896	1.843
2.255	2.183	2.108	2.027	1.984	1.939	1.892	1.842	1.790	1.733
2.165	2.092	2.015	1.932	1.887	1.841	1.792	1.740	1.684	1.622
2.077	2.004	1.925	1.839	1.793	1.744	1.693	1.637	1.577	1.509
1.993	1.917	1.836	1.748	1.700	1.649	1.594	1.534	1.467	1.389
1.911	1.834	1.751	1.659	1.608	1.554	1.495	1.429	1.352	1.254
1.831	1.752	1.666	1.517	1.571	1.459	1.394	1.318	1.221	1.000

TABLE D (continued)

D.2. Two-Tailed Test, 5 Percent Critical Region

$k_2 =$ DE-GREES OF FREEDOM IN DE-NOMINATOR	$k_1 =$ DEGREES OF								
	1	2	3	4	5	6	7	8	9
1	647.79	799.50	864.16	899.58	921.85	937.11	948.22	956.66	963.28
2	38.506	39.000	39.165	39.248	39.298	39.331	39.355	39.373	39.387
3	17.443	16.044	15.439	15.101	14.885	14.735	14.624	14.540	14.473
4	12.218	10.649	9.979	9.605	9.365	9.1973	9.074	8.980	8.905
5	10.007	8.434	7.764	7.388	7.146	6.978	6.853	6.757	6.681
6	8.813	7.260	6.599	6.227	5.988	5.820	5.696	5.600	5.523
7	8.073	6.542	5.890	5.523	5.285	5.119	4.995	4.899	4.823
8	7.571	6.060	5.416	5.053	4.817	4.652	4.529	4.433	4.357
9	7.209	5.715	5.078	4.718	4.484	4.320	4.197	4.102	4.026
10	6.937	5.456	4.826	4.468	4.236	4.072	3.950	3.855	3.779
12	6.554	5.096	4.474	4.121	3.891	3.728	3.607	3.512	3.436
15	6.200	4.765	4.153	3.804	3.576	3.415	3.293	3.199	3.123
20	5.872	4.461	3.859	3.515	3.289	3.128	3.007	2.913	2.837
24	5.717	4.319	3.721	3.379	3.155	2.995	2.874	2.780	2.703
30	5.568	4.182	3.589	3.250	3.027	2.867	2.746	2.651	2.575
40	5.424	4.051	3.463	3.126	2.904	2.744	2.624	2.529	2.452
60	5.286	3.925	3.343	3.008	2.786	2.627	2.507	2.412	2.334
120	5.152	3.805	3.227	2.894	2.674	2.515	2.395	2.299	2.222
∞	5.024	3.689	3.116	2.786	2.567	2.408	2.288	2.192	2.114

10	12	15	20	24	30	40	60	120	∞
968.63	976.71	984.87	993.10	997.25	1001.4	1005.6	1009.8	1014.0	1018.8
39.398	39.415	39.431	39.448	39.456	39.465	39.473	39.481	39.490	39.498
14.419	14.337	14.253	14.167	14.124	14.081	14.037	13.992	13.947	13.902
8.844	8.751	8.657	8.560	8.511	8.461	8.411	8.360	8.309	8.257
6.619	6.525	6.428	6.329	6.278	6.227	6.175	6.123	6.069	6.015
5.461	5.366	5.269	5.168	5.117	5.065	5.013	4.959	4.905	4.849
4.761	4.666	4.568	4.467	4.415	4.362	4.309	4.254	4.199	4.142
4.295	4.200	4.101	4.000	3.947	3.894	3.840	3.784	3.728	3.670
3.964	3.868	3.769	3.667	3.614	3.560	3.560	3.449	3.392	3.333
3.717	3.621	3.522	3.419	3.365	3.311	3.255	3.198	3.140	3.080
3.374	3.277	3.177	3.073	3.019	2.963	2.906	2.848	2.787	2.725
3.060	2.963	2.862	2.756	2.701	2.644	2.585	2.524	2.461	2.395
2.774	2.676	2.573	2.465	2.408	2.349	2.287	2.223	2.156	2.085
2.640	2.541	2.437	2.327	2.269	2.209	2.146	2.080	2.010	1.935
2.511	2.412	2.307	2.195	2.136	2.074	2.009	1.940	1.866	1.787
2.388	2.288	2.182	2.068	2.007	1.943	1.875	1.803	1.724	1.637
2.270	2.169	2.061	1.945	1.882	1.815	1.744	1.667	1.581	1.482
2.157	2.055	1.945	1.825	1.760	1.690	1.614	1.530	1.433	1.310
2.048	1.945	1.833	1.709	1.640	1.566	1.484	1.388	1.263	1.000

TABLE D   **(continued)**

D.3.   One-Tailed Test, 1 Percent Critical Region

$k_2 =$ DE-GREES OF FREEDOM IN DE-NOMINATOR	$k_1 =$ DEGREES OF								
	1	2	3	4	5	6	7	8	9
1	4052.2	4999.5	5403.3	562.46	5763.7	5895.0	5928.3	5981.6	6022.5
2	98.503	99.000	99.166	99.249	99.299	99.332	99.356	99.374	99.388
3	34.116	30.817	29.457	28.710	28.237	27.911	27.672	27.489	27.345
4	21.198	18.000	16.694	15.977	15.522	15.207	14.976	14.799	14.659
5	16.258	13.274	12.060	11.392	10.967	10.672	10.456	10.289	10.158
6	13.745	10.925	9.780	9.148	8.746	8.466	8.260	8.102	7.976
7	12.246	9.547	8.451	7.847	7.460	7.191	6.993	6.840	6.719
8	11.259	8.649	7.591	7.006	6.632	6.371	6.178	6.029	5.911
9	10.561	8.022	6.992	6.422	6.057	5.802	5.613	5.467	5.351
10	10.044	7.559	6.552	5.994	5.636	5.386	5.200	5.057	4.942
12	9.330	6.927	5.953	5.412	5.064	4.821	4.640	4.499	4.388
15	8.683	6.359	5.417	4.893	4.556	4.318	4.142	4.005	3.895
20	8.096	5.849	4.938	4.431	4.103	3.871	3.699	3.564	3.457
24	7.823	5.614	4.718	4.218	3.895	3.667	3.496	3.363	3.256
30	7.563	5.390	4.510	4.018	3.699	3.474	3.305	3.173	3.067
40	7.314	5.179	4.313	3.828	3.514	3.291	3.124	2.993	2.888
60	7.077	4.977	4.126	3.649	3.339	3.119	2.953	2.823	2.719
120	6.851	4.787	3.949	3.480	3.174	2.956	2.792	2.663	2.559
∞	6.635	4.605	3.782	3.319	3.017	2.802	2.639	2.511	2.407

FREEDOM IN NUMERATOR

10	12	15	20	24	30	40	60	120	∞
6055.8	6106.3	6157.3	6208.7	6234.6	6260.7	6286.8	6313.0	6339.4	6366.0
99.399	99.416	99.432	99.449	99.458	99.466	99.474	99.483	99.491	99.501
27.229	27.052	26.872	26.690	26.598	26.505	26.411	26.316	26.221	26.125
14.546	14.374	14.198	14.020	13.929	13.838	13.745	13.652	13.558	13.463
10.051	9.888	9.722	9.553	9.467	9.379	9.291	9.202	9.112	9.020
7.874	7.718	7.559	7.396	7.313	7.229	7.143	7.057	6.969	6.880
6.620	6.469	6.314	6.155	6.074	5.992	5.908	5.824	5.737	5.650
5.814	5.667	5.515	5.359	5.279	5.198	5.116	5.032	4.946	4.859
5.257	5.111	4.962	4.808	4.729	4.649	4.567	4.483	4.398	4.311
4.849	4.706	4.558	4.405	4.327	4.247	4.165	4.082	3.997	3.909
4.296	4.155	4.010	3.858	3.781	3.701	3.619	3.536	3.449	3.361
3.805	3.666	3.522	3.372	3.294	3.214	3.132	3.047	2.960	2.868
3.368	3.231	3.088	2.938	2.859	2.779	2.695	2.608	2.517	2.421
3.168	3.032	2.889	2.738	2.659	2.577	2.492	2.404	2.310	2.211
2.980	2.843	2.700	2.549	2.469	2.386	2.299	2.208	2.111	2.006
2.801	2.665	2.522	2.369	2.288	2.203	2.114	2.019	1.917	1.805
2.632	2.496	2.352	2.198	2.115	2.029	1.936	1.836	1.726	1.601
2.472	2.336	2.192	2.035	1.950	1.860	1.763	1.656	1.533	1.381
2.321	2.185	2.039	1.878	1.791	1.696	1.592	1.473	1.325	1.000

TABLE D (continued)

D.4.   Two-Tailed Test, 1 Percent Critical Region

k_2 = DE- GREES OF FREEDOM IN DE- NOMINATOR						k_1 = DEGREES OF			
	1	2	3	4	5	6	7	8	9
1	16211.0	20000.0	21615.0	22500.0	23056.0	23437.0	23715.0	23925.0	24091.0
2	198.50	199.00	199.17	199.25	199.30	199.33	199.36	19.937	199.39
3	55.552	49.799	47.467	46.195	45.392	44.838	44.434	44.126	43.882
4	31.333	26.284	24.259	23.155	22.456	21.975	21.622	21.352	21.139
5	22.785	18.314	16.530	15.556	14.940	14.513	14.200	13.961	13.772
6	18.635	14.544	12.917	12.028	11.464	11.073	10.786	10.566	10.391
7	16.236	12.404	10.882	10.050	9.522	9.155	8.885	8.678	8.514
8	14.688	11.042	9.597	8.805	8.302	7.952	7.694	7.496	7.339
9	13.614	10.107	8.717	7.956	7.471	7.134	6.885	6.693	6.541
10	12.826	9.427	8.081	7.343	6.872	6.545	6.303	6.116	5.968
12	11.754	8.510	7.226	6.521	6.071	5.757	5.525	5.345	5.202
15	10.798	7.701	6.476	5.803	5.372	5.071	4.847	4.674	4.536
20	9.944	6.987	5.818	5.174	4.762	4.472	4.257	4.090	3.956
24	9.551	6.661	5.519	4.890	4.486	4.202	3.991	3.826	3.695
30	9.180	6.355	5.239	4.623	4.228	3.949	3.742	3.580	3.451
40	8.828	6.066	4.976	4.374	3.986	3.713	3.509	3.350	3.222
60	8.495	5.795	4.729	4.140	3.760	3.492	3.291	3.134	3.008
120	8.179	5.539	4.497	3.921	3.548	3.285	3.087	2.933	2.808
∞	7.879	5.298	4.279	3.715	3.350	3.091	2.897	2.744	2.621

10	12	15	20	24	30	40	60	120	∞
24224.0	24426.0	24630.0	24836.0	24940.0	25044.0	25148.0	25253.0	25359.0	25465.0
199.40	199.42	199.43	199.45	199.46	199.47	199.47	199.48	199.49	199.51
43.686	43.387	43.085	42.778	42.622	42.466	42.308	42.149	41.989	41.829
20.697	20.705	20.438	20.167	20.030	19.892	19.752	19.611	19.468	19.325
13.618	13.384	13.146	12.903	12.780	12.656	12.530	12.402	12.274	12.144
10.250	10.034	9.814	9.589	9.474	9.358	9.241	9.122	9.002	8.879
8.380	8.176	7.968	7.754	7.645	7.535	7.423	7.309	7.193	7.076
7.211	7.015	6.814	6.608	6.503	6.396	6.288	6.177	6.065	5.951
6.417	6.227	6.033	5.832	5.729	5.625	5.519	5.410	5.300	5.188
5.847	5.661	5.471	5.274	5.173	5.071	4.966	4.860	4.750	4.639
5.086	4.906	4.721	4.530	4.432	4.331	4.228	4.123	4.015	3.904
4.424	4.250	4.070	3.883	3.786	3.687	3.585	3.480	3.372	3.260
3.847	3.678	3.502	3.318	3.222	3.123	3.022	2.916	2.806	2.690
3.587	3.420	3.246	3.062	2.967	2.868	2.765	2.659	2.546	2.428
3.344	3.179	3.006	2.823	2.727	2.628	2.524	2.415	2.300	2.176
3.117	2.953	2.781	2.598	2.502	2.402	2.296	2.184	2.064	1.932
2.904	2.742	2.571	2.387	2.290	2.187	2.079	1.962	1.834	1.689
2.705	2.544	2.373	2.188	2.089	1.984	1.871	1.747	1.606	1.431
2.519	2.358	2.187	2.000	1.898	1.789	1.669	1.533	1.364	1.000

TABLE E   Durbin-Watson statistic—d_1 and d_2 for two-tailed, 5 percent critical region

N	$K = 1$		$K = 2$		$K = 3$		$K = 4$		$K = 5$	
	d_1	d_2	d_1	d_2	d_1	d_2	d_1	d_2	d_1	d_2
15	0.95	1.23	0.83	1.40	0.71	1.61	0.59	1.84	0.48	2.09
16	0.98	1.24	0.86	1.40	0.75	1.59	0.64	1.80	0.53	2.03
17	1.01	1.25	0.90	1.40	0.79	1.58	0.68	1.77	0.57	1.98
18	1.03	1.26	0.93	1.40	0.82	1.56	0.72	1.74	0.62	1.93
19	1.06	1.28	0.96	1.41	0.86	1.55	0.76	1.72	0.66	1.90
20	1.08	1.28	0.99	1.41	0.89	1.55	0.79	1.70	0.70	1.87
21	1.10	1.30	1.01	1.41	0.92	1.54	0.83	1.69	0.73	1.84
22	1.12	1.31	1.04	1.42	0.95	1.54	0.86	1.68	0.77	1.82
23	1.14	1.32	1.06	1.42	0.97	1.54	0.89	1.67	0.80	1.80
24	1.16	1.33	1.08	1.43	1.00	1.54	0.91	1.66	0.83	1.79
25	1.18	1.34	1.10	1.43	1.02	1.54	0.94	1.65	0.86	1.77
26	1.19	1.35	1.12	1.44	1.04	1.54	0.96	1.65	0.88	1.76
27	1.21	1.36	1.13	1.44	1.06	1.54	0.99	1.64	0.91	1.75
28	1.22	1.37	1.15	1.45	1.08	1.54	1.01	1.64	0.93	1.74
29	1.24	1.38	1.17	1.45	1.10	1.54	1.03	1.63	0.96	1.73
30	1.25	1.38	1.18	1.46	1.12	1.54	1.05	1.63	0.98	1.73
31	1.26	1.39	1.20	1.47	1.13	1.55	1.07	1.63	1.00	1.72
32	1.27	1.40	1.21	1.47	1.15	1.55	1.08	1.63	1.02	1.71
33	1.28	1.41	1.22	1.48	1.16	1.55	1.10	1.63	1.04	1.71
34	1.29	1.41	1.24	1.48	1.17	1.55	1.12	1.63	1.06	1.70
35	1.30	1.42	1.25	1.48	1.19	1.55	1.13	1.63	1.07	1.70
36	1.31	1.43	1.26	1.49	1.20	1.56	1.15	1.63	1.09	1.70
37	1.32	1.43	1.27	1.49	1.21	1.56	1.16	1.62	1.10	1.70
38	1.33	1.44	1.28	1.50	1.23	1.56	1.17	1.62	1.12	1.70
39	1.34	1.44	1.29	1.50	1.24	1.56	1.19	1.63	1.13	1.69
40	1.35	1.45	1.30	1.51	1.25	1.57	1.20	1.63	1.15	1.69
45	1.39	1.48	1.34	1.53	1.30	1.58	1.25	1.63	1.21	1.69
50	1.42	1.50	1.38	1.54	1.34	1.59	1.30	1.64	1.26	1.69
55	1.45	1.52	1.41	1.56	1.37	1.60	1.33	1.64	1.30	1.69
60	1.47	1.54	1.44	1.57	1.40	1.61	1.37	1.65	1.33	1.69
65	1.49	1.55	1.46	1.59	1.43	1.62	1.40	1.66	1.36	1.69
70	1.51	1.57	1.48	1.60	1.45	1.63	1.42	1.66	1.39	1.70
75	1.53	1.58	1.50	1.61	1.47	1.64	1.45	1.67	1.42	1.70
80	1.54	1.59	1.52	1.62	1.49	1.65	1.47	1.67	1.44	1.70
85	1.56	1.60	1.53	1.63	1.51	1.65	1.49	1.68	1.46	1.71
90	1.57	1.61	1.55	1.64	1.53	1.66	1.50	1.69	1.48	1.71
95	1.58	1.62	1.56	1.65	1.54	1.67	1.52	1.69	1.50	1.71
100	1.59	1.63	1.57	1.65	1.55	1.67	1.53	1.70	1.51	1.72

K = number of independent variables
N = number of observations

Table F is reprinted from J. Durbin and G. S. Watson, "Testing for Serial Correlation in Least Squares Regression." *Biometrika*, Vol. 38 (1951), pp. 159–177, with the permission of the authors and the Trustees of Biometrika.

References

Adelman, Irma [1965]. "Long Cycles—Fact or Artifact?" *American Economic Review*, vol. 55, no. 3 (June 1965), pp. 444–463.

Aitken, A.C. [1934–1935]. "On Least Squares and Linear Combination of Observations," *Proceedings of the Royal Society of Edinburgh*, vol. 55, Part I (1934–1935), pp. 42–48.

Allen, R.G.D. [1966]. *Statistics for Economists*, revised edition, London: Hutchinson & Co., Ltd., 1966.

Almon, Shirley [1965]. "The Distributed Lag between Capital Appropriations and Expenditures," *Econometrica*, vol. 33, no. 1 (January 1965), pp. 444–463.

Anderson, Richard L., and Theodore A. Bancroft [1952]. *Statistical Theory in Research*, New York: McGraw-Hill Book Company, 1952.

Anderson, Theodore W. [1958]. *An Introduction to Multivariate Statistical Analysis*, New York: John Wiley & Sons, Inc., 1958.

Anderson, Theodore W., and R.R. Bahadur [1962]. "Classification into Two Multivariate Normal Distributions with Different Covariance Matrices," *Annals of Mathematical Statistics*, vol. 33, no. 2 (June 1962), pp. 420–431.

Anderson, Theodore W., and Herman Rubin [1956]. "Statistical Inference in Factor Analysis," *Proceedings of the Third Berkeley Symposium on Mathematical Statistics and Probability*, vol. V, Berkeley: University of California, 1966.

Basmann, Robert L. [1960]. "On the Asymptotic Distribution of Generalized Linear Estimators," *Econometrica*, vol. 28, no. 1 (January 1960), pp. 97–107.

Blackwell, David, and M.A. Girshick [1954]. *Theory of Games and Statistical Decisions*, New York: John Wiley & Sons, Inc., 1954.

Box, G.E.P. [1954]. "Some Theorems on Quadratic Forms Applied in the Study of Analysis of Variance Problems, I and II," *Annals of Mathematical Statistics*, vol. 25, nos. 2 and 3 (January and September 1954), pp. 290–302 and pp. 484–498.

Bryant, Edward C. [1966]. *Statistical Analysis*, 2nd edition, New York: McGraw-Hill Book Co., 1966.

Chernoff, Herman, and Nathan Divinsky [1953]. "The Computation of Maximum-Likelihood Estimates of Linear Structural Equations," in Hood and Koopmans [1953], pp. 236–271.

Chernoff, Herman, and Leon E. Moses [1959]. *Elementary Decision Theory*, New York: John Wiley & Sons, Inc., 1959.

Chernoff, Herman, and Herman Rubin [1953]. "Asymptotic Properties of Limited Information Estimates under Generalized Conditions," in Hood and Koopmans [1953], pp. 200–212.

Chipman, John S., and M.M. Rao [1964]. "The Treatment of Linear Restrictions in Regression Analysis, *Econometrica*, vol. 32, nos. 1 and 2 (January–April 1964), pp. 198–209.

Chou, Ya-Lun [1969]. *Statistical Analysis*, New York: Holt, Rinehart and Winston, Inc., 1969.

Chow, Gregory C. [1960]. "Tests of Equality between Sets of Coefficients in Two Linear Regressions," *Econometrica*, vol. 28, no. 3 (July 1960), pp. 591–605.

Christ, Carl F. [1966]. *Econometric Models and Methods*, New York: John Wiley & Sons, Inc., 1966.

Christ, Carl F. [1960]. "Simultaneous Equation Estimation: Any Verdict Yet?," *Econometrica*, vol. 28, no. 4 (October 1960), pp. 835–845.

Cochran, William G. [1947]. "Some Consequences when the Assumptions for the A.O.V. Are Not Satisfied," *Biometrics*, vol. 3, no. 1 (March 1947), pp. 22–38.

Cochran, William G. [1963]. *Sampling Techniques*, 2nd edition, New York: John Wiley & Sons, Inc., 1963.

Cochran, William G., and Gertrude M. Cox [1956]. *Experimental Designs*, 2nd edition, New York: John Wiley & Sons, Inc., 1956.

Cochrane, Donald, and Guy H. Orcutt [1949]. "Application of Least Squares Regression to Relationships Containing Auto-Correlated Error Terms," *Journal of the American Statistical Association*, vol. 44, no. 245 (March 1949) pp. 32–61.

Cramér, Harald [1955]. *The Elements of Probability Theory and Some of Its Applications*, New York: John Wiley & Sons, Inc., 1955.

Cramér, Harald [1946]. *Mathematical Methods of Statistics*, Princeton: Princeton University Press, 1946.

Croxton, Frederick E., and Dudley J. Cowden [1960]. *Applied General Statistics*, 2nd edition, Englewood Cliffs: Prentice-Hall, Inc., 1960.

David, F.N., and N.L. Johnson [1951]. "A Method of Investigating the Effect of Non-Normality and Heterogeneity of Variance on Tests of the General Linear Hypothesis," *Annals of Mathematical Statistics*, vol. 22, no. 3 (September 1951), pp. 382–392.

Dempster, Arthur P. [1969]. *Elements of Continuous Multivariate Analysis*, Reading, Mass.: Addison-Wesley Publishing Company, Inc., 1969.

Dhrymes, Phoebus, J. [1970]. *Econometrics*, New York: Harper & Row, 1970.

Durbin, James, and G.S. Watson [1950]. "Testing for Serial Correlation in Least Squares Regression I," *Biometrika*, vol. 37, parts 3–4 (December 1950), pp. 409–428.

Durbin, James, and G.S. Watson [1951]. "Testing for Serial Correlation in Least Squares Regression II," *Biometrika*, vol. 38, parts 1–2 (June 1951), pp. 159–178.

Dwyer, Paul [1941]. "The Doolittle Technique," *Annals of Mathematical Statistics*, vol. 12, no. 4 (December 1941), pp. 449–458.

Eisner, Robert [1960]. "A Distributed Lag Investment Function," *Econometrica*, vol. 28, no. 1 (January 1960), pp. 1–29.

Ezekiel, Mordecai [1941]. *Methods of Correlation Analysis*, 2nd edition, New York: John Wiley & Sons, Inc., 1941.

Feller, William [1957]. *An Introduction to Probability Theory and Its Applications*, vol. I, 2nd edition, New York: John Wiley & Sons, Inc., 1957.

Feller, William [1966]. *An Introduction to Probability Theory and Its Applications*, vol. II, New York: John Wiley & Sons, Inc., 1966.

Fisher, Franklin M. [1966]. *The Identification Problem in Econometrics*, New York: McGraw-Hill Book Co., 1966.

Fishman, George S. [1968]. *Spectral Methods in Econometrics*, Santa Monica: RAND, April 1968.

Fox, Karl A. [1958]. *Econometric Analysis for Public Policy*, Ames, Iowa: Iowa State College Press, 1958.

Frank, Charles R., Jr. "Measurement of Debt Servicing Capacity: An Application of Discriminant Analysis when Covariance Matrices Are Unequal," unpublished paper.

Freund, John E. [1962]. *Mathematical Statistics*, Englewood Cliffs, N.J.: Prentice-Hall, Inc., 1962.

Freund, John E. [1967]. *Modern Elementary Statistics*, 3rd edition, New York: Prentice-Hall, Inc., 1967.

Freund, John E. *Statistics: A First Course*, Englewood Cliffs, N.J.: Prentice-Hall, Inc., 1970.

Freund, John E., and Frank J. Williams [1964]. *Elementary Business Statistics: The Modern Approach*, Englewood Cliffs, N.J.: Prentice-Hall, Inc., 1964.

Friedman, Milton [1957]. *A Theory of the Consumption Function*, New York: National Bureau of Economic Research, 1957.

Godfrey, Michael D. [1967]. "The Statistical Analysis of Stochastic Processes in Economics," *Kyklos*, vol. 20, no. 1 (January 1967), pp. 373–386.

Godfrey, Michael D., Clive W.J. Granger, and O. Morgenstern [1964]. "The Random-Walk Hypothesis of Stock Market Behavior," *Kyklos*, vol. 17, no. 1 (January 1964), pp. 1–30.

Goldberg, Samuel [1960]. *Probability: An Introduction*, Englewood Cliffs, N.J.: Prentice-Hall, Inc., 1960.

Goldberger, Arthur S. [1964]. *Econometric Theory*, New York: John Wiley & Sons, Inc., 1964.

Granger, Clive W.J. [1966]. "The Typical Spectral Shape of an Economic Variable," *Econometrica*, vol. 34, no. 1 (January 1966), pp. 150–161.

Granger, Clive W.J., and M. Hatanaka [1961]. *Spectral Analysis of Economic Time Series*, Princeton: Princeton University Press, 1961.

Granger, Clive W.J., and H.J.B. Rees [1968]. "Spectral Analysis of the Term Structure of Interest Rates," *Review of Economic Studies*, vol. 35(1), no. 101 (January 1968), pp. 67–76.

Graybill, Franklin A. [1961]. *An Introduction to Linear Statistical Models*, vol. I, New York: McGraw-Hill Book Co., 1961.

Graybill, Franklin A. [1969]. *Introduction to Matrices with Applications in Statistics*, Belmont, Calif.: Wadsworth Publishing Company, 1969.

Griliches, Zvi [1961]. "A Note on the Serial Correlation Bias in Estimates of Distributed Lags," *Econometrica*, vol. 29, no. 1 (January 1961), pp. 65–73.

Griliches, Zvi [1967]. "Distributed Lags: A survey," *Econometrica*, vol. 35, no. 1 (January 1967), pp. 16–49.

Haavelmo, Trygve [1943]. "The Statistical Implications of a System of Simultaneous Equations," *Econometrica*, vol. 11, no. 1 (January 1943), pp. 1–12.

Hadley, G. [1961]. *Linear Algebra*, Reading, Mass.: Addison-Wesley Publishing Company, Inc., 1961.

Halperin, Max [1961]. "Fitting of Straight Lines and Prediction when Both Variables Are Subject to Error," *Journal of the American Statistical Association*, vol. 56, no. 295 (September 1961), pp. 657–669.

Hannan, Edward J. [1965]. "The Estimation of Relationships Involving Distributed Lags, *Econometrica*, vol. 33, no. 1 (January 1965), pp. 206–224.

Harman, Harry Horace [1967]. *Modern Factor Analysis*, 2nd edition revised, Chicago: University of Chicago Press, 1967.

Hoel, Paul G. [1966]. *Elementary Statistics*, 2nd edition, New York: John Wiley & Sons, Inc., 1966.

Hoel, Paul G. [1962]. *Introduction to Mathematical Statistics*, 3rd edition, New York: John Wiley & Sons, Inc., 1962.

Hogg, Robert V., and Allen T. Craig [1970]. *Introduction to Mathematical Statistics*, 3rd edition, New York: The Macmillan Company, 1970.

Hood, William C., and Tjalling C. Koopmans, eds. [1953]. "Studies in Econometric Method," Cowles Commision Monograph, no. 14, New York: John Wiley & Sons, Inc.

Hooper, John W. [1959]. "Simultaneous Equations and Canonical Correlation Theory," *Econometrica*, vol. 27, no. 2 (April 1959), pp. 245–256.

Hope, Keith [1968]. *Methods of Multivariate Analysis with Handbook of Multivariate Methods Programmed in Atlas Autocode*, London: University of London Press, 1968.

Horsnell, G. [1953]. "The Effect of Unequal Group Variances on the F-Test for the Homogeneity of Group Means. *Biometrika*, vol. 40, parts 1–2 (June 1953), pp. 128–136.

Horst, Paul [1965]. *Factor Analysis of Data Matrices*, New York: Holt, Rinehart and Winston, Inc., 1965.

Hotelling, Harold [1936]. "Relations between Two Sets of Variates," *Biometrika*, vol. 28, parts 3–4 (December 1936), pp. 321–377.

Howrey, E. Phillip, "A Spectrum Analysis of the Long-Swing Hypothesis," *International Economic Review*, vol. 9, no. 2 (June 1968), pp. 228–252.

Huff, Darrell [1954]. *How to Lie with Statistics*, New York: W.W. Norton & Company, Inc., 1954.

Jenkins, Gwilym M., and Donald G. Jenkins [1968]. *Spectral Analysis and Its Applications*, San Francisco: Holden Day, 1968.

Johnston, J. [1960]. *Econometric Methods*, New York: McGraw-Hill Book Co., 1960.

Jorgenson, Dale W. [1966]. "Rational Distributed Lag Functions," *Econometrica*, vol. 34, no. 1 (January 1966), pp. 135–149.

Kane, Edward J. [1968]. *Economic Statistics and Econometrics: An Introduction to Quantitative Economics*, New York: Harper & Row, 1968.

Kendall, Maurice G., and Allan Stuart [1969]. *The Advanced Theory of Statistics*, vol. 1 (Distribution Theory), London: Charles Griffin and Company, 1969.

Kendall, Maurice G., and Allan Stuart [1967]. *The Advanced Theory of Statistics*, vol. 2 (Inference and Relationship) London: Charles Griffin and Company, Ltd., 1967.

Kendall, Maurice G., and Allan Stuart [1966]. *The Advanced Theory of Statistics*, vol. 3 (Design and Analysis, and Time Series), London: Charles Griffin and Company, Ltd., 1966.

Klein, Lawrence R. [1953]. *A Textbook of Econometrics*, Evanston, Ill.: Row, Peterson & Co., 1953.

Klein, Lawrence R. [1958]. "The Estimation of Distributed Lags," *Econometrica*, vol. 26, no. 4 (October 1958), pp. 553–565.

Klein, Lawrence R., and Nakamura Mitsugu [1962]. "Singularity in the Equation Systems of Econometrics: Some Aspects of the Problem of Multicollinearity," *International Economic Review*, vol. 3, no. 3 (September 1962), pp. 274–299.

Kloek, T., and L.B.M. Mennes [1960]. "Simultaneous Equations Estimation Based on Principal Components of Predetermined Variables," *Econometrica*, vol. 28, no. 1 (January 1960), pp. 45–61.

Kolmogorov, Andrei N. [1950]. *Foundations of the Theory of Probability*, New York: Chelsea Publishing Company, 1950.

Koopmans, Tjalling, C. [1953]. "Identification Problems in Economic Model Construction," in Hood and Koopmans [1953], pp. 27–48.

Koopmans, Tjalling C., ed. [1950]. *Statistical Inference in Dynamic Economic Models*, Cowles Commission Monograph, no. 10, John Wiley & Sons, Inc., 1950.

Koopmans, Tjalling C., and William C. Hood [1953]. "The Estimation of

Simultaneous Linear Economic Relationships,"in Hood and Koopmans [1953], pp. 112–119.

Koyck, L.M. [1954]. *Distributed Lags and Investment Analysis*, Amsterdam: North-Holland Publishing Company, 1954.

Lawley, D.N., and A.E. Maxwell [1963]. *Factor Analysis as a Statistical Method*, London: Butterworth & Co., Ltd., 1963.

Levinson, Horace C., *Chance, Luck, and Statistics*, 2nd edition, New York: Dover Publications, Inc., 1963 (first edition published under the title *The Science of Chance*).

Madansky, Albert [1964]. "On the Efficiency of Three-Stage Least-Squares Estimation," *Econometrica*, vol. 32, nos. 1–2 (January–April 1964), pp. 51–56.

Madansky, Albert [1959]. "The Fitting of Straight Lines when Both Variables Are Subject to Error," *Journal of the American Statistical Association*, vol. 54, no. 285 (March 1959), pp. 173–205.

Malinvaud, Edmond [1966]. *Statistical Methods of Econometrics*, Skokie, Ill.: Rand McNally, 1966.

Malinvaud, Edmond [1961]. "The Estimation of Distributed Lags: A Comment," *Econometrica*, vol. 29, no. 3 (July 1961), pp. 430–433.

Mills, Frederick C. [1955]. *Statistical Methods*, 3rd edition, New York: Holt, Rinehart and Winston, Inc., 1955.

Mood, Alexander M., and Franklin A. Graybill [1963]. *Introduction to the Theory of Statistics*, 2nd edition, New York: McGraw-Hill Book Co., 1963.

Nagar, A.L. [1959]. "The Bias and Moment Matrix of the General k-Class Estimators of the Parameters in Simultaneous Equations," *Econometrica*, vol. 27, no. 4 (October 1959), pp. 575–595.

Nerlove, Marc [1958a]. *Distributed Lags and Demand Analysis for Agricultural and Other Commodities*, Agricultural Handbook, no. 141, Washington, D.C.: U.S. Department of Agriculture, 1958.

Nerlove, Marc [1964]. "Spectral Analysis of Seasonal Adjustment Procedures," *Econometrica*, vol. 32, no. 3 (July 1964), pp. 241–286.

Nerlove, Marc [1958b]. *The Dynamics of Supply: Estimation of Farmers' Response to Price*, Baltimore: The Johns Hopkins Press, 1958.

Neter, John, and William Wassermann [1966]. *Fundamental Statistics for Business and Economics*, 3rd edition, Boston: Allyn and Bacon, Inc., 1966.

Neyman, Jerzy [1950]. *First Course in Probability and Statistics*, New York: Holt, Rinehart and Winston, Inc., 1950.

Orcutt, Guy H. [1950]. "Measurement of Price Elasticities in International Trade," *Review of Economics and Statistics*, vol. 32, no. 2 (May 1950), pp. 117–132.

Parzen, Emanuel [1960]. *Modern Probability Theory and Its Applications*, New York: John Wiley & Sons, Inc., 1960.

Perlis, Sam [1958]. *Theory of Matrices*, Reading, Mass.: Addison-Wesley Publishing Company, Inc., 1958.

Raiffa, Howard, and Robert Schlaifer [1961]. *Applied Statistical Decision Theory*, Cambridge: Graduate School of Business Administration, Harvard University, 1961.

Raj, Des [1968]. *Sampling Theory*, New York: McGraw-Hill Book Co., 1968.

Richmond, Samuel B. [1964]. *Statistical Analysis*, 2nd edition, New York: The Ronald Press Company, 1964.

Rosett, Richard N. [1959]. "A Statistical Model of Friction in Economics," *Econometrica* vol. 27, no. 2 (April 1959), pp. 263–267.

Rummel, R.J. [1970]. *Applied Factor Analysis*, Evanston, Ill.: Northwestern University Press, 1970.

Sargan, J.D. [1958]. "The Estimation of Economic Relationships Using Instrumental Variables," *Econometrica*, vol. 26, no. 3 (July 1958), pp. 393–415.

Sargan, J.D. [1961]. "The Maximum Likelihood Estimation of Economic Relationships with Autoregressive Residuals," *Econometrica*, vol. 29, no. 3 (July 1961), pp. 414–426.

Savage, L.J. [1962]. *The Foundations of Statistical Inference*, New York: John Wiley & Sons, Inc., and London, Methuen and Co., Ltd., 1962.

Scheffé, Henry [1959]. *The Analysis of Variance*, New York: John Wiley & Sons, Inc., 1959.

Solow, Robert [1960]. "On a Family of Lag Distributions," *Econometrica*, vol. 28, no. 2 (April 1960), pp. 393–406.

Stein, Max [1967]. *Introduction to Matrices and Determinants*, Reading, Mass.: Addison-Wesley Publishing Company, Inc., 1967.

Suits, Daniel B. [1963]. *Statistics: An Introduction to Quantitative Economic Research*, Skokie, Ill.: Rand McNally, 1963.

Suits, Daniel B. [1957]. "Use of Dummy Variables in Regression Equations," *Journal of the American Statistical Association*, vol. 52, no. 280 (December 1957), pp. 548–551.

Summers, Robert [1965]. "A Capital Intensive Approach to the Small Sample Properties of Various Simultaneous Equation Estimators," *Econometrica*, vol. 35, no. 1 (January 1965), pp. 1–41.

Theil, Henri [1965]. *Economic Forecasts and Policy*, Amsterdam: North-Holland Publishing Company, 1965.

Theil, Henri, and Robert M. Stern [1960]. "A Simple Unimodal Lag Distribution," *Metroeconomica*, vol. 12, nos. 2–3 (August–December 1960), pp. 111–119.

Tintner, Gerhard [1952]. *Econometrics*, New York: John Wiley & Sons, Inc., 1952.

Tobin, James [1958]. "Estimation of Relationships for Limited Dependent Variables," *Econometrica*, vol. 26, no. 1 (January 1958), pp. 24–36.

Tobin, James [1955]. "The Application of Multivariate Probit Analysis to Economic Survey Data," Cowles Foundation Discussion Paper 1, 1955.

Valavanis, Stefan [1959]. *Econometrics: An Introduction to Maximum Likelihood Methods*, New York: McGraw-Hill Book Co., 1959.

Wald, Abraham [1950]. *Statistical Decision Functions*, New York: John Wiley & Sons, Inc., 1950.

Walker, Helen M., and Joseph Lev [1953]. *Statistical Inference*, New York: Holt, Rinehart and Winston, Inc., 1953.

Wallis, W. Allen, and Harvey V. Roberts [1956]. *Statistics, A New Approach*, Glencoe, Ill.: The Free Press, 1956.

Wallis, W. Allen, and Harry V. Roberts [1962]. *The Nature of Statistics*, New York: P.F. Collier & Son Corp., 1962 (a new, revised edition of the basic first section of *Statistics: A New Approach*).

Warner, Stanley L. [1962]. *Stochastic Choice of Mode in Urban Travel: A Study in Binary Choice*, Evanston, Ill.: Northwestern University Press, 1962.

Weiss, Lionel [1961]. *Statistical Decision Theory*, New York: McGraw-Hill Book Co., 1961.

Wilks, Samuel S. [1962]. *Mathematical Statistics*, New York: John Wiley & Sons, Inc., 1962.

Wonnacott, Ronald J., and Thomas H. Wonnacott [1970]. *Econometrics*, New York: John Wiley & Sons, Inc., 1970.

Wonnacott, Thomas H., and Ronald J. Wonnacott [1969]. *Introductory Statistics*, New York: John Wiley & Sons, Inc., 1969.

Working, Elmer J. [1927]. "What Do 'Statistical Demand Curves' Show?," *Quarterly Journal of Economics*, vol. 4, no. 1 (February 1927), pp. 212–235.

Yamane, Taro [1967]. *Statistics: An Introductory Analysis*, New York: Harper & Row, 1967.

Yule, George U., and Maurice G. Kendall [1950]. *An Introduction to the Theory of Statistics*, 14th ed., London: Charles Griffin & Co., Ltd., 1950.

Zellner, Arnold [1961a]. "Econometric Estimation with Temporally Dependent Disturbance Terms," *International Economic Review*, vol. 2, no. 2 (May 1961), pp. 164–178.

Zellner, Arnold [1961b]. "Linear Regression with Inequality Constraints on the Coefficients: An Application of Quadratic Programming and Linear Decisions Rules," International Center for Management Science, Report 6109 (MS No. 9), 1961.

Zellner, Arnold, and Henri Theil [1962]. "Three-Stage Least Squares: Simultaneous Estimation of Simultaneous Equations," *Econometrica*, vol. 30, no. 1 (January 1962), pp. 54–78.

Subject Index